ADVANCES IN SECOND MESSENGER
AND PHOSPHOPROTEIN RESEARCH

Volume 31

Signal Transduction in Health and Disease

Advances in Second Messenger and Phosphoprotein Research

ADVANCES IN SECOND MESSENGER
AND PHOSPHOPROTEIN RESEARCH

Volume 31

Signal Transduction in Health and Disease

Editors

Jackie D. Corbin, Ph.D.
Department of Molecular Physiology and Biophysics
Vanderbilt University School of Medicine
Nashville, Tennessee

and

Sharron H. Francis, Ph.D.
Department of Molecular Physiology and Biophysics
Vanderbilt University School of Medicine
Nashville, Tennessee

Lippincott - Raven
PUBLISHERS
Philadelphia • New York

Acquisitions Editor: Mark Placito
Manufacturing Manager: Dennis Teston
Associate Managing Editor: Kathleen Bubbeo
Production Services: Colophon
Cover Designer: Patricia Gast
Indexer: Indexing Research
Compositor: Eastern Compositon
Printer: Maple Press

Printed and bound in the United Kingdom

International Standard Book Number 0-397-51685-1

Transferred to Digital Printing, 2011

To the memory of Earl W. Sutherland (1915–1974) of the Department of Physiology, Vanderbilt Medical School. Dr. Sutherland won the Nobel Prize in Physiology or Medicine in 1971 for his work which led to the second messenger hypothesis for signal transduction.

Contents

Contributing Authors

A. Allgeier, M.S.
Faculte de Medecine
Institute of Interdisciplinary Research (I.R.I.B.H.N.)
Free University of Brussels
Campus Erasme
808 Route de Lennik (Bldg C)
B-1070 Brussels, Belgium

M. Baptist, Ph.D.
Faculte de Medecine
Institute of Interdisciplinary Research (I.R.I.B.H.N.)
Free University of Brussels
Campus Erasme
808 Route de Lennik (Bldg C)
B-1070 Brussels, Belgium

David Barford, B.Sc., D.Phil.
David Phillips Professor of Molecular Biophysics
Laboratory of Molecular Biophysics
University of Oxford
Rex Richards Building
South Parks Road
Oxford OX1 3QU, England

Joseph A. Beavo, Ph.D.
Professor
Department of Pharmacology
University of Washington School of Medicine
Box 357280
Seattle, Washington 98195

Karin E. Bornfeldt, Ph.D.
Department of Pathology
University of Washington
Box 357470
Seattle, Washington 98195

David L. Brautigan, Ph.D.
Professor of Microbiology and Medicine
Center for Cell Signaling
University of Virginia
Box 577-HSC-Hospital West
Charlottesville, Virginia 22908

Lewis C. Cantley, Ph.D.
Professor of Medicine
Department of Cell Biology
Harvard Medical School
and
Chief
Division of Signal Transduction
Beth Israel Deaconness Medical Center
330 Brookline Avenue
Boston, Massachusetts 02115

William A. Catterall, Ph.D.
Professor and Chair
Department of Pharmacology
University of Washington School of Medicine
Box 357280, F-427, Health Sciences Center
Seattle, Washington 98195-7280

D. Christophe, Ph.D.
Faculte de Medecine
Institute of Interdisciplinary Research (I.R.I.B.H.N.)
Free University of Brussels
Campus Erasme
808 Route de Lennik (Bldg C)
B-1070 Brussels, Belgium

F. Coppee, M.S.
Faculte de Medecine
Institute of Interdisciplinary Research (I.R.I.B.H.N.)
Free University of Brussels
Campus Erasme
808 Route de Lennik (Bldg C)
B-1070 Brussels, Belgium

Jackie D. Corbin, Ph.D.
Professor
Department of Molecular Physiology and Biophysics
Vanderbilt University School of Medicine
702 Light Hall
Nashville, Tennessee 37232

K. Coulonval, M.S.
Faculte de Medecine
Institute of Interdisciplinary Research (I.R.I.B.H.N.)
Free University of Brussels
Campus Erasme
808 Route de Lennik (Bldg C)
B-1070 Brussels, Belgium

S. Deleu, M.S.
Faculte de Medecine
Institute of Interdisciplinary Research (I.R.I.B.H.N.)
Free University of Brussels
Campus Erasme
808 Route de Lennik (Bldg C)
B-1070 Brussels, Belgium

F. Depoortere, M.S.
Faculte de Medecine
Institute of Interdisciplinary Research (I.R.I.B.H.N.)
Free University of Brussels
Campus Erasme
808 Route de Lennik (Bldg C)
B-1070 Brussels, Belgium

Peter N. Devreotes, Ph.D.
Department of Biological Chemistry
The Johns Hopkins University School of Medicine
725 North Wolfe Street
Baltimore, Maryland 21205-2185

Edgar Dippel, M.D.
Postdoctoral Fellow
Institute of Pharmacology
Free University of Berlin
Thielallee 69-73
D-14195 Berlin, Germany

Julian Downward, Ph.D.
Signal Transduction Laboratory
Imperial Cancer Research Fund
44 Lincoln's Inn Fields
London WC2A 3PX, England

S. Dremier, M.S.
Faculte de Medecine
Institute of Interdisciplinary Research (I.R.I.B.H.N.)
Free University of Brussels
Campus Erasme
808 Route de Lennik (Bldg C)
B-1070 Brussels, Belgium

E. Dumont, M.D., Ph.D.
Professor of Biochemistry and Head
Institute of Interdisciplinary Research (I.R.I.B.H.N.)
Free University of Brussels
Campus Erasme, Building C
808 Route de Lennik (Bldg C)
B-1070 Brussels, Belgium

Nicholas S. Foulkes, Ph.D.
Institut de Génétique et de Biologie Moléculaire et Cellulaire
B.P. 163, 67404 Illkirch-Cédex
Strasbourg, France

Sharron H. Francis, Ph.D.
Research Professor
Department of Molecular Physiology and Biophysics
Vanderbilt University School of Medicine
701 Light Hall
Nashville, Tennessee 37232

Stephan Frings, M.D.
Institut für Biologische Informationsverarbeitung
Forschungszentrum Jülich GmbH
D-52425 Jülich, Germany

Zeren Gao, M.S.
Investigator, Howard Hughes Medical Institute
Department of Pharmacology
University of Texas Southwestern Medical Center at Dallas
5323 Harry Hines Boulevard
Dallas, Texas 75235-9041

David L. Garbers, Ph.D.
Investigator, Howard Hughes Medical Institute
and
Professor
Department of Pharmacology
University of Texas Southwestern Medical Center at Dallas
5323 Harry Hines Boulevard
Dallas, Texas 75235-9041

Elspeth F. Garman, M.A., D.Phil.
David Phillips Professor of Molecular Biophysics
Laboratory of Molecular Biophysics
University of Oxford
Rex Richards Building
South Parks Road
Oxford OX1 3QU, England

Lee M. Graves, Ph.D.
Assistant Professor
Department of Pharmacology
University of North Carolina at Chapel Hill
The School of Medicine
Manning Drive
Chapel Hill, North Carolina 27599-7365

Thomas Gudermann, M.D.
Research Assistant
Institute of Pharmacology
Free University of Berlin
Thielallee 69-73
D-14195 Berlin, Germany

Vidar Hansson, M.D., Ph.D.
Professor
Institute of Medical Biochemistry
University of Oslo
P.O. Box 1112, Blindern
N-0317 Oslo, Norway

Robert A. Harris, Ph.D.
Showalter Professor and Chairman
Department of Biochemistry and Molecular Biology
Indiana University School of Medicine
635 Barnhill Drive
Indianapolis, Indiana 46202-5122

Brian E. Hawes, Ph.D.
Senior Scientist
CNS/Cardiovascular Department
Schering Plough Research Institute
2015 Galloping Hill Road
Kenilworth, New Jersey 07033

John W. Hawes, Ph.D.
Department of Biochemistry and Molecular Biology
Indiana University School of Medicine
Van Nuys Medical Science Building 410
635 Barnhill Drive
Indianapolis, Indiana 46202-5122

Jörg Heierhorst, M.D.
St. Vincent's Institute of Medical Research
41 Victoria Parade
Fitzroy 3065, Victoria, Australia

Hiroyoshi Hidaka, M.D., Ph.D
Chairman
Department of Pharmacology
Nagoya University School of Medicine
Tsurumai-cho 65, Showa-ku
Nagoya 466, Japan

Tore Jahnsen, M.D., Ph.D.
Professor
Institute of Medical Biochemistry
University of Oslo
P.O. Box 1112, Blindern
N-0317 Oslo, Norway

Ann Kirsti Johansen, M.Sc.
Institute of Medical Biochemistry
University of Oslo
P.O. Box 1112, Blindern
N-0317 Oslo, Norway

Louise N. Johnson, B.Sc (Hons), Ph.D.
Professor
Laboratory of Molecular Biophysics
University of Oxford
Rex Richards Building
South Parks Road
Oxford OX1 3QU, England

Frank Kalkbrenner, M.D.
Research Assistant
Institute of Pharmacology
Free University of Berlin
Thielallee 69-73
D-14195 Berlin, Germany

Kristine E. Kamm, Ph.D.
Associate Professor
Department of Physiology
University of Texas Southwestern Medical Center at Dallas
5323 Harry Hines Boulevard
Dallas, Texas 75235-9040

Bruce E. Kemp, Ph.D.
St. Vincent's Institute of Medical Research
41 Victoria Parade
Fitzroy 3065, Victoria, Australia

Ji-Yun Kim, Ph.D.
Department of Biological Chemistry
The Johns Hopkins University School of Medicine
725 North Wolfe Street
Baltimore, Maryland 21205-2185

Helle K. Knutsen, Ph.D.
Postdoctoral Fellow
Institute of Medical Biochemistry
University of Oslo
P.O. Box 1112, Blindern
N-0317 Oslo, Norway

Boštjan Kobe, Ph.D.
St. Vincent's Institute of Medical Research
41 Victoria Parade
Fitzroy 3065, Victoria, Australia

Walter J. Koch, Ph.D.
Department of Medicine
Duke University Medical Center
Post Office Box 3821
Room 468 CARL Building
Research Drive
Durham, North Carolina 27710

Edwin G. Krebs, M.D.
Sr. Investigator Emeritus
Department of Pharmacology, SL-15
Howard Hughes Medical Institute
School of Medicine
J681 University of Washington Medical Center
Seattle, Washington 98195

Joanna K. Krueger
Department of Physiology
University of Texas Southwestern Medical Center at Dallas
5323 Harry Hines Boulevard
Dallas, Texas 75235-9040

Kathleen M. Krueger, Ph.D.
Postdoctoral Fellow
Bioscience and Biotechnology Group
Chemical Science and Technology Division
Los Alamos National Laboratory
Los Alamos, New Mexico 87545

Enzo Lalli, M.D.
Institut de Génétique et de Biologie Moléculaire et Cellulaire
B.P. 163, 67404 Illkirch-Cédex
Strasbourg, France

Monica Lamas, Ph.D.
Institut de Génétique et de Biologie Moléculaire et Cellulaire
B.P. 163, 67404 Illkirch-Cédex
Strasbourg, France

F. Lamy, Ph.D.
Faculte de Medecine
Institute of Interdisciplinary Research (I.R.I.B.H.N.)
Free University of Brussels
Campus Erasme
8-08 Route de Lennik (Bldg C)
B-1070 Brussels, Belgium

Turid Larsen, M.Sc.
Institute of Medical Biochemistry
University of Oslo
P.O. Box 1112, Blindern
N-0317 Oslo, Norway

Karl-Ludwig Laugwitz, M.D.
Postdoctoral Fellow
Institute of Pharmacology
Free University of Berlin
Thielallee 69-73
D-14195 Berlin, Germany

C. Ledent, Ph.D.
Faculte de Medecine
Institute of Interdisciplinary Research (I.R.I.B.H.N.)
Free University of Brussels
Campus Erasme
808 Route de Lennik (Bldg C)
B-1070 Brussels, Belgium

Robert J. Lefkowitz, M.D.
James B. Duke Professor of Medicine
Department of Medicine
Duke University Medical Center
Room 468, CARL Building
Research Drive
Durham, North Carolina 27710

Finn Olav Levy, M.D., Ph.D.
Research Fellow
Institute of Medical Biochemistry
University of Oslo
P.O. Box 1112, Blindern
N-0317 Oslo, Norway

Pei-ju Lin
Department of Physiology
University of Texas Southwestern Medical Center at Dallas
5323 Harry Hines Boulevard
Dallas, Texas 75235-9040

Suzanne M. Lohmann, M.D.
Department of Clinical Biochemistry
Medical University Clinic
Wurzburg 8700, Germany

Katherine Luby-Phelps, Ph.D.
Assistant Professor
Department of Physiology
University of Texas Southwestern Medical Center at Dallas
5323 Harry Hines Boulevard
Dallas, Texas 75235-9040

Louis M. Luttrell, M.D., Ph.D.
Associate in Medicine
Department of Medicine
Duke University Medical Center
Box 3821
Erwin Road
Durham, North Carolina 27710

C. Maenhaut, Ph.D.
Faculte de Medecine
Institute of Interdisciplinary Research (I.R.I.B.H.N.)
Free University of Brussels
Campus Erasme
808 Route de Lennik (Bldg C)
B-1070 Brussels, Belgium

Cristina Mazzucchelli, Ph.D.
Institut de Génétique et de Biologie Moléculaire et Cellulaire
B.P. 163, 67404 Illkirch-Cédex
Strasbourg, France

Jacqueline L. S. Milne, Ph.D.
Department of Biological Chemistry
The Johns Hopkins University School of Medicine
725 North Wolfe Steet
Baltimore, Maryland 21205-2185

F. Miot, Ph.D.
Faculte de Medecine
Institute of Interdisciplinary Research (I.R.I.B.H.N.)
Free University of Brussels
Campus Erasme
808 Route de Lennik (Bldg C)
B-1070 Brussels, Belgium

Lucia Monaco, Ph.D.
Institut de Génétique et de Biologie Moléculaire et Cellulaire
B.P. 163, 67404 Illkirch-Cédex
Strasbourg, France

Yasuhito Naito, B.S.
Department of Pharmacology
Nagoya University School of Medicine
Tsurumai-cho 65, Showa-ku
Nagoya 466, Japan

Hiroyuki Nakanishi, M.D., Ph.D.
Group Leader
Takai Biotimer Project
ERATO Research Development Corporation of Japan
2-2-10 Murotani, Nishi-ku
Kobe 651-22, Japan

François Nantel, Ph.D.
Institut de Génétique et de Biologie Moléculaire et Cellulaire
B.P. 163, 67404 Illkirch-Cédex
Strasbourg, France

Martin E. M. Noble, B.Sc., Ph.D.
David Phillips Professor of Molecular Biophysics
Laboratory of Molecular Biophysics
University of Oxford
Rex Richards Building
South Parks Road
Oxford OX1 3QU, England

Sigurd Ørstavik, M.D.
Institute of Medical Biochemistry
University of Oslo
P.O. Box 1112, Blindern
N-0317 Oslo, Norway

David J. Owen, M.A., D.Phil.
David Phillips Professor of Molecular Biophysics
Laboratory of Molecular Biophysics
University of Oxford
Rex Richards Building
South Parks Road
Oxford OX1 3QU, England

V. Panneels, Ph.D.
Faculte de Medecine
Institute of Interdisciplinary Research (I.R.I.B.H.N.)
Free University of Brussels
Campus Erasme
808 Route de Lennik (Bldg C)
B-1070 Brussels, Belgium

J. Parma, Ph.D.
Faculte de Medecine
Institute of Interdisciplinary Research (I.R.I.B.H.N.)
Free University of Brussels
Campus Erasme
808 Route de Lennik (Bldg C)
B-1070 Brussels, Belgium

M. Parmentier, M.D., Ph.D.
Faculte de Medecine
Institute of Interdisciplinary Research (I.R.I.B.H.N.)
Free University of Brussels
Campus Erasme
808 Route de Lennik (Bldg C)
B-1070 Brussels, Belgium

Lucia Penna, B.S.
Institut de Génétique et de Biologie Moléculaire et Cellulaire
B.P. 163, 67404 Illkirch-Cédex
Strasbourg, France

I. Pirson, Ph.D.
Faculte de Medecine
Institute of Interdisciplinary Research (I.R.I.B.H.N.)
Free University of Brussels
Campus Erasme
808 Route de Lennik (Bldg C)
B-1070 Brussels, Belgium

V. Pohl, Ph.D.
Faculte de Medecine
Institute of Interdisciplinary Research (I.R.I.B.H.N.)
Free University of Brussels
Campus Erasme
808 Route de Lennik (Bldg C)
B-1070 Brussels, Belgium

Kirill M. Popov, Ph.D.
Assistant Scientist
Department of Biochemistry and\ Molecular Biology
Indiana University School of Medicine
635 Barnhill Drive
Indianapolis, Indiana 46202-5122

Celeste E. Poteet-Smith
Research Fellow
Department of Molecular Physiology and Biophysics
Vanderbilt University School of Medicine
702 Light Hall
Nashville, Tennessee 37232

Robin B. Reed, Ph.D.
Research Fellow
Department of Molecular Physiology and Biophysics
Vanderbilt University School of Medicine
701 Light Hall
Nashville, Tennessee 37232

Nils T. K. Reinton, M.Sc.
Institute of Medical Biochemistry
University of Oslo
P.O. Box 1112, Blindern
N-0317 Oslo, Norway

P. Roger, Ph.D.
Faculte de Medecine
Institute of Interdisciplinary Research (I.R.I.B.H.N.)
Free University of Brussels
Campus Erasme
808 Route de Lennik (Bldg C)
B-1070 Brussels, Belgium

Mårten Sandberg, M.D., Ph.D.
Postdoctoral Fellow
Institute of Medical Biochemistry
University of Oslo
P.O. Box 1112, Blindern
N-0317 Oslo, Norway

Takuya Sasaki, M.D., Ph.D.
Assistant Professor
Department of Molecular Biology and Biochemistry
Osaka University Medical School
2-2, Yamada-oka
Suita 565, Osaka
Japan

Paolo Sassone-Corsi, Ph.D.
Directeur de Recherche
Institut de Génétique et de Biologie Moléculaire et Cellulaire
1, Rue Laurent Fries
B.P. 163
67404 Illkirch
Strasbourg, France

V. Savonet, M.S.
Faculte de Medecine
Institute of Interdisciplinary Research (I.R.I.B.H.N.)
Free University of Brussels
Campus Erasme
808 Route de Lennik (Bldg C)
B-1070 Brussels, Belgium

Hans Petter Schrader
Institute of Medical Biochemistry
University of Oslo
P.O. Box 1112, Blindern
N-0317 Oslo, Norway

Günter Schultz, M.D.
Professor
Institute of Pharmacology
Free University of Berlin
Thielallee 69-73
D-14195 Berlin, Germany

Hiromichi Shirataki, M.D., Ph.D.
Assistant Professor of Biochemistry
Department of Cell Physiology
National Institute for Physiological Sciences
Okasazi 444, Japan

Bjørn S. Skålhegg, Ph.D.
Research Fellow
Institute of Medical Biochemistry
University of Oslo
P.O. Box 1112, Blindern
N-0317 Oslo, Norway

Rigmor Solberg, Ph.D.
Postdoctoral Fellow
Institute of Medical Biochemistry
University of Oslo
P.O. Box 1112, Blindern
N-0317 Oslo, Norway

Zhou Songyang, Ph.D.
Research Fellow
Department of Biology
Massachusetts Institute of Technology
31 Ames Street
Cambridge, Massachusetts 02139

William K. Sonnenburg, Ph.D.
Research Assistant Professor
Department of Pharmacology
University of Washington School of Medicine
Box 357280
Seattle, Washington 98195

James T. Stull, Ph.D.
Professor and Chairman
Department of Physiology
University of Texas Southwestern Medical Center at Dallas
5323 Harry Hines Boulevard
Dallas, Texas 75235-9040

Ryotaro Sugita, M.D., Ph.D.
Department of Pharmacology
Nagoya University School of Medicine
Tsurumai-cho 65, Showa-ku
Nagoya 466, Japan

Yoshimi Takai, M.D., Ph.D.
Professor
Department of Molecular Biology and Biochemistry
Osaka University Medical School
2-2 Yamada-oka
Suita 565, Osaka
Japan

Katherine Tamai, Ph.D.
Institut de Génétique et de Biologie Moléculaire et Cellulaire
B.P. 163, 67404 Illkirch-Cédex
Strasbourg, France

Kjetil Taskén M.D., Ph.D.
Research Fellow
Institute of Medical Biochemistry
University of Oslo
P.O. Box 1112, Blindern
N-0317 Oslo, Norway

Kristin Austlid Taskén, Ph.D.
Postdoctoral Fellow
Institute of Medical Biochemistry
University of Oslo
P.O. Box 1112, Blindern
N-0317 Oslo, Norway

M. Taton, Ph.D.
Faculte de Medecine
Institute of Interdisciplinary Research (I.R.I.B.H.N.)
Free University of Brussels
Campus Erasme
808 Route de Lennik (Bldg C)
B-1070 Brussels, Belgium

Osamu Terada, B.S.
Department of Pharmacology
Nagoya University School of Medicine
Tsurumai-cho 65, Showa-ku
Nagoya 466, Japan

M. Tonacchera, M.D., Ph.D.
Faculte de Medecine
Institute of Interdisciplinary Research (I.R.I.B.H.N.)
Free University of Brussels
Campus Erasme
808 Route de Lennik (Bldg C)
B-1070 Brussels, Belgium

Knut Martin Torgersen, M.Sc.
Institute of Medical Biochemistry
University of Oslo
P.O. Box 1112, Blindern
N-0317 Oslo, Norway

Kazushige Touhara, Ph.D.
Assistant Professor
Department of Neurobiochemistry
Faculty of Medicine
The University of Tokyo
7-3-1 Hongo, Bunkyo-ku
Tokyo 113 Japan

N. Uyttersprot, M.S.
Faculte de Medecine
Institute of Interdisciplinary Research (I.R.I.B.H.N.)
Free University of Brussels
Campus Erasme
808 Route de Lennik (Bldg C)
B-1070 Brussels, Belgium

Tim van Biesen, Ph.D.
Research Pharmacologist
Department of Neurological and Urological Diseases Research
Abbott Laboratories
100 Abbott Park Road, D-4PM AP10
Abbott Park, Illinois 60064-3500

J. van Sande, Ph.D.
Faculte de Medecine
Institute of Interdisciplinary Research (I.R.I.B.H.N.)
Free University of Brussels
Campus Erasme
808 Route de Lennik (Bldg C)
B-1070 Brussels, Belgium

Torkel Vang
Institute of Medical Biochemistry
University of Oslo
P.O. Box 1112, Blindern
N-0317 Oslo, Norway

G. Vassart, M.D., Ph.D.
Faculte de Medecine
Institute of Interdisciplinary Research (I.R.I.B.H.N.)
Free University of Brussels
Campus Erasme
808 Route de Lennik (Bldg C)
B-1070 Brussels, Belgium

F. Wilkin, Ph.D.
Faculte de Medecine
Institute of Interdisciplinary Research (I.R.I.B.H.N.)
Free University of Brussels
Campus Erasme
808 Route de Lennik (Bldg C)
B-1070 Brussels, Belgium

Chen Yan, Ph.D.
Postdoctoral Fellow
Department of Pathology
University of Washington School of Medicine
Box 357710
Seattle, Washington 98195

Hisayuki Yokokura
Department of Pharmacology
Nagoya University School of Medicine
Tsurumai-cho 65, Showa-ku
Nagoya 466, Japan

Peter S. T. Yuen, Ph.D.
Assistant Professor
University of Tennessee, Memphis
858 Madison Ave. Suite G01
Memphis, Tennessee 38163

Emmanuel Zazopoulos, Ph.D.
Institut de Génétique et de Biologie Moléculaire et Cellulaire
B.P. 163, 67404 Illkirch-Cédex
Strasbourg, France

Allan Z. Zhao, Ph.D.
Instructor
Department of Pharmacology
University of Washington School of Medicine
Box 357280
Seattle, Washington 98195

Gang Zhi, Ph.D.
Assistant Professor
Department of Physiology
University of Texas Southwestern Medical Center at Dallas
5323 Harry Hines Boulevard
Dallas, Texas 75235-9040

Preface

Signal transduction in health and disease has emerged as a primary focus of biomedical research, and most of the salient aspects of this subject are covered in this book. Biologic signals are diverse and provide the basis for communication between or within organisms. Many drugs and medications simply mimic natural biologic signals to bring about their palliative effects. Signaling molecules include simple molecules such as nitric oxide, catecholamines, and glutamate, as well as complex molecules such as proteins. Signal transduction encompasses the entire process of signal communication, including signal synthesis, transport, translation, and amplification, into a physiologic response. Some of the best-known signals are derived from the environment and include odorants and light, which are detected and processed in order to elicit a given physiologic response. In mammals, hormonal signals are transported by the bloodstream, allowing communication between different tissues. Some signals, such as neurotransmitters and paracrine factors, may travel only short distances in the intercellular spaces to elicit effects on their target cells. Signals such as steroids or nitric oxide physically traverse the membranes of cells to cause intracellular actions. Other signals interact with specific receptors on the membrane surface, which are then coupled with G protein systems or other processing systems within the membrane. The membrane-bound receptor is sometimes a tyrosine-specific protein kinase or is closely coupled to such a kinase, and the signal can be transferred downstream in this type of pathway by protein structures termed SH2 domains. Second messengers such as cyclic nucleotides, calcium, or phospholipids, may provide the next step in communicating a signal to the cell interior in a process that can involve protein phosphorylation catalyzed by serine/threonine-specific, tyrosine-specific, or dual-specificity protein kinases. Protein dephosphorylation catalyzed by protein phosphatases also communicates signals, and specific phosphatases remove phosphate from serine/threonine, tyrosine, or both types of residues. Ion channels are also frequent targets of particular signal transduction pathways. Almost all body processes are regulated at least in part by the aforementioned pathways. These pathways may involve acute regulation of cell processes or involve long-term developmental changes that utilize gene regulation. Studies of a variety of simplified biologic systems allow more careful delineation of these mechanisms, which have commonly gone awry in pathologies such as cancer, cardiovascular disease, diabetes, and most other diseases.

Signal Transduction in Health and Disease contains the edited proceedings of the Ninth International Conference on Second Messengers and Phosphoproteins, held at Vanderbilt University in Nashville, Tennessee, from October 27 through November 1, 1995.

Acknowledgments

We want to thank all of the symposium speakers and workshop organizers at the Ninth International Conference on Second Messengers and Phosphoproteins. We also thank the members of the International Scientific Advisory Board, Local Organizing Committee, symposium session chairs, and workshop participants, as well as the many following organizations, for their support of the conference.

Symposium Speakers

J. Beavo
D. Brautigan
L. Cantley
W. Catterall
S. Cohen
P. Devreotes
J. Dixon
J. Downward
J. Dumont
S. Frings
D. Garbers
A. Gilman
H. Hamm
R. Harris
W. Hendrickson
H. Hidaka
R. Hilgenfeld
T. Jahnsen
L. Johnson
M. Karin
B. Kemp
E. Krebs
R. Lefkowitz
M. Marletta
A. Pawson
P. Sassone-Corsi
J. Schlessinger
G. Schultz
J. Stull
Y. Takai
S. Taylor
M. Welsh

Workshop Organizers

A. Aitken
J. Beavo
S. Beebe
B. Brown
E. Carafoli
R. Colbran
M. Conti
D. Cooper
M. Currie
P. Dobson
J. Exton
S. Fleischer
L. Forte
J. Franks
K. Gould
D. Granner
S. Hanks
A. Harmon
J. Hawiger
M. Hosey
R. Iyengar
R. Johnson
L. Limbird
V. Manganiello
D. Mochly-Rosen
M. Moran
H. Moses
F. Murad
A. Nairn
E. Neer
L. Pinna
M. Rasenick
S. Shenolikar
J. Scott
B.-H. Shieh
T. Soderling
M. Thomas
J. Thompson
B. Wadzinski
T. Woodford-Thomas

Local Organizing Committee

J. Corbin
S. Francis
G. Carpenter
A. Cherrington
S. Cohen
J. Exton
S. Fleischer
D. Granner
J. Hardman
T. Inagami
L. Limbird
H. Moses
J. Wells

Symposium Session Chairs

G. Carpenter
A. Cherrington
J. Exton
S. Fleischer
D. Granner
J. Hardman
J. Hawiger
T. Inagami
L. Limbird
H. Moses
J. Oates
C. Park
J. Staros
M. Waterman
J. Wells

Workshop Participants

K. Ahmed
J. Allende
K. Al-Sakkaf
W. Anderson
B. Andrews
F. Antoni
D. Armstrong
V. Arshavsky
N. Artemyev
D. Ballard
T. Blanck
A. Bleecker
G. Bolger
M. Bollen
E. Butt
J. Campbell
P. Casey
E. Chambaz
M. Chinkers
J. Clark
M. Cohen
J. Corbin
D. Corda
S. Cox
J. Crawley
Z. Damuni
S. Davis
H. De Jonge
A. De Paoli-Roach
S. Döskeland
A. Edelman
R. Ferl
M. Fonteles
S. Francis
J. Frearson
E. Freed
H. Fu
N. Gautam
C. Glover
J. Goldberg
T. Goldkorn
J. Goris
M. Goy
P. Grondin
F. Guesdon
K. Hamra
L. Hein
M. Houslay
R. Huganir
R. Idzerda
Y. Ishikawa
O.-G. Issinger
S. Jaken
G. Kammer
Y. Kanaho
R. Kessin
J. Kieber
U. Kikkawa
Y. Kimura
T. Kitagawa
J. Knopf
W. Koch
M. Kon-Kozlowski
D. Kosk-Kosicka
J. Krupinski
M. Kuhn
D. Lawrence
J. Lawrence
E. Lee
L. Levin
I. Levitan
T. Lincoln
F. Liu
P.-M. Lledo
D. MacLennan

L. MacMillan
S. Marcus
T. Martins
A. McDermott
S. Meyer
E. Miyamoto
C. Moxham
M. Mumby
A. Muslin
R. Neubig
D. Nicholson
B. Niemeyer
C. Nobes
R. O'Brien
M. Okumura
H. Ostergaard
R. Pack
J. Parry
T. Patel
J. Penniston
K. Penta
E. Peralta
C. Proud
W. Pyerin
H. Rasmussen
E. Reimann
F. Rendu
M. Rizzo
P. Roach
L. Robinson
J. Robishaw
C. Rommel
C. Rubin
J. Ruderman
A. Saltiel
S. Sazer
D. Seldin
C. Sette
J. Shabb
M. Siletta
N. Skiba
E. Skoulakis
C. Smith
T. Smith
W. Sonnenburg
J. Stamler
R. Stevenson
C. Sutherland
D. Sweatt
M. Tada
W.-J. Tang
K. Taskén
R. Taussig
S. Thomas
T. Torgerson
T. Torphy
J. Traugh
M. Uhler
L. Van Kaer
T. Vanaman
T. Voyno-Yasenetskaya
J. Walker
D. Walsh
A. Yamazaki
A. Young

Sponsor: Vanderbilt University Medical Center

Distinguished Lecture Sponsors: Cadus Pharmaceutical Corporation
Vanderbilt University Research Council

Donors

Abbott Laboratories; Amgen Incorporated; Avanti Polar Lipids, Incorporated; BASF Bioresearch Corporation; Berlex Biosciences; BioLog Life Science Institute; Boehringer-Ingelheim Pharmaceuticals, Incorporated; Boehringer Mannheim GmbH; Ciba Pharmaceuticals; Cystic Fibrosis Foundation; Daiichi Pharmaceutical Company, Ltd.; Jack Daniel Distillery; Eisai Company, Ltd.; Glaxo Wellcome Research and Development; Green Cross Corporation; Hoffman La Roche, Incorporated; ICOS Corporation; Juvenile Diabetes Foundation International; March of Dimes Birth Defects Foundation; Mead-Johnson/Bristol-Myers Squibb Company; Merck and Company Incorporated; Monsanto; National Institutes of Health (NIH)/National Cancer Institute; NIH/National Institute for Environmental Health Sciences; NIH/National Institute of Diabetes and Digestive and Kidney Diseases; NIH/National Institute of General Medical Sciences; Otsuka Pharmaceutical Company, Ltd., Osaka; Otsuka Pharmaceutical Company, Ltd., Tokushima; Oxford Biomedical Research, Incorporated; Parke-Davis Pharmaceutical Research, Warner Lambert Company; Peptides International, Incorporated; Peterbilt Motors Company; Pfizer Limited, UK; Pfizer Incorporated, USA; Pierce Chemical Company; Pioneer Hi-Bred International, Incorporated; Procter and Gamble Pharmaceuticals; Rhône-Poulenc Rorer Foundation; Sankyo Company, Ltd.; Sanofi Winthrop, Incorporated; Schering-Plough Research Institute; G. D. Searle and Company; Shionogi Research Laboratories; Shoney's Incorporated; SmithKline Beecham, UK; SmithKline Beecham Pharmaceuticals, USA; Solvay Duphar, B.V.; South Central Bell; Takeda Chemical Industries, Ltd.; Tanabe Seiyaku Company, Ltd.; Teijin Institute for Biomedical Research; Terrapin Technologies, Incorporated; Council for Tobacco Research-U.S.A., Incorporated; Upstate Biotechnology Incorporated/Argonex; Vanderbilt University Cancer Center; Carl Zeiss, Incorporated; Zeneca Pharmaceuticals.

Exhibitors

Alexis Corporation; Amersham Life Science Incorporated; BioLog Life Science Institute; BIOMOL Research Laboratories, Inc.; Calbiochem-Novabiochem International; DNA Proscan Incorporated; DuPont New England Nuclear Products; Elsevier Science Incorporated; Transduction Laboratories, Incorporated.

Signal Transduction in Health and Disease,
Advances in Second Messenger and Phosphoprotein
Research, Vol. 31, edited by J. Corbin and S. Francis.
Lippincott–Raven Publishers, Philadelphia © 1997.

1

Role of Phosphoinositide-3-OH Kinase in Ras Signaling

Julian Downward

Signal Transduction Laboratory, Imperial Cancer Research Fund, London WC2A 3PX, England

Ras proteins are protooncogene products that are critical components of signaling pathways leading from cell-surface receptors to the control of cellular proliferation, morphology, and differentiation. The ability of Ras to activate the mitogen activated protein (MAP) kinase pathway through interaction with the serine/threonine kinase Raf is now well established. However, recent work has shown that Ras can also interact directly with a number of other potential effectors, including the catalytic subunit of phosphoinositide-3-OH kinase, and is involved in control of this lipid kinase in intact cells. A model is presented in which both tyrosine phosphoprotein interaction with the regulatory p85 subunit and Ras.guanosine triphosphate (GTP) interaction with the catalytic p110 subunit is required for optimal activation of phosphoinositide-3-OH kinase in response to extracellular stimuli. The ability of Ras to regulate phosphoinositide-3-OH kinase may be important in both Ras control of cellular morphology through the actin cytoskeleton and in Ras control of deoxyribonucleic acid (DNA) synthesis.

REGULATION OF PHOSPHOINOSITIDE-3-OH KINASE

Phosphoinositide-3-OH kinase is stimulated in response to treatment of cells with a very wide variety of stimuli. It is subject to regulation by a number of known mechanisms, the best characterized of which is the interaction of the p85 regulatory subunit bound to p110α and β with tyrosine-phosphorylated sequences in other proteins. This interaction leads to a modest stimulation of lipid kinase activity. The interaction of p85 with tyrosine phosphoproteins such as growth-factor receptors may contribute to increased phosphoinositide-3-OH kinase activity within the cell through a combination of translocation of the enzyme to the plasma membrane, where its substrate lipid is located, and allosteric regulation of the kinase activity, possibly by alteration of the relative orientation of the two SH2 domains.

In addition, it is now clear that there are two proline-rich motifs in p85 that are capable of binding to SH3 domains both in other proteins ($p145^{abl}$, $p56^{lck}$, $p59^{fyn}$, $p55^{blk}$ and $p60^{src}$) and in p85 itself (1–5). *In vitro* work suggests that the interaction of p85 with the SH3 domain of $p56^{lck}$ may result in activation of the lipid kinase activity of phosphoinositide-3-OH kinase (6). It is possible that normal regulation of phosphoinositide 3-OH kinase involves competition between exogenous SH3 domains and the SH3 domain of p85 itself for binding to the proline-rich motifs in p85.

Another possible regulatory mechanism for phosphoinositide-3-OH kinase may involve the Bcr-like domain of p85. Proteins with homology to this sequence are known to act as guanosine triphosphatase (GTPase)-activating proteins for members of the Rho family of Ras-related proteins, which includes Rho, Rac, and CDC42 (7). It has not been possible to demonstrate that the Bcr region of p85 acts as a GTPase-activating protein. However, Rho does appear to be involved, either directly or indirectly, in the regulation of phosphoinositide-3-OH kinase in platelets (8). Furthermore, *in vitro* evidence has been obtained for an interaction of the p85 subunit of phosphoinositide-3-OH kinase with guanosine triphosphate (GTP)-bound Rac and CDC42, with subsequent stimulation of lipid kinase activity (9). Because other investigators have been unsuccessful in finding interactions of Rho family proteins with phosphoinositide-3-OH kinase either *in vitro* or *in vivo* (10), the significance of such mechanisms in the regulation of phosphoinositide-3-OH kinase remains uncertain.

The family of phosphoinositide-3-OH kinases consists of many members other than those such as p110α and β that bind to p85 regulatory subunits. Other members of the family are likely to be regulated in very different ways, since they are not associated with subunits containing SH2, SH3, SH3-binding, and Bcr-like domains. The best characterized control mechanism operating independently of p85 is the regulation of certain phosphoinositide-3-OH kinases by the βγ subunits of heterotrimeric GTP-binding proteins (11), although this may also be capable of acting on some p85-associated phosphoinositide-3-OH kinases (12). It therefore appears that several different regulatory mechanisms can act simultaneously on these lipid kinases to control their activity in response to extracellular signals. This chapter will consider in more detail whether the small GTP-binding protein Ras is also involved directly in the control of phosphoinositide-3-OH kinase, and the possible significance of such an interaction in the function of Ras in the control of cellular behavior.

RAS PROTEINS AND THEIR EFFECTORS

The three protooncogenes H-, K-, and N-*ras* encode very closely related 21,000-kDa monomeric GTP-binding proteins. These proteins are the prototypic members of the large Ras superfamily of low-molecular-weight GTPases (13). The superfamily is made up of a number of families, the best characterized of which is the Ras family, whose members play a role in the control of cellular proliferation; the Rho family, whose members are involved in regulation of the cytoskeleton, among other functions; and the Rab family, which is involved in the regulation of intracellular membrane trafficking. All of these proteins bind to GTP and catalyze its hydrolysis to guanosine diphosphate (GDP). When in the GTP-bound state, they are biologically active, whereas in the GDP-bound conformation they are inactive. The nucleotide binding of Ras proteins is controlled by two sets of proteins, the GTPase-activating proteins (GAPs), which stimulate the rate of hydrolysis of GTP bound to Ras, thus switching off the Ras proteins, and the guanine nucleotide exchange factors, which stimulate the rate at which nucleotide exchanges on and off Ras. Since the predominant guanine nucleotide in the cytosol is GTP, exchange factors act within the cell to activate Ras proteins.

The activation state of Ras itself is regulated in response to a very wide variety of extracellular stimuli. Most factors that interact with cell-surface receptors of the tyrosine kinase or seven membrane-spanning families have been shown to be capable of activating Ras in at least some cell types. In addition, a great many hematologic cytokines activate Ras. The list of factors that will activate Ras is remarkably similar to those that will activate phosphoinositide-3-OH kinase (14). Over the past few years, enormous advances have been made in understanding the mechanisms involved in the regulation of Ras, particularly the formation of complexes of tyrosine-phosphorylated proteins, such as the epidermal growth factor (EGF) receptor or Shc, with SH2- and SH3-domain-containing adaptor proteins, such as Grb2, and guanine-nucleotide-exchange factors, such as Sos (15).

Once in the active, GTP-bound state, Ras influences cell behavior by interacting with target proteins, generally known as effectors. The best characterized effectors for Ras are the serine/threonine kinases of the Raf family, A-Raf, B-Raf, and c-Raf-1 (16). These kinases interact with Ras.GTP (but not Ras.GDP) *in vitro* (17–20), in the yeast two-hybrid system (17,19,21), and in intact cells (22,23). The interaction occurs through the regulatory CR1 region in the amino-terminal part of Raf. As a result of this interaction, Raf is brought to the plasma membrane, where it is localized as a result of its posttranslational modification with farnesyl and palmityl moieties (24,25). Although the interaction of Ras.GTP with Raf does not appear to directly cause activation of its kinase activity, this interaction does appear to be responsible for presenting Raf to another signaling component, probably localized in the plasma membrane, which is able to further modify it, resulting in increased kinase activity. Once Raf has been activated, it phosphorylates and activates MEK, the MAP kinase kinase, which in turn phosphorylates and activates MAP kinase. Through this kinase cascade, Ras is able to activate the MAP kinase pathway, leading to phosphorylation of a number of transcription factors and hence to stimulation of the expression of immediate early genes (26).

Many of the effects of Ras on the cell may be mediated through the MAP kinase pathway. However, it is clear that the induction of cellular proliferation caused by expression of activated mutants of MEK in fibroblasts is distinct from that caused by expression of activated mutants of Ras (27). The MAP kinase pathway may be capable of causing many of the transcriptional effects of Ras and its stimulation of DNA synthesis, but it does not induce the profound changes in cell morphology that Ras does. This raises the likelihood of there being other effectors of Ras in addition to Raf. It has long been considered a possibility that the GTPase-activating proteins $p120^{GAP}$ and neurofibromin might act as Ras effectors, since they interact only with the GTP- and not with the GDP-bound form of Ras. Some evidence exists to support this hypothesis for $p120^{GAP}$ in the case of Ras-induced germinal-vesicle breakdown in *Xenopus* oocytes (28); however, the ability of fibroblasts from mouse embryos that lack neurofibromin or $p120^{GAP}$ due to homologous recombination to grow apparently normally in culture argues against a major role for these proteins downstream of Ras. In contrast, blocking Ras function in most cell types, through the use of dominant negative mutants of Ras or anti-Ras antibodies, results in arrest of cell growth.

Another family of proteins that may have Ras effector function was identified in the yeast two-hybrid screen (29). This family consists of two closely related members that interact with Ras.GTP but not with Ras.GDP. One of these proteins is identical to a previously characterized protein known as RalGDS, which was originally identified as a guanine-nucleotide-exchange factor (GDS stands for *GDP dissociation stimulator*) for the Ras family member Ral. The function of Ral is unknown. Despite having about 50% identity with Ras, it cannot transform cells. RalGDS does not have exchange activity toward Ras; in fact, Ras does not interact with the exchange-factor domain (CDC25-like) of RalGDS, but rather with the amino-terminal regulatory region. One might imagine that activated Ras controls the guanine-nucleotide-exchange rate of Ral through RalGDS; however, it has not yet been possible to show such regulation. The role of RalGDS and RalGDS-related protein in the effects of Ras on cellular function remains entirely unknown.

The other family of proteins that are candidate effectors for Ras are the phosphoinositide-3-OH kinases. The evidence for believing that Ras effects its signals at least in part through phosphoinositide-3-OH kinase is discussed below.

Evidence that Ras Regulates Phosphoinositide 3-OH Kinase

The first indication that Ras might interact with phosphoinositide 3-OH kinase came from the observation that some phosphoinositide 3-OH kinase activity could be found in immunoprecipitates of Ras from Ras-transformed cells (30). At the time, the significance of this observation was questioned, because the monoclonal antibody used to immunoprecipitate Ras was Y13-259, which blocks the biologic activity of Ras by binding close to the region that undergoes a conformational change upon GTP hydrolysis, and which is implicated as the site of interaction of Ras with biologic targets from mutational analysis (13). The antibody would therefore not be expected to allow coimmunoprecipitation of Ras with an effector.

These observations prompted study in my laboratory of the possible interaction of Ras with phosphoinositide 3-OH kinase (10). Ras proteins were expressed using a baculovirus system, and were purified to homogeneity. Phosphoinositide-3-OH kinase p85α or β and p110α subunits were expressed in a similar manner. Untagged p85 and p110 were purified, and a form in which the p110 subunit was glutathione-S-transferase (GST)-tagged to allow easier purification. It was found that purified Ras immobilized on agarose beads would bind to p85 and p110 when it was GTP-bound but not when it was GDP-bound; this interaction could be observed either by measurement of lipid kinase activity or by immunoblotting for p85. The interaction could be shown to be direct and not to involve other proteins. This indicated that phosphoinositide 3-OH kinase might act as a direct effector of Ras, a hypothesis that was further supported by the observation that mutations in the "effector loop" of Ras, which are known to destroy its biologic activity, prevented it from interacting with p85 and/or p110. The interaction between Ras and phosphoinositide 3-OH kinase was found to be very heavily, though possibly incompletely, inhibited by the antibody Y13-259, but not by the nonneutralizing antibody Y13-238. It is possible that the original observation of phosphoinositide 3-OH kinase activity in Y13-259 immunoprecipitates was due to heterogeneity in the preparations of antibody used in different laboratories.

Although Ras interacted well with phosphoinositide 3-OH kinase, we could find no evidence for association of any Rho family proteins with p85 and/or p110. However, the closely related Ras-family members Rap1a, Rap1b, R-Ras, and TC21 all interacted with phosphoinositide 3-OH kinase, although Rap2 and Ral did not. A similar pattern of binding is observed for Raf, although the significance of these Ras-related proteins is not well understood. The interaction only of Ras and not Rho proteins with phosphoinositide 3-OH kinase suggested that the interaction did not occur through the Bcr-like domain in p85. When binding of Ras to p85 or p110 was assessed separately, it was found that Ras interacted with p110, the catalytic subunit of phosphoinositide 3-OH kinase, but not with p85, the regulatory subunit. It therefore seems likely that possible regulation of phosphoinositide 3-OH kinase by Ras would occur through a different subunit than would regulation by tyrosine phosphoproteins.

The convincing interaction of Ras with the catalytic subunit of phosphoinositide 3-OH kinase in a strictly GTP-dependent manner *in vitro* suggested that Ras might be controlling phosphoinositide-3-OH kinase activity in intact cells. To investigate this possibility, expression plasmids coding for activated mutants of Ras were transfected into COS cells. The levels of 3′-phosphorylated phosphoinositides were measured in the cells 48 hours after transfection. Activated Ras was found to cause a large increase in the amount of phosphatidylinositol-3,4-bisphosphate (PI[3,4]P_2) and phosphatidylinositol-3,4,5-trisphosphate (PI[3,4,5]P_3) in the cells. A similar increase was observed when the catalytic and regulatory subunits of phosphoinositide-3-OH kinase were transfected into the cells. When Ras and p85α and/or p110α were transfected together into COS cells, there was a synergistic increase in PI(3,4,5)P_3 levels. Effector mutants of Ras were inactive. Interestingly, activated mutants of Raf were also inactive in this assay, suggesting that the effect of Ras on phosphoinositide-3-OH kinase activity was the result of a bifurcation in Ras effector pathways. Rho-family proteins were inactive in this assay.

It therefore appears that Ras is capable of stimulating phosphoinositide-3-OH kinase activity in intact cells, at least when it is overexpressed. However, there remains the question of the importance of the function of endogenous Ras in the activation of phosphoinositide-3-OH kinase in response to growth-factor treatment of cells. To investigate this, PC12 rat pheochromocytoma cells were created that expressed a dominant negative mutant of Ras, N17. Unlike most untransformed cells in culture, PC12 cells do not require Ras function for continued proliferation, but instead differentiate into neuron-like cells in response to activated Ras. Normally, treatment of PC12 cells with nerve growth factor (NGF) or EGF results in strong activation of phosphoinositide-3-OH kinase activity, as can be seen from measuring the levels of 3′-phosphorylated phosphoinositides produced by these cells (31). When PC12 cells are expressing dominant negative Ras, the ability of NGF and EGF to increase levels of these lipids is greatly reduced, to about 25% of the levels achieved in the absence of N17 Ras. These data indicate that Ras contributes significantly to the activation of phosphoinositide-3-OH kinase activity in response to growth-factor activation of receptor tyrosine kinases. It is likely that interaction of Ras.GTP with the catalytic subunit of phosphoinositide-3-OH cooperates with tyrosine phosphoprotein interaction with the regulatory subunit of the enzyme to give optimal activation of phosphoinositide 3-OH kinase activity.

What other data exist to support the hypothesis that phosphoinositide-3-OH kinase acts as an effector of Ras? It has been known for some time that the amount of phosphoinositide-3-OH kinase found in antiphosphotyrosine immunoprecipitates following cellular stimulation does not always correlate well with the levels of $PI(3,4)P_2$ and $PI(3,4,5)P_3$ in the cells. For example, when mast cells were incubated with a range of cytokines, including interleukins-3, -4, and -5 (IL-3, -4, and -5), granulocyte-macrophage colony stimulating factor (GM-CSF), and Steel factor, it was found that IL-4 caused by far the weakest increase in 3′-phosphorylated lipid levels, but the largest increase in phosphoinositide-3-OH kinase activity associated with antiphosphotyrosine immunoprecipitates, indicating that the *in vitro* assay is a poor reflector of the *in vivo* activation state of the lipid kinase (32). This is similar to what was found in the N17 Ras-expressing PC12 cells, in which 3′-phosphorylated lipid levels were greatly reduced without affecting the amount of phosphoinositide-3-OH kinase activity in antiphosphotyrosine immunoprecipitates. Interestingly, IL-4 is incapable of activating Ras in mast cells (33), whereas the other cytokines do activate it, suggesting that IL-4 is a poor activator of phosphoinositide-3-OH kinase activity *in vivo* because it activates only one of the two pathways leading to phosphoinositide-3-OH kinase, namely tyrosine phosphoprotein interaction with p85.

Another related case is that of the polyoma virus middle-T antigen, which is phosphorylated at multiple tyrosine residues by $p60^{c\text{-}src}$. A nontransforming mutant of middle-T does not bind to Shc/Grb2 and therefore does not activate guanine nucleotide exchange on Ras (34,35). This mutant binds the same amount of phosphoinositide-3-OH kinase as the nonmutated antigen through its tyrosine-phosphorylated p85 binding site; however, it does not cause elevation of 3′-phosphorylated phosphoinositides in whole cells (36). This may be due to failure of the mutant antigen to activate both phosphotyrosine-mediated p85 binding and Ras.GTP interaction with p110 at the same time.

One other system has been reported in which there is good evidence that Ras controls the activity of phosphoinositide-3-OH kinase. This is an artificially created strain of the yeast *Schizosaccharomyces pombe* that expresses mammalian p110 on an inducible promoter. Normally, yeast contains only phosphatidylinositol-3-phosphate, but not more heavily phosphorylated 3′-phosphoinositides. On expression of p110, $PI(3,4)P_2$ and $PI(3,4,5)P_3$ are found in the cells, and their growth is profoundly inhibited (38). When p85 is also expressed in the same cells, the amount of $PI(3,4)P_2$ and $PI(3,4,5)P_3$ is greatly reduced and the inhibition of cell growth is relieved. This suggests that the basal function of p85 is to inhibit the activity of p110; presumably, binding to tyrosine phosphoproteins normally overcomes this inhibition. When activated mutant Ras, but not effector mutants, are put into yeast expressing p85 and p110, growth is once again inhibited and cellular levels of $PI(3,4)P_2$ and $PI(3,4,5)P_3$ are increased. Thus, activated Ras opposes the effect of unstimulated p85 on the activity of the catalytic subunit of phosphoinositide-3-OH kinase.

POSSIBLE ROLES OF PHOSPHOINOSITIDE-3-OH KINASE IN MEDIATING RAS EFFECTS ON THE CELL

As discussed above, there is good evidence that the interaction of Ras.GTP with p110 is one of a number of pathways involved in regulating the activity of phosphoinositide-3-OH kinase. It is likely that this pathway synergizes with the interaction of tyrosine phosphoproteins with p85, and that both inputs are required to achieve good activation of lipid kinase activity in response to growth factors and other stimuli. Several forms of phosphoinositide-3-OH kinase do not interact with p85 regulatory subunits. Some of these enzymes do, however, interact with Ras.GTP (J. Downward and R. Wetzker, unpublished observations); in these cases Ras may synergize with other signaling pathways. The question remains, however, of how important phosphoinositide-3-OH kinase activity is to the effects of Ras on cellular function. Although little is now definitively known on this topic, the final section of this review provides some speculations about it.

The importance of the Raf/MAP kinase pathway in mediating the effects of Ras on activation of gene transcription is now well established. However, is the activation of this pathway alone sufficient to produce cellular proliferation, or, in cases of constitutive activation, transformation? Clearly, activated Raf mutants are potent oncogenes (16), and activated mutants of MEK can induce proliferation of NIH 3T3 fibroblasts (27). However, it is possible that some of these effects are secondary to the activation of expression of genes encoding autocrine growth factors, such as transforming growth factor-α (TGF-α). In the case of activated MEK mutants, morphologic transformation is inhibited by neutralizing anti-Ras antibodies, suggesting that the cytoskeletal changes accompanying transformation are caused by factors that the cells secreted and which act through receptors coupled to Ras along pathways distinct from the Raf/MAP kinase pathway. Of the possible alternative effectors to Ras other than Raf, phosphoinositide-3-OH kinase is probably the best candidate to be involved in inducing cytoskeletal changes.

Activated Ras has been known for 10 yr to be able to induce changes in cellular morphology: when injected into fibroblasts it causes membrane ruffling within 30 min (39). The relatively long period required for this effect is probably due to the need for microinjected Ras to become posttranslationally modified after it enters the cells. Similar effects are seen when activated mutants of the Ras-related protein Rac are microinjected into Swiss 3T3 cells (40). Dominant negative mutants of Rac inhibit the effects of Ras, suggesting that Ras acts upstream of Rac in a pathway that controls the organization of the cortical actin cytoskeleton. It is known that platelet-derived growth factor (PDGF) induces membrane ruffling in a manner that is sensitive to the phosphoinositide-3-OH kinase inhibitor wortmannin; this effect is also blocked by mutations that remove the p85-binding phosphotyrosine residues from the PDGF receptor (Y740 and 751), and by the overexpression of a mutant p85 that does not interact with p110 (41,42). Thus, both Rac and phosphoinositide-3-OH kinase lie upstream of actin rearrangement. It is possible that Ras partly mediates the activation of phosphoinositide-3-OH kinase in response to growth factors, which in turn induces the activation of Rac, which in its turn induces effects on the actin cytoskeleton. Since it is not possible to activate phosphoinositide-3-OH kinase by transfecting activated Rac into COS cells (10), it is likely that Rac acts downstream of the lipid kinase products.

The effect of inhibiting Ras function on the ability of growth factors to induce membrane ruffling is not currently clear. The ability of dominant negative Ras mutants to inhibit short-term effects of growth factors on a number of pathways is not straightforward: N17 Ras has no effect on the ability of EGF to activate MAP kinase, presumably because there are at least two alternative pathways to MAP kinase activation from the EGF receptor, in addition to that involving Ras (43). Similar redundancy may exist in the control of phosphoinositide-3-OH kinase and the cytoskeleton. We have, however, found that the effects of Ras on membrane ruffling are sensitive to the phosphoinositide-3-OH kinase inhibitor wortmannin (P. Rodriguez-Viciana and J. Downward, unpublished observations). It is very likely that Ras mediates a major component of the effects of growth factors on cell morphology, and that phosphoinositide-3-OH kinase mediates a large proportion of the effects of Ras on cell morphology.

Through the use of inhibitors such as wortmannin, phosphoinositide-3-OH kinase has been implicated in the regulation of a number of other cellular activities, including glucose transport and nicotinamide adenine dinucleotide phosphate (NADP)-oxidase function in neutrophils. The possible importance of these specialized functions to the cellular effects of Ras is not clear. A more general effect of inhibiting phosphoinositide-3-OH kinase function has been found with the microinjection of antibodies against p110 into fibroblasts: DNA synthesis in response to a number of growth factors is inhibited (44). Although the pathway that might be involved in this is not immediately obvious, it may reflect the regulation of certain protein kinase C (PKC) isotypes by phosphatidylinositol-3,4,5-trisphosphate (45). It has been known for some time that the ability of activated Ras to induce DNA synthesis in fibroblasts depends on the presence of functional PKC (46), whereas Ras-induced morphologic changes are independent of PKC. It is conceivable that the PKC-dependent pathway leading from Ras to DNA synthesis involves phosphoinositide-3-OH kinase-mediated PKC activation, whereas the PKC-independent pathway from Ras to morphologic transformation involves targets of phosphoinositide-3-OH kinase other than PKC.

REFERENCES

1. Prasad KVS, Kapeller R, Janssen O, et al. Phosphatidylinositol (PI) 3-kinase and PI 4-kinase binding to the CD4-p56^{lck} complex: the p56^{lck} SH3 binds to PI 3-kinase but not PI 4-kinase. *Mol Cell Biol* 1993;13:7708–17.
2. Prasad KVS, Janssen O, Kapeller R, Cantley LC, Rudd CE. The p59fyn SH3 domain mediates binding to phosphatidylinositol 3-kinase in T cells. *Proc Natl Acad Sci USA* 1993;90:7366–70.
3. Pleiman CM, Clark MR, Timson Gauen LK, et al. Mapping of sites on the src family protein tyrosine kinases p55^{blk}, p59^{fyn} and p56^{lyn} which interact with the effector molecules phospholipase c-γ2, MAP kinase, GAP and phosphatidylinositol 3-kinase. *Mol Cell Biol* 1993;13:5877–87.
4. Vogel LB, Fujita DJ. The SH3 domain of p56^{lck} is involved in binding to phosphatidylinositol 3′-kinase from T lymphocytes. *Mol Cell Biol* 1993;13:7408–17.
5. Kapeller R, Prasad KVS, Janssen O, et al. Identification of two SH3-binding motifs in the regulatory subunit of phosphatidylinositol 3-kinase. *J Biol Chem* 1994;269:1927–33.
6. Pleiman CM, Hertz WM, Cambier JC. Activation of phosphatidylinositol 3-kinase by src family kinase SH3 binding to the p85 subunit. *Science* 1994;263:1609–12.
7. Diekman D, Brill S, Garrett M, et al. Bcr encodes a GTPase activating protein for p21rac. *Nature* 1991;351:400–2.
8. Zhang J, King WG, Dillon S, Hall A, Feig L, Rittenhouse S. Activation of platelet phosphatidylinositol 3-kinase requires the small GTP binding protein Rho. *J Biol Chem* 1993;268:22251–4.
9. Zheng Y, Bagrodia S, Cerione RA. Activation of phosphoinositide 3-kinase activity by Cdc42Hs binding to p85. *J Biol Chem* 1994;269:18727–30.
10. Rodriguez-Viciana P, Warne PH, Dhand R, et al. Phosphatidylinositol-3-OH kinase as a direct target of Ras. *Nature* 1994;370:527–32.
11. Stephens L, Smrcka A, Cooke FT, Jackson TR, Sternweis PC, Hawkins PT. A novel phosphoinositide 3 kinase activity in myeloid-derived cells is activated by G protein βγ subunits. *Cell* 1994; 77:83–93.
12. Thomason PA, James SR, Casey PJ, Downes CP. A G-protein beta gamma-subunit-responsive phosphoinositide 3-kinase activity in human platelet cytosol. *J Biol Chem* 1994;269:16525–8.
13. Lowy DR, Willumsen BM. Function and regulation of RAS. *Annu Rev Biochem* 1993;62:851–91.
14. Stephens LR, Jackson TR, Hawkins PT. Agonist-stimulated synthesis of phosphatidylinositol-3,4,5-trisphosphate: a new intracellular signalling system. *Biochim Biophys Acta* 1993;1179:27–75.
15. McCormick F. Signal transduction: how receptors turn Ras on. *Nature* 1993;363:15–6.
16. Rapp UR. The Raf serine/threonine kinases. *Oncogene* 1991;6:495–500.
17. Vojtek AB, Hollenberg SM, Cooper JA. Mammalian Ras interacts directly with the serine/threonine kinase Raf. *Cell* 1993;74:205–14.
18. Warne PH, Vicinia PR, Downward J. Direct interaction of Ras and the amino-terminal region of Raf-1 *in vitro*. *Nature* 1993;364:352–5.
19. Zhang X-F, Settleman J, Kyriakis JM, et al. Normal and Oncogenic p21ras proteins bind to the amino-terminal regulatory domain of c-Raf-1. *Nature* 1993;364:308–13.
20. Moodie SA, Willumsen BM, Weber MJ, Wolfman A. Complexes of Ras.GTP with Raf-1 and mitogen activated protein kinase kinase. *Science* 1993;260:1658–61.
21 Van Aelst L, Barr M, Marcus S, Polverino A, Wigler M. Complex formation between Ras and Raf and other protein kinases. *Proc Natl Acad Sci USA* 1993;90:6213–7.
22. Finney RE, Robbins SM, Bishop JM. Association of pRas and pRaf-1 in a complex correlates with activation of a signal transduction pathway. *Curr Biol* 1993;3:805–12.
23. Hallberg B, Rayter SI, Downward J. Interaction of Ras and Raf in intact mammalian cells upon extracellular stimulation. *J Biol Chem* 1994;270:3913–6.
24. Stokoe D, MacDonald SG, Cadwallader K, Symons M, Hancock JF. Activation of raf as a result of recruitment to the plasma-membrane. *Science* 1994;264:1463–7.
25. Leevers SJ, Paterson HF, C.J. M. Requirement for ras in raf activation is overcome by targeting raf to the plasma-membrane. *Nature* 1994;369:411–4.
26. Egan SE, Weinberg RA. The pathway to signal achievement. *Nature* 1993;365:781–3.
27. Cowley S, Paterson H, Kemp P, Marshall CJ. Activation of map kinase kinase is necessary and sufficient for pc12 differentiation and for transformation of NIH 3T3 cells. *Cell* 1994;77:841–52.
28. Duchesne M, Schweighoffer F, Parker F, et al. Identification of the SH3 domain of GAP as an essential sequence for Ras-GAP mediated signaling. *Science* 1993;259:525–8.
29. Albright CF, Giddings BW, Liu J, Vito M, Weinberg RA. Characterization of a guanine nucleotide dissociation stimulator for a ras-related GTPase. *EMBO J* 1993;12:339–47.

30. Sjölander A, Yamamoto K, Huber BE, Lapetina EC. Association of p21^{ras} with phosphatidylinositol 3-kinase. *Proc Natl Acad Sci USA* 1991;88:7908–12.
31. Carter AN, Downes CP. Phosphatidylinositol 3-kinase is activated by nerve growth factor and epidermal growth factor in PC12 cells. *J Biol Chem* 1992;267:14563–7.
32. Gold MR, Duronio V, Saxena SP, Schrader JW, Aebersold R. Multiple cytokines activate phosphatidylinositol 3-kinase in hemopoietic cells. Association of the enzyme with various tyrosine phosphorylated proteins. *J Biol Chem* 1994;269:5403–12.
33. Duronio V, Welham MJ, Abraham S, Dryden P, Schrader JW. p21ras activation via hemopoietin receptors and c-kit requires tyrosine kinase activity but not tyrosine phosphorylation of GAP. *Proc Natl Acad Sci USA* 1992;89:1587–91.
34. Campbell KS, Ogris E, Burke B, et al. Polyoma middle tumor antigen interacts with SHC protein via the NPTY (Asn-Pro-Thr-Tyr) motif in middle tumor antigen. *Proc Natl Acad Sci USA* 1994;91:6344-8.
35. Dilworth SM, Brewster CE, Jones MD, Lanfrancone L, Pelicci G, Pelicci PG. Transformation by polyoma virus middle T-antigen involves the binding and tyrosine phosphorylation of Shc. *Nature* 1994;367:87–90.
36. Ling LE, Druker BJ, Cantley LC, Roberts TM. Transformation-defective mutants of polyomavirus middle T antigen associate with phosphatidylinositol 3-kinase (PI 3-kinase) but are unable to maintain wild-type levels of PI 3-kinase products in intact cells. *J Virol* 1992;66:1702–8.
37. Stephens RM, Loeb DM, Copeland T, Pawson T, Greene LA, Kaplan DR. Trk receptors use redundant signal transduction pathways involving Shc and PLCγ1 to mediate NGF responses. *Neuron* 1994;12:691–705.
38. Kodaki T, Wosholski R, Hallberg B, Rodiriguez-Viciana P, Downward J, Parker PJ. The activation of phosphatidylinositol 3-kinase by Ras. *Curr Biol* 1994;4:798–806.
39. Bar Sagi D, Feramisco JR. Induction of membrane ruffling and fluid-phase pinocytosis in quiescent fibroblasts by ras proteins. *Science* 1986;233:1061–8.
40. Ridley A, Paterson HF, Johnston CL, Diekmann D, Hall A. The small GTP-binding protein rac regulates growth factor-induced membrane ruffling. *Cell* 1992;70:401–10.
41. Kotani K, Yonezawa K, Hara K, et al. Involvement of phosphoinositide 3-kinase in insulin or IGF-1 induced membrane ruffling. *EMBO J* 1994;13:2313–21.
42. Wennström S, Hawkins P, Cooke F, et al. Activation of phosphoinositide 3-kinase is required for PDGF-stimulated membrane ruffling. *Curr Biol* 1994;4:385–93.
43. Burgering BMT, de Vries-Smits AMM, Medema RH, van Weeren PC, Tertoolen LGJ, Bos JL. Epidermal growth factor induces phosphorylation of extracellular signal-regulated kinase 2 via multiple .pathways. *Mol Cell Biol* 1993;13:7248–56.
44 Roche S, Koegl M, Courtneidge SA. The phosphatidylinositol 3-kinase α is required for DNA synthesis induced by some, but not all, growth factors. *Proc Natl Acad Sci USA* 1994;91:9185–9.
45. Toker A, Meyer M, Reddy KK, et al. Activation of protein kinase C family members by the novel phosphoinositides PI-3,4-P2 and PI-3,4,5-P3. *J Biol Chem* 1994;269:32358–67.
46. Morris JD, Price B, Lloyd AC, Self AJ, Marshall CJ, Hall A. Scrape-loading of Swiss 3T3 cells with ras protein rapidly activates protein kinase C in the absence of phosphoinositide hydrolysis. *Oncogene* 1989;4:27–31.

Signal Transduction in Health and Disease, Advances in Second Messenger and Phosphoprotein Research, Vol. 31, edited by J. Corbin and S. Francis.
Lippincott–Raven Publishers, Philadelphia © 1997.

2

From Phosphorylase to Phosphorylase Kinase

Louise N. Johnson, David Barford, David J. Owen, Martin E. M. Noble, and Elspeth F. Garman

Laboratory of Molecular Biophysics, University of Oxford, Oxford OX1 3QU, England

Control by protein phosphorylation almost always implies significant conformational change in the target protein and the creation of phosphate recognition sites. Tight phosphate-binding sites most commonly involve interactions with the guanidinium groups of arginine residues and/or interactions with the peptide nitrogens of residues at the amino-termini of α-helices. Structural studies of glycogen phosphorylase have shown that conversion of the inactive to the active protein on phosphorylation results in a shift by about 50 Å of the serine (Ser^{14}) after its phosphorylation, coupled with substantial tertiary and quaternary structural changes. The Ser^{14}-phosphate contacts two arginine residues. The main incentive for the shift in the serine on phosphorylation appears to be electrostatic.

The catalytic subunit core of phosphorylase kinase is a constitutively active kinase that requires no posttranslational modification for activity. Determination of the x-ray crystal structure of the kinase has shown that the arginine, which is adjacent to the catalytic base, is compensated by interaction with a glutamate from the activation segment of the kinase. In many other kinases the activation segment carries a phosphorylatable residue whose phosphorylation is essential for activation. In these kinases there is invariably a charge cluster whose neutralization can best be effected by a dianionic phosphate group. In phosphorylase kinase there is no charge cluster, and the active conformation of the activation segment can be promoted by a carboxylate group.

INTRODUCTION

The control of glycogen metabolism by reversible phosphorylation–dephosphorylation events is one of the best understood pathways of signal transduction. Following the pioneering work of Krebs et al. (1,2), glycogen phosphorylase (GP) was the first protein for which activation by phosphorylation was established, phosphorylase kinase was the first protein kinase to be characterized (1), and the concept of inactivation by protein phosphatases was established (2). The discovery of cyclic adenosine monophosphate (cAMP)-dependent protein kinase in 1968 (3) provided the important link for the hormone-stimulated cascade that uses cAMP as the second messenger. Extracellular signals from hormones such as glucagon and catecholamines such as adrenaline lead to increased intracellular levels of cAMP and the activation of the cAMP-dependent protein kinase (cAPK). cAPK in turn activates phosphorylase kinase. Phosphorylase kinase is also sensitive to calcium, released in response to neuronal signals, and thus signals used to initiate muscle contraction also initiate glycogen degradation. GP, the major physiologic substrate for phosphorylase kinase, is phosphorylated on a single serine residue, Ser^{14}, resulting in the conversion of inactive GPb to active GPa. Coincidentally, the activated cAPK, together with phosphorylase kinase and other kinases such as glycogen synthase kinase-3, casein kinase-1, and casein kinase-2, phosphorylate glycogen synthase (GS) and convert GS from its active, nonphosphorylated state to the inactive phosphorylated state. The signal-transduction pathway promotes activation of glycogen degradation through GP and inhibition of glycogen synthesis by GS. The reversal of these effects is catalyzed by the catalytic subunit of protein phosphatase-1 in complex with a glycogen binding subunit (PP1G). PP1G is controlled by an inhibitor protein which, when phosphorylated by cAPK, strongly inhibits the free catalytic subunit. In the muscle cell, PP1G is activated in response to insulin by a mechanism that involves phosphorylation of its glycogen-binding subunit by the insulin-stimulated protein kinase (ISPK-1/RSK-2) (4,5). Activated PP1G dephosphorylates GP, phosphorylase kinase, and GS, and thus inactivates glycogen degradation and promotes glycogen synthesis. The integrated control mechanisms for phosphorylation and dephosphorylation provide a remarkable example of control processes that are now recognized in many other signaling pathways. The kinases and phosphatases are controlled by concerted mechanisms that allow differential responses to different stimuli.

PROPERTIES CONFERRED BY ADDITION OF A PHOSPHATE GROUP

Phosphorylation on single or multiple sites can result in activation or inhibition of proteins. What special properties may be conferred by a phosphate group that cannot be achieved by the naturally occurring amino acids? Phosphorylation provides a reversible process in which the forward and the backward reactions are catalyzed by different enzymes operating with different specificities. Thus, reactions can be turned on or off in response to different stimuli. The phosphate group (pK ~ 6.7), with four oxygen atoms, can participate in extensive hydrogen-bond interactions, and these can link different parts of the polypeptide chain. In most of these interactions the phosphate oxygens participate as hydrogen-bond acceptors, although, depending on pH and environment, participation of the phosphate group as a donor for one hydrogen bond is possible. Most often the phosphate group is dianionic at physiologic pH. The double negative charge of the group is a property that is not available to the naturally occurring amino acids, and we may expect electrostatic effects to be important in control by phosphorylation.

Analysis of protein–phosphate interactions in existing protein structures (6,7) has shown that the most common interaction is between the phosphate oxygens and the main-chain nitrogens at the start of a helix (8), where the most frequently found residue is glycine. In nonhelix interactions the most commonly found residue is arginine. The guanidinium group is suited for interactions with phosphate by virtue of its planar structure and its ability to form multiple hydrogen bonds. Because of its resonance stabilization, the guanidinium group is a poor proton donor ($pK_a>12$) and cannot function as a general acid catalyst in the hydrolysis of phosphorylated amino acids. Electrostatic interactions between arginine and phosphate groups provide tight binding sites that appear to play a dominant role in recognition and the stabilization of protein conformations. An exception is found in the bacterial enzyme isocitrate dehydrogenase, whose inhibition is achieved by phosphorylation. The phosphate group does not make strong interactions with the protein, but operates through an electrostatic blocking mechanism that prevents binding of the charged substrate isocitrate (9). For this enzyme, there is no change in conformation on phosphorylation.

GLYCOGEN PHOSPHORYLASE

In GP, there is a major change in conformation on phosphorylation that results in a shift of more than 50 Å of the phosphorylated serine residue (Ser^{14}). The structural changes that occur on conversion of inactive GPb to active GPa have been described (10–15) and reviewed (16,17). To a first approximation, the enzyme can be viewed in terms of the Monod–Wyman–Changeux model for control by allosteric effectors (Fig. 1). The protein exists in two conformational states, a T state that is less active and has low affinity for substrate, and an R state that is active and has high affinity for substrate. The equilibrium between the T and R states can be affected either by phosphorylation or by noncovalent association of metabolites such as AMP. The structural results (14) have shown that the active R-state GPa promoted by Ser^{14} phosphorylation and active R-state GPb promoted by binding of AMP are essentially the same. Likewise, the T-state GPb and the T-state GPa induced by glucose have similar quaternary structures and a similar constellation of residues at the catalytic site, although they differ in the region surrounding the site of phosphorylation (10).

In the inactive T state of GPb, N-terminal residues 1 to 10 have not been located. The following residues, numbers 10 to 20, are some of the least-well-defined residues in the large protein structure, but they have been located (11,12) and make suitable contacts against the protein surface within the same subunit (Fig. 2). On activation to GPa there is a conformational change in the region of residues 22 to 23, so that the N-terminal tail rotates through 120° and reaches up to make intersubunit contacts. The Ser^{14} shifts 50 Å. The serine phosphate binds between two arginine residues, $Arg^{43'}$ from the other subunit and Arg^{69} from its own subunit (Fig. 3). Both

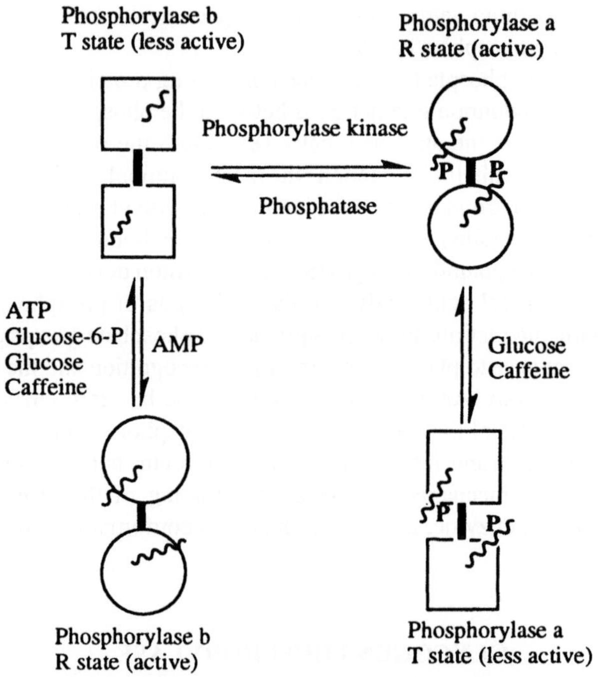

FIG. 1. Schematic representation of the regulation of glycogen phosphorylase between the inactive T-state conformation (*squares*) to the active R-state conformation (*circles*) by either reversible phosphorylation or by noncovalent allosteric effectors.

these arginines shift in order to make these contacts, and these movements are part of a global response in both tertiary and quaternary structure that results in significant changes at the subunit–subunit interface and elsewhere in the GP molecule. In GPb, C-terminal residues 836 to 842 are located in the intersubunit region, and on conversion to GPa these residues are displaced. Thus, phosphorylation results in an ordering of the N-terminal tail and a disordering of the C-terminal tail, together with other major changes.

The site of phosphorylation is located in the sequence at residues 10 to 16 (Arg-Lys-Gln-Ile-SerP-Val-Arg), which contains three basic residues and no acidic residues. This basic character around the site of phosphorylation was recognized in 1959 by Fischer et al. (18), who speculated that "the accumulation of positive charges on this exposed segment of the phosphorylase molecule" could be significant for control. The structural results indicate that, in the first instance, electrostatic effects do appear to dominate in directing the N-terminal tail to its new location (19). The electrostatic potential surface of the GPb molecule, displayed with the computer program GRASP (20), shows that the surface of the enzyme against which the N-terminal tail docks exhibits a negative charge resulting from acidic residues that include Asp^{109}, Glu^{110}, Glu^{120}, Glu^{501}, Glu^{505}, and Glu^{509}. The N-terminal tail is lo-

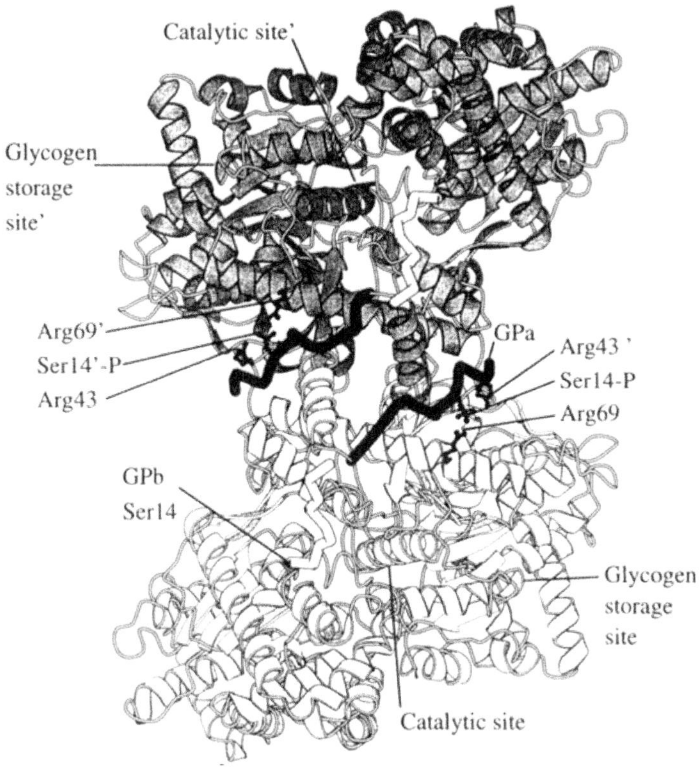

FIG. 2. Schematic diagram of the phosphorylase dimer viewed down the 2-fold axis with the phosphorylation (Ser^{14}-P) toward the viewer. Access to the catalytic site is on the far side of the molecule. One subunit is shown with *light shading* and the other with *darker shading*. The conformation of residues 14 to 23 in the inactive T state of GPb is shown as a *solid white line*. The conformation of residues 10 to 23 in the active R state of GPa is shown as a *dark line*. The Ser^{14}-P and the two arginine residues are indicated. This figure and Figs. 3, 4, and 6 were prepared with XOBJECTS (M. E. M. Noble, unpublished work).

cated with both Ser^{14} and Arg^{16} directed into this acidic site. The high concentration of negative charge makes this a favorable surface against which the basic peptide can dock, but it provides an inhospitable site for a negatively charged phosphate group. In order for the kinase to have access to the serine, a conformational change in the tail is necessary, but this can be accomplished readily, since these residues are already somewhat mobile. Indeed, those ligands that promote changes to the R state (such as AMP) also make GP a better substrate for phosphorylase kinase. The structure of R-state GPb in the presence of AMP showed that the N-terminal tail adopts a similar conformation to that in R-state GPa (15).

In R-state GPa, as a result of the changes in tertiary and quaternary structure, a positively charged region is created by $Arg^{43'}$ and Arg^{69} and by the displacement of the C-terminal tail. The serine phosphate binds into the site between the two

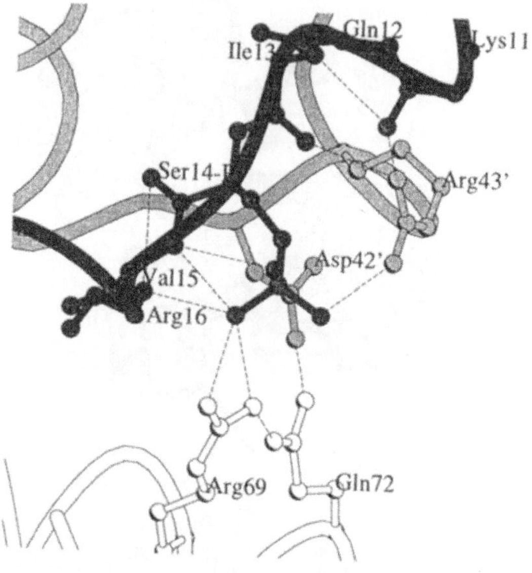

FIG. 3. The contacts of the Ser[14]-P in glycogen phosphorylase. The N-terminal residues are shown in *dark grey*; one subunit is in *white* and the other in *light grey*. The Ser[14]-P phosphate group is hydrogen bonded to two arginine residues, one from its own subunit, Arg[69], and the other from the symmetry related subunit, Arg[43′]. It is also hydrogen bonded to the main-chain nitrogen of Val[15] and Arg[16] at the start of a 3_{10} helix. Each of the arginines is involved in a network of hydrogen bonds: Arg[69] to Gln[72] to Asp[42′] and Arg[43′] to main-chain carbonyl oxygens from Lys[11] and Ile[13].

arginines; the nonpolar residues Ile[13] and Val[15] dock into adjacent nonpolar pockets; Arg[10] contacts the C-terminal end of an α-helix in the other subunit; and Arg[16] is exposed to the solvent. The changes in tertiary and quaternary structure allow events at the phosphorylation site located at the subunit–subunit interface to be communicated to the catalytic site, which is situated more than 30 Å away, in the center of the subunit. The changes have been described elsewhere (14,21). In brief, they result in a displacement of an acidic residue at the catalytic site and its replacement by an arginine residue to create the substrate-phosphate recognition site. This site is adjacent to the 5′-phosphate of the cofactor pyridoxal phosphate. Here again an arginine residue provides contacts to create a phosphate–substrate binding site. In the catalytic mechanism, the cofactor 5′-phosphate acts as a general acid to promote attack by the substrate phosphate on the glycosidic oxygen of the glucosyl substrate (21). The cofactor 5′-phosphate plays a dynamic role in catalysis, in which it switches from a monoanionic to a dianionic state. The 5′-phosphate group is localized by interactions with a lysine residue and main-chain NH groups at the start of an α-helix. There are no strong ionic interactions with arginine residues that would favor the dianionic state.

PHOSPHORYLASE KINASE

Phosphorylase kinase represents one of the largest and most complex of the protein kinases. The enzyme exists as a heterotetramer with stoichiometry $(\alpha\beta\gamma\delta)_4$ and a total molecular weight of approximately 1.3×10^6 Da (reviewed in [22]). A model for the intact holoenzyme has been derived from high-resolution electron microscopy studies (23). The α and β subunits are regulatory and the targets for phosphorylation by cAPK (and by autophosphorylation). The δ subunit is an integral calmodulin subunit and provides the calcium sensitivity. The γ subunit is the catalytic subunit and is composed of an N-terminal kinase domain and a C-terminal regulatory domain. Peptides from two distinct regions in the regulatory domain have been shown to exhibit a high affinity for calmodulin, the most potent being peptides Phk13 (residues 302 to 326; $K_D = 5$ nM) and Phk5 (residues 342 to 366; $K_D = 20$ nM) (24). Both peptides inhibit the Ca^{2+}-activated enzyme, with $K_i = 0.3$ μM and $K_i = 20$ μM, respectively (25,26). These results support the notion that the regulatory domain provides peptide regions that operate through an autoinhibitory mechanism but are able to transduce Ca^{2+}-induced conformational changes in the δ subunit to the catalytic γ subunit to allow activation of the enzyme.

Bacterial expression of the intact γ subunit (27) resulted in a product that was insoluble and did not refold. A stop codon was introduced in the gene between the kinase and the regulatory domains, and residues 1 to 298 were expressed. This fragment (termed here PhK) was purified, shown to have very similar enzyme kinetics to the activated holoenzyme (27), and crystallized (28). The enzyme represents a constitutively active protein kinase that requires no posttranslational modification for activity. All the regulatory machinery for this particular kinase is located outside the kinase core domain. The binary structures of this kinase core of the catalytic γ subunit, in complex with Mn^{2+}/AMPPNP (a nonhydrolyzable analogue of ATP) and with the product Mg^{2+}/ADP, have been solved at 2.6 Å and at 3.0 Å resolution, respectively (29).

Structure

The kinase core domain PhK, residues 1 to 298, represents a minimal kinase structure. The structure is composed of two lobes: an N-terminal lobe composed mostly of β-sheet and a C-terminal lobe that is mostly α-helix (Fig. 4). The structure is similar to that of the kinase core of cAPK (30,31), and the root-mean-square difference in Cα coordinates between the two structures is 1.3 Å, a value that is typical for two proteins that share 33% sequence identity. There are some differences in the relative orientations of the G and H helices in the C-terminal lobe and in some surface loops. The link region between the lobes, extending from residues 103 to 112, is well located, and part of the chain in this region contributes to a lattice contact. Comparison of the relative lobe orientations between the binary PhK structure and the ternary

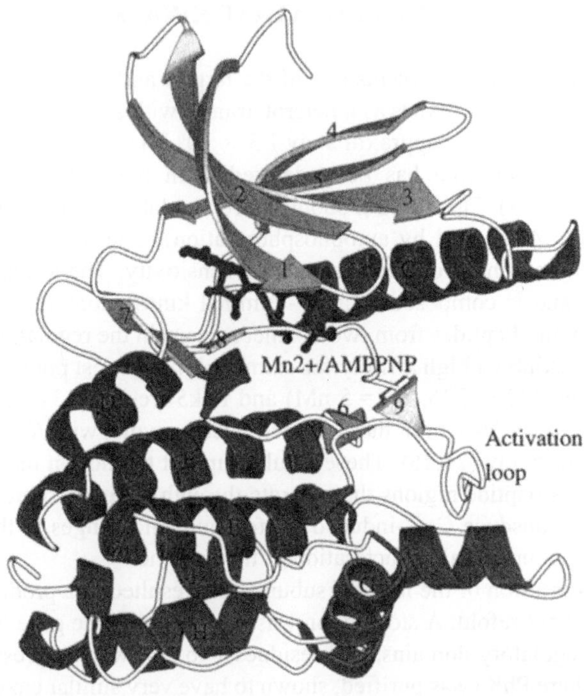

FIG. 4. The structure of PhK. α helices and β strands are represented as spirals and arrows, respectively. The secondary structural elements are labeled with β strands numbered from 1 to 9 and α helices lettered from C to I, as in cAPK. Mn^{2+}/AMPPNP is shown bound at the catalytic site. The activation segment starts just before β strand 9 and the catalytic loop is between strands 6 and 7.

complex of cAPK with inhibitor and ATP shows that PhK adopts a slightly more open structure and that the shift can be represented as a rotation of about 5° of the N-terminal lobe with respect to the C-terminal lobe. This small rotation means that in the nomenclature of open and closed conformations for protein kinases (32), the PhK structure can be classed as closed. The key catalytic aspartate (Asp^{149} in PhK) is located in the loop between β6 and β7. This aspartate is conserved in all protein kinases and promotes catalysis through its action as a general base.

Nucleotide Binding

The nucleotide AMPPNP is associated mainly with the N-terminal domain of phosphorylase kinase, but also makes important links to residues in the C-terminal domain. The polar contacts are shown in Fig. 5. The adenine is docked into a nonpolar pocket comprising Leu^{25}, Val^{33}, and Ala^{46} from strands β1, β2, and β3, respec-

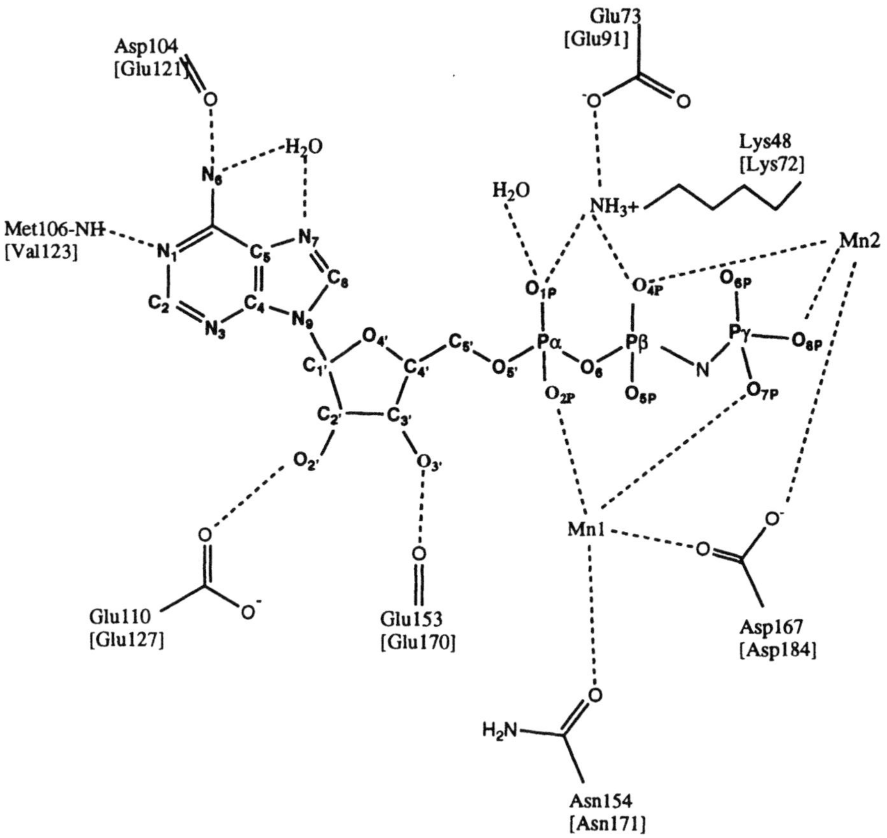

FIG. 5. Schematic diagram of the polar contacts between Mn^{2+}/AMPPNP and PhK. Residues making potential contacts shorter than 3.3 Å have been included. Equivalent residues in cAPK are shown in brackets (29).

tively, and Ile^{87}, Phe^{103}, Met^{105}, and Leu^{156}. The ribose makes two hydrogen bonds through its O2′ and O3′ hydroxyls, to Glu^{110} and the main-chain carbonyl oxygen of Glu^{153}, respectively. The triphosphate moiety is stabilized by contacts to Lys^{48} and two metal ions. The Mn^{2+} in site 2 chelates the β and γ phosphates, and the Mn^{2+} in site 1 chelates the α and γ phosphates. The metal ions make contact to residues Asp^{167} and Asn^{154}, as shown in Fig. 5. The major difference in the nucleotide contacts between PhK and cAPK is in the contacts to the glycine-rich loop between strands β1 and β2 (Fig. 4). Because of the more open structure in PhK, the loop is just too far away to contribute direct hydrogen-bond interactions from the main-chain NH groups. The sequence of residues 26 to 32 in PhK is Gly-Arg-Gly-Val-Ser-Ser-Val. The serine at position 31 replaces the third glycine that is normally found in the glycine-rich loops of nucleotide-binding sites, and the side chain of Ser^{31} may be responsible for some of the displacement. In addition, there may be further closure of the loop upon formation of the ternary complex. In the binary product complex with Mg^{2+}/ADP, the magnesium ions occupy the same positions as the manganese ions in the AMPPNP complex, but there is a shift in the phosphate positions so that Mg2 contacts only the β phosphate and Mg1 contacts the α and β phosphates. The loop closes down slightly so that its position is intermediate between that of the PhK–AMPPNP complex and the cAPK ternary complex.

The holoenzyme of phosphorylase kinase exhibits significant ATPase activity, with a k_{cat} of approximately 0.06 s^{-1} (33), a value that is approximately 1/1,000 of the k_{cat} for kinase activity (60 to 90/s). The characteristics of ATPase activity parallel the activation properties of the kinase, and the K_m values for ATPase (140 μM) are of the same order of magnitude as those that have been reported for ATP in the kinase assay (about 60 μM). The structure of PhK suggests that the catalytic aspartate (Asp^{149}) would be poised to promote attack by a water molecule on the γ position of the ATP after minimal conformational changes. With the crystals obtained in the presence of 3 mM ATP and 10 mM magnesium acetate, the electron-density map showed no density for the γ phosphate. Since the γ phosphate was apparent in the electron-density maps with the nonhydrolyzable analogue AMPPNP, it is assumed that PhK had hydrolyzed ATP to give the product adenosine diphosphate (ADP) in the crystals that were grown with ATP.

The kinase activity of phosphorylase kinase is sensitive to metal ions. PhK activity is increased by free Mg^{2+} and inhibited by free Mn^{2+} (27,34). With Mg^{2+}, activity increases as Mg^{2+} concentrations are increased until they are equimolar with ATP, and there is a further increase in activity as Mg^{2+} ions are increased to about 10 times those of ATP. With Mn^{2+} there is a similar activation as concentrations are raised to equimolarity with ATP; above this concentration, Mn^{2+} ions are inhibitory. This behavior is different from that observed with cAPK for which Mg^{2+} ions above equimolarity with ATP are inhibitory and there is no activation at all with Mn^{2+}. The activatory and inhibitory Mg sites have been established for cAPK. Metal site 2 is the activatory site and metal site 1 is the inhibitory site (as discussed in [35]). The role of the metals in catalysis is crucial. It is assumed that they serve to bind and stabilize the triphosphate group of the ATP substrate and to orient the phosphate group for the favored transfer of the γ-phosphate to the seryl or threonyl OH group of the protein substrate. The metals neutralize the charge on the phosphates and facilitate the developing negative charge during the phosphoryl transfer. Clearly there is a subtle balance between the strengths of these interactions, which are required for binding and neutralization, and those that are required to allow product dissociation after the reaction is completed. In the activity of cAPK the rate-limiting step is product (ADP) release (36). If we assume a similar mechanism for PhK, it may be that product release, although still rate limiting, is slightly easier because of the more open structure of the glycine-rich loop that results from changes in amino-acid sequence in this region. Although we do not have an entirely satisfactory explanation for the metal dependence of kinase activity, the results indicate the exquisite sensitivity of the kinase reaction and indicate some properties that may be exploited in the control of kinase activity. In the inactive kinase structures of CDK2 (37), MAPK

(38), and insulin-receptor kinases (39), the ATP-recognition sites are either blocked or incorrectly formed for locating the phosphates.

Substrate Recognition

We have not yet succeeded in studying the binding of peptide substrates to the catalytic site of PhK. However, some indication of the likely interactions can be obtained by superimposing the inhibitor peptide bound to cAPK on the PhK structure with residues changed so that the side chains correspond to those around the Ser^{14} recognition site of GP (29). Each of the sites in the GP sequence Lys-Gln-Ile-Ser-Val-Arg can make satisfactory contacts to residues in PhK. In particular, the basic residue in the P^{-3} position could contact Glu^{110} and the polar residue at the P^{-2} position could contact Glu^{153}, both of which residues have been shown to be important in site-directed mutagenesis experiments (25). Further, the arginine at the P^{+2} site, which ensures that this peptide is not a substrate for cAPK, could make satisfactory contacts to acidic residues.

If we assume that the conformation adopted by the inhibitor of cAPK as it lies in the catalytic site of cAPK is the conformation adopted for substrates of PhK, we note that the conformation differs from that of the peptide around Ser^{14} in GPb and GPa. This suggests that protein kinases may recognize a part of the peptide chain that, in the intact protein substrate may have a different conformation than that required at the catalytic site of the kinase. Thornton et al. have noted that the nick sites in proteins that are attacked by serine proteinases such as trypsin also tend to have a different conformation from that adopted by, for example, the trypsin inhibitor when bound to trypsin (40). The tetradecapeptide substrate (residues 5 to 18 of phosphorylase) exhibits a significantly greater K_m value (K_m = 1.49 μM) in the phosphorylase kinase assay than that observed with phosphorylase b as substrate (K_m = 20 μM) (41). This result may indicate that the phosphorylase molecule contains additional recognition sites that may help locate the substrate. Interestingly, there are some similarities between one of the calmodulin-binding peptides of the regulatory domain, Phk13, with the sequence around the phosphorylatable serine in phosphorylase, and between another calmodulin-binding peptide, Phk5, and part of the α-2 helix in phosphorylase (26).

Activation Segment

Almost all protein kinases (with a few exceptions such as phosphorylase kinase, twitchin kinase, casein kinase I, calmodulin-dependent protein kinase II, epidermal growth factor [EGF] receptor tyrosine kinase) are regulated by phosphorylation on a loop of chain that links the β9 strand and the F α-helix (Fig. 4). We propose, after discussions with Susan Taylor, to call the region from residues 184 to 208 in cAPK the "activation segment." This segment includes the motif known as DFG, beginning in cAPK at residue 184, that is conserved in all kinases. Asp^{184} chelates the activatory metal, and the positions of the equivalent residues in all kinase structures so far elucidated are similar. The segment, which includes the β9 strand in cAPK and the active kinases, continues with a region that is remarkably different in the active and inactive kinase structures. This region carries the residue(s) that is(are) phosphorylated (Thr^{197} in cAPK). The segment ends in a region that has a 3_{10} helix structure and which includes residues (Leu^{198}, Gly^{200}, Pro^{202}, and Leu^{203}) that are important for the P^{+1} peptide-substrate recognition site in cAPK. Residue 208 is a glutamate and is part of the APE motif that is conserved in all kinases. The activation segment therefore includes, at the beginning and the end, two highly conserved motifs, while the middle region is variable among the different kinases. Comparison of the activation segments of PhK and cAPK shows that they are similar in these two active protein kinases (the root-mean-square difference in Cα positions is 0.9 Å for 25 residues), apart from a two-residue insertion in PhK.

In PhK, the residue corresponding to Thr^{197} of cAPK is a glutamate, Glu^{182}. The glutamate turns its side chain inward and makes an ionic link with Arg^{148}, the arginine adjacent to the catalytic aspartate Asp^{149} (Fig. 6a). All protein kinases that are regulated by phosphorylation in the activation segment have an arginine residue that immediately precedes the catalytic aspartate. In PhK the arginine is also hydrogen bonded to Tyr^{206} (as in cAPK), but is otherwise open to the solvent. In cAPK, Thr^{197}-P is involved in interactions through its phosphate group not only to the arginine (Arg^{165}) but also to Lys^{189}, His^{87}, and Thr^{195} (Fig. 6b; Table 1). Kinetic studies of the His^{87}Ala mutant of cAPK have shown that although the mutant is impaired in its recognition of peptide substrate, especially at the P^{+2} site, it exhibits a 2- to 3-fold higher k_{cat} than does the native enzyme (42). On the other hand, mutants of cAPK in which the threonine was changed to an aspartate (Thr^{197}Asp) showed K_M values for peptide substrate that were increased by two orders of magnitude. Analysis showed that the changes were associated with reduced rates for phosphoryl transfer of about 500-fold, and not with reduced substrate affinities (43). These results imply that the Thr^{197}-P group is important as a molecular switch for enzymatic function. In cAPK the phosphate is located so as to neutralize the cluster of positively charged residues. In PhK the residue corresponding to Lys^{189} is a cysteine (Cys^{172}), and the residue corresponding to His^{87} is an alanine (Ala^{69}) (Table 1). Clearly there is no charge cluster, and the positive charge on Arg^{148} can be neutralized successfully by the single negative charge of the glutamate. This appears to explain why this kinase core domain is constitutively active. All the elaborate control machinery provided by the rest of the molecule, which includes the regulatory domain of the γ subunit and the α, β, and δ subunits, is designed to keep the kinase inactive until its action is required.

Implications for Other Kinases

As noted by Taylor and Radzio-Andzelm (44), both CDK2 (37) and MAP kinase (38) have equivalent charge clusters to those observed in cAPK. We can now extend their structural observations and include PhK (29), twitchin kinase (45), insulin

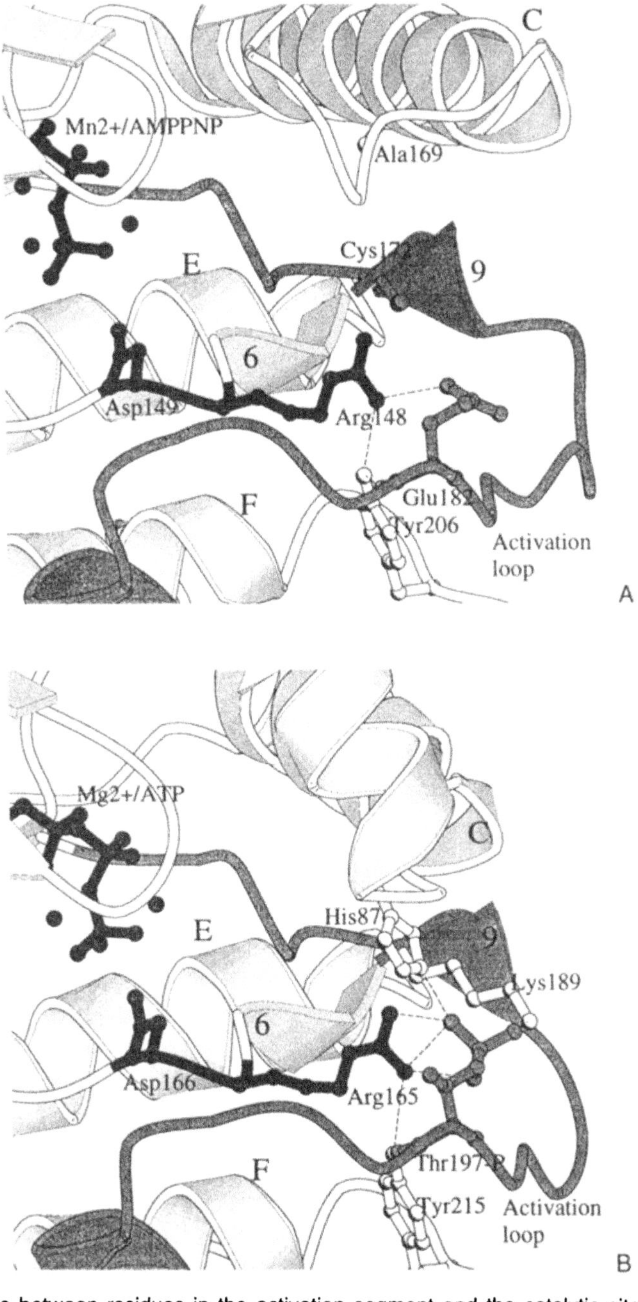

FIG. 6. Contacts between residues in the activation segment and the catalytic site. **A:** PhK. Residue Glu^{182} from the activation segment is hydrogen bonded to Arg^{148}, the arginine adjacent to the catalytic base, Asp^{149}. **B:** cAPK. Thr^{197}-P of the activation segment is hydrogen bonded to Arg^{165}, the arginine adjacent to the catalytic base, Asp^{166}, and to Lys^{189} of the activation segment and His^{87} of the N-terminal domain.

TABLE 1. *Possible Contact Residues to Phosphorylated Amino Acids and Corresponding Nonphosphorylated Amino Acids in the Activation Segment of Protein Kinases*

			Contact residues to Thr^{197}-P in cAPK and equivalent residues in other kinases				
Kinase	Conformational state in crystal	Phosphoamino acid(s) or equivalent	Residue preceding catalytic aspartate	Residue in DFG motif loop	Contact from N-terminal domain	Residue in vicinity but not in contact in cAPK	Reference
Kinases controlled by phosphorylation on residues in the activation segment							
cAPK	Active Closed Phosphorylated	Thr^{197}	Arg^{165}	Lys^{189}	His^{87}	Asn^{90}	30,31
CDK2	Inactive Closed Nonphosphorylated	Thr^{160}	Arg^{126}	Arg^{150}	Thr^{47}	Arg^{50}	37
CDK2/Cyclin A	Partially active Closed Nonphosphorylated	Glu^{162} mimics Thr^{160}-P	Arg^{126}	Arg^{150}	Thr^{47}	Arg^{50}	48
MAP kinase	Inactive Open Nonphosphorylated	Thr^{183} Tyr^{185}	Arg^{146}	Arg^{170}	Arg^{65}	Arg^{68}	38
IRK	Inactive Open Nonphosphorylated	$Tyr1^{158}$ $Tyr1^{162}$ $Tyr1^{163}$	Arg^{1131}	Arg^{1155}	Glu^{1043}	Asn^{1046}	39
Kinases not controlled by phosphoryation on residues in the activation segment							
Twitchin	Inactive Open	Val^{6098}	Leu^{6062}	Ala^{6088}	Thr^{5983}	Lys^{5986}	45
CK1	Active Closed	Asn^{177}	Arg^{130}	Lys^{159}	Deletion	Asp^{54}	47
PhK	Active Closed	Glu^{182}	Arg^{148}	Cys^{162}	Ala^{69}	Lys^{72}	29

receptor tyrosine kinase (IRK) (46) and CK1 (47) (Table 1). In the inactive nonphosphorylated conformations of CDK2 and the insulin receptor tyrosine kinase domain, the activation segments adopt very different conformations and partially block the ATP and substrate binding sites. In MAP kinase, the inactive structure adopts an open conformation so that the residues across the domains are not in the appropriate relative positions to promote catalysis. It is reasonable to assume that on activation, the activation segment undergoes significant shifts that allow both the correct substrate-recognition sites to form and realignment of catalytic residues.

In the recently solved complex of CDK2 with cyclin A (48), just such conformational changes are seen coupled with movement of the C helix that includes the sequence motif PSTAIRE (using the single-letter amino acid code), which is known to be important in binding cyclin. Comparison of the structure of free cyclin A, recently determined in this laboratory (49), with the CDK2-bound cyclin A structure shows no significant change between the free and bound forms. Cyclin A activates CDK2 through docking of a pliable kinase against a rigid cyclin. As anticipated, the changes in the CDK2 structure do bring together a charge cluster, and in the complex this cluster is partly neutralized by Glu162 turning inward to hydrogen bond to the arginine adjacent to the catalytic aspartate, resembling the contact made by the glutamate in PhK, and partly through the use of main-chain carbonyl oxygens on the cyclin to hydrogen bond to the arginines. The CDK2–cyclin A complex exhibits approximately 1/16 the full activity achieved when the kinase is activated by phosphorylation on Thr^{160}. It seems likely that on phosphorylation, the Thr^{160}, which is exposed in the complex, turns in and interacts with the arginine residues, possibly displacing Glu^{162}.

In the inactive twitchin kinase structure, the C-terminal 60 residues partially block the ATP site as well as the peptide substrate site in an autoinhibitory mechanism. The mechanism of activation must allow relief of this autoinhibition. It is interesting to note that this kinase has a nonpolar residue preceding the catalytic base (Leu^{6062}), and that the residue equivalent to Thr^{197} in cAPK is also nonpolar (Val^{6098}) (45). Evidently, no charge compensation is required. Casein kinase I is seemingly anomalous (Table 1). It has no phosphorylatable group equivalent to Thr^{197} in cAPK, but it does contain a charge cluster of Arg^{130}, adjacent to the catalytic base, and Lys^{159}, equivalent to Lys^{189} in cAPK. The crystal structure of the *Schizosaccharomyces pombe* CKI contains a sulfate ion that stabilizes this cluster (47). Thus, the ion appears to promote the correct conformation of the activation loop, but it is not known whether this is essential for activity of the enzyme *in vivo*.

The extent to which kinases that are normally activated by phosphorylation on the activation segment can be activated by mutagenesis of the phosphorylated residue to glutamate or aspartate varies considerably. Both cAPK and MAP kinase are not appreciably activated by the mutations of Thr^{197}Asp or Thr^{183}Glu, Tyr^{185}Glu, respectively (43,50). In *S. pombe* cells, a mutant p34^{cdc2} kinase in which a glutamate replaces the phosphorylatable Thr^{167} gives rise to cells that arrest in a pseudomitotic state characterized by an abundance of mitotic spindles. This suggests that the kinase may have been active enough to drive the cells into mitosis, but that dephosphorylation is important for exit (51). In MEK1 (the activatory kinase of MAP kinase), and in protein kinase C (PKC), significant biologic activity has been reported for the mutants $Ser^{218}Asp$ and $Ser^{222}Glu$, and $Thr^{500}Glu$, respectively (52,53).

The present structural results lead to the notion that phosphorylation of residue(s) in the activation segment results in conformational changes that bring the segment into its correct conformation for substrate binding and promote the correct orientation and electrostatic environment for catalytic groups. In those kinases that are activated by phosphorylation, a phosphorylated residue is necessary to neutralize a charge cluster that would otherwise persist. Phosphorylase kinase requires no such phosphorylation because there is only one charged group requiring neutralization, and neutralization is achieved with a glutamate residue. Although it is also possible to explain the other kinases whose structures are known according to this hypothesis (Table 1), there are likely to be exceptions to these simple proposals. A complete understanding will come only when the structures of both dephospho and phospho forms of several kinases are available.

REFERENCES

1. Fischer EH, Krebs EG. Conversion of phosphorylase b to phosphorylase a in muscle extracts. *J Biol Chem* 1955;216:121–32.
2. Sutherland EW, Wosilait WD. Inactivation and activation of liver phosphorylase. *Nature* 1955;175: 169–71.
3. Walsh DA, Perkins JP, Krebs EG. An adenosine 3′,5′-monophosphate dependent protein kinase from rabbit muscle. *J Biol Chem* 1968;243:3763–74.
4. Bollen M, Stalmans W. The structure, role and regulation of Type I protein phosphatases. *Crit Rev Biochem Mol Biol* 1992;27:227–81.
5. Hubbard MJ, Cohen P. On target with a new mechanism for the regulation of protein phosphorylation. *Trends Biochem Sci* 1993;18:172–7.
6. Chakrabarti P. Anion binding sites in protein structures. *J Mol Biol* 1993;234:463–82.
7. Copley RR, Barton GJ. A structural analysis of phosphate and sulphate binding sites in proteins. *J Mol Biol* 1994;242:321–9.
8. Hol WGJ. The role of the alpha helix dipole in protein structure and function. *Prog Biophys Mol Biol* 1985;45:149–95.
9. Stroud RM. Mechanisms of biological control by phosphorylation. *Curr Opin Struct Biol* 1991; 1:826–35.
10. Sprang SR, Acharya KR, Goldsmith EJ, et al. Sructural changes in glycogen phosphorylase induced by phosphorylation. *Nature* 1988;336:215–21.
11. Martin JL, Withers SG, Johnson LN. Comparison of the binding of glucose and glucose-1-phosphate derivatives to T-state glycogen phosphorylase b. *Biochemistry* 1990;29:10745–57.
12. Acharya KR, Stuart DI, Varvill KM, Johnson LN, *Glycogen phosphorylase: description of the protein structure*. London and Singapore: World Scientific; 1991: 123.
13. Barford D, Johnson LN. The allosteric transition of glycogen phosphorylase. *Nature* 1989;340:609–14.
14. Barford D, Hu S-H, Johnson LN. The structural mechanism for glycogen phosphorylase control by phosphorylation and by AMP. *J Mol Biol* 1991;218:233–60.
15. Sprang SR, Withers SG, Goldsmith EJ, Fletterick RJ, Madsen NB. The structural basis for the association of phosphorylase b with AMP. *Science* 1991;254:1367–71.
16. Johnson LN. Glycogen phosphorylase: control by phosphorylation and allosteric effectors. *FASEB J* 1992;6:2274–82.
17. Johnson LN, Barford D. The effects of phosphorylation on the structure and function of proteins. *Annu Rev Biophys Biomol Struct* 1993;22:199–232.
18. Fischer EH, Graves DJ, Sydney-Crittenden EB, Krebs EG. Structure of the site phosphorylated in the phosphorylase b to a reaction. *J Biol Chem* 1959;234:1698–1704.

19. Johnson LN, Barford D. Electrostatic effects in the control of glycogen phosphorylase by phosphorylation. *Protein Sci* 1994;3:1726–30.
20. Nicholls A, Honig B. A rapid finite difference algorithm, utilising successive over relaxation to solve the Poisson Boltzmann equation. *J Comp Chem* 1991;12:435–45.
21. Johnson LN, Acharya KR, Jordan MD, McLaughlin PJ. The refined crystal structure of the phosphorylase-heptulose 2–phosphate-oligosaccharide-AMP complex. *J Mol Biol* 1990;211:645–61.
22. Pickett-Gies CA, Walsh DA. *Glycogen phosphorylase kinase.* In: Boyer PD, Krebs EG, eds. *The enzymes.* Orlando, Florida: Academic Press; 1986: 396–459.
23. Norcum MT, Wilkinson DA, Carlson MC, Hainfeld JF, Carlson GM. The structure of phosphorylase kinase: a three dimensional model derived from stained and unstained electron micrographs. *J Mol Biol* 1994;241:94–102.
24. Dasgupta M, Honeycutt T, Blumenthal DK. The γ subunit of skeletal muscle phosphorylase kinase contains two non-contiguous domains that act in concert to bind calmodulin. *J Biol Chem* 1989; 264:17156–63.
25. Huang C-YF, Yuan C-J, Blumenthal DK, Graves DJ. Identification of the substrate and pseudosubstrate binding sites of phosphorylase kinase γ. *J Biol Chem* 1995;270:7183–8.
26. Dasgupta M, Blumenthal DK. Characterisation of the regulatory domain of the gamma subunit of phosphorylase kinase. *J Biol Chem* 1995;270:22283–9.
27. Cox S, Johnson LN. Expression of the phosphorylase kinase γ subunit catalytic domain in *E. coli*. *Protein Engineering* 1992;5:811–9.
28. Owen DJ, Papageorgiou AC, Garman EF, Noble MEM, Johnson LN. Expression, purification and crystallisation of phosphorylase kinase catalytic domain. *J Mol Biol* 1995;246:376–83.
29. Owen DJ, Noble ME, Garman EF, Papageorgiou AC, Johnson LN. Two structures of the catalytic domain of phosphorylase kinase: an active protein kinase complexed with substrate analogue and product. *Structure* 1995;3:467–82.
30. Knighton DR, Zheng J, Ten Eyck LF, et al. Crystal structure of the catalytic subunit of cyclic adenosinemonophosphate-dependent protein kinase. *Science* 1991;253:407–13.
31. Bossmeyer D, Engh RA, Kinzel V, Ponstingl H, Huber R. Phosphotransferase and substrate binding mechanism of the cAMP dependent protein kinase subunit from porcine heart as deduced from the 2.0 Å structure of the complex with Mn^{2+} adenylylimidophosphate and inhibitor peptide PKI(5-24). *EMBO J* 1993;12:849–59.
32. Cox S, Radzio-Andzelm E, Taylor SS. Domain movements in protein kinases. *Curr Opin Struct Biol* 1994;4:893–901.
33. Paudel HK, Carlson GM. The ATPase activity of phosphorylase kinase is regulated in parallel with its protein kinase activity. *J Biol Chem* 1991;266:16524–9.
34. Kee SM, Graves DJ. Properties of the γ subunit of phosphorylase kinase. *J Biol Chem* 1987;262: 9448–53.
35. Adams JA, Taylor SS. Divalent metal ions influence catalysis and active site accessibility in the cAMP-dependent protein kinase. *Protein Sci* 1993;2:2177–86.
36. Adams JA, Taylor SS. Energetic limits of phosphotransfer in the catalytic subunit of cAMP-dependent protein kinase as measured by viscosity experiments. *Biochemistry* 1992;31:8516–22.
37. De Bondt HL, Rosenblatt J, Jancarik J, Jones HD, Morgan DO, Kim S-H. Crystal structure of cyclin dependent kinase 2. *Nature* 1993;363:592–602.
38. Zhang F, Strand A, Robbins D, Cobb MH, Goldsmith EJ. Atomic structure of the MAP kinase ERK2 at 2.3 Å resolution. *Nature* 1994;367:704–11.
39. Hubbard SR, Wei L, Ellis L, Hendrickson WA. Crystal structure of the tyrosine kinase domain of the human insulin receptor. *Nature* 1994;372:746–54.
40. Hubbard S, Eisenmenger F, Thornton JM. Modelling studies of the change in conformation required for cleavage of limited proteolytic sites. *Protein Sci* 1994;3:757–68.
41. Tabatabai LB, Graves DJ. Kinetic mechanism and specificity of the phosphorylase kinase reaction. *J Biol Chem* 1978;253:2196–202.
42. Cox S, Taylor SS. Kinetic analysis of cAMP-dependent protein kinase mutations at histidine 87 effect peptide binding and pH dependence. *Biochemistry* 1995;34:16203–16209.
43. Adams JA, McGlone ML, Gibson R, Taylor SS. Phosphorylation modulates catalytic function and regulation in the cAMP-dependent protein kinase. *Biochemistry* 1995;34:2447–54.
44. Taylor SS, Radzio-Andzelm E. Three protein kinase structures define a common motif. *Structure* 1994;2:345–55.
45. Hu SH, Parker MW, Lei JY, Wilce MCJ, Benian GM, Kemp BE. Insights into autoregulation from the crystal structure of twitchin kinase. *Nature* 1994;369:581–4.

46. Hubbard SR, Wei L, Ellis L, Hendrickson WA. Crystal structure of the tyrosine kinase domain of the human insulin receptor. *Nature* 1994;372:746–54.
47. Xu R-M, Carmel G, Sweet RM, Kuret J, Cheng X. Crystal structure of casein kinase-1, a phosphate directed protein kinase. *EMBO J* 1995;14:1015–23.
48. Jeffrey PD, Russo AA, Polyak K, et al. Mechanism of CDK activation revealed by the structure of a cyclinA-CDK2 complex. *Nature* 1995;376:313–20.
49. Brown NR, Noble MEM, Endicott JA, et al. The crystal structure of cyclin A. *Structure* 1995;3: 1235–47.
50. Zhang J, Zhang F, Ebert D, Cobb MH, Goldsmith EJ. Activity of the MAP kinase ERK2 is controlled by a flexible surface loop. *Structure* 1995;3:299–307.
51. Gould KL, Moreno S, Owen DJ, Sazer S, Nurse P. Phosphorylation at Thr167 is required for *Schizosaccharomyces pombe* $p34^{cdc2}$ function. *EMBO J* 1991;10:3297–3309.
52. Huang W, Erickson RL. Constitutive activation of MEK1 by mutation of serine phosphorylation sites. *Proc Natl Acad Sci USA* 1994;91:8960–3.
53. Orr JW, Newton AC. Requirement for negative charge on activation loop of protein kinase C. *J Biol Chem* 1994;269:27715–8.

Signal Transduction in Health and Disease, Advances in Second Messenger and Phosphoprotein Research, Vol. 31, edited by J. Corbin and S. Francis.
Lippincott–Raven Publishers, Philadelphia © 1997.

3

Intrasteric Regulation of Protein Kinases

Boštjan Kobe, Jörg Heierhorst, and Bruce E. Kemp

St. Vincent's Institute of Medical Research, Fitzroy 3065, Victoria, Australia

INTRODUCTION

Most protein kinases exist in latent inactive forms requiring specific signals for activation. The first such kinase identified was phosphorylase kinase, which could be interconverted between an inactive autoregulated form and an activated form, which differed in their kinetics of phosphorylation of phosphorylase b (1). Phosphorylase kinase can be activated by calmodulin binding, phosphorylation by the cyclic adenosine monophosphate (cAMP)-dependent protein kinase (cAPK), or partial proteolysis with chymotrypsin. This theme of protein kinase autoregulation and activation by binding of allosteric regulators and phosphorylation has since been observed repeatedly. Some examples of second-messenger-dependent autoregulated protein kinases are listed in Table 1.

The cAPK regulatory subunits and the inhibitory protein PKI contain substrate-like mimics (pseudosubstrate sequences) of the local phosphorylation-site sequences that are responsible for inhibition; this led to the idea that the inhibitory structure was a pseudosubstrate (reviewed in [2]). The concept that protein kinases can be regulated by pseudosubstrates evolved over a number of years from studies of cAPK, in which the regulatory subunit RI could be displaced by Arg-rich substrates or polyarginine (3,4), implying that the regulatory subunit functioned by masking the active site (5,6). The demonstration that PKI acted as a competitive substrate antagonist through the presence of arginine residues analogous to those in protein substrates (7) further reinforced this concept. Furthermore, studies of the regulation and substrate specificity of the myosin light-chain kinase (MLCK), protein kinase C (PKC), phosphorylase kinase, calmodulin-dependent protein kinases I (CaMK-I) and II (CaMK-II), and the plant calcium-regulated protein kinases (CDK) indicated that the pseudosubstrate model of regulation is a widely exploited regulatory mechanism (8).

The crystal structure of cAPK with a bound inhibitory peptide provided direct evidence that inhibition occurs by a pseudosubstrate mechanism (9). Subsequently, the crystal structure of twitchin expanded this theme and showed that autoinhibition could be mediated by a complex array of contacts, not limited to pseudosubstrate

TABLE 1. *Second-Messenger-Dependent Autoregulated Protein Kinases*

cAMP-dependent protein kinase subfamily
cGMP-dependent protein kinase subfamily
Protein kinase C (PKC) subfamily
Myosin light chain kinase (MLCK) subfamily
Calmodulin-dependent protein kinase (CaMK) subfamily
Plant calcium-dependent protein kinase subfamily

contacts, within the active site (10). In order to more appropriately describe these types of regulation, we have applied the term *intrasteric* regulation (see Table 2) to specify that the regulatory events involve the active site directly. This contrasts with *allosteric* regulation, in which the regulatory ligand is structurally unrelated to the substrate and binds at a site distinct from the substrate-binding site. Intrasteric regulation can be achieved by pseudosubstrate binding or by phosphorylation events within the active site that are typically catalyzed by distinct protein kinases (the phosphorylation events within the active site can thus be termed *intrasteric phosphorylation*). In some cases, such as that of CaMK-I, both types of control operate simultaneously, with calmodulin binding being a prerequisite for phosphorylation of the activation loop. The cAPK, the insulin receptor, and CaMK-I and CaMK-IV are all activated by phosphorylation in their activation loops. In addition, the cyclin-dependent kinase cdc2 is also regulated by intrasteric phosphorylation by wee-1, but at the glycine-rich phosphate-anchoring loop, resulting in an inhibition of enzyme activity. *Extrasteric phosphorylation* refers to phosphorylation that occurs distal to the active site and is thus the covalent modification equivalent to allosteric regulation, which is classically a noncovalent binding event. Smooth-muscle MLCK and CaMK-II are regulated by extrasteric phosphorylation that occurs in proximity to their respective calmodulin-binding sequences and modulates the calmodulin-dependence of enzyme activation.

It is now clear that there is considerable diversity of protein kinase regulation, with a variety of combinations of intrasteric and extrasteric regulation. These involve intrasteric regulation by sequences outside the catalytic core or by loops within the catalytic core, such as the substrate-anchoring loop. Superimposed on this

TABLE 2. *Mechanisms of Enzyme Regulation*

Extrasteric regulation
Allosteric regulation (cAPK, cGPK, PKC)
Extrasteric phosphorylation (smMLCK, CaMK-II)
Intrasteric regulation
Intrasteric autoinhibition (PKC, cAPK, MLCK)
Intrasteric phosphorylation (CDK, IRPK)

Abbreviations: cAPK, cAMP-dependent protein kinase; cGPK, cGMP-dependent protein kinase; PKC, protein kinase C; MLCK, myosin light-chain kinase; smMLCK, smooth-muscle MLCK; CaMK-II, calmodulin-dependent protein kinase II; CDK, cyclin-dependent protein kinase; IRPK, insulin receptor kinase.

TABLE 3. *Myosin Light-Chain Kinase Subfamily*

Kinase	Mass	Additional domains
Titin	3 MDa	Ig and Fn-III
Twitchin	700 kDa	Ig and Fn-III
smMLCK	130–152 kDa	Ig and Fn-III
skMLCK	65 kDa	

is a variety of activation mechanisms including allosteric regulation by Ca^{2+}/calmodulin or intrasteric phosphorylation that reverses the autoinhibition.

MYOSIN LIGHT-CHAIN KINASE SUBFAMILY

Myosin light-chain kinases (MLCKs) constitute an important subfamily of autoregulated protein kinases (Table 3). This subfamily includes the classical MLCKs, such as smooth-muscle MLCK (130 to 155 kDa) and skeletal muscle MLCK (87 kDa), and the giant protein kinases titin (at 3 MDa the largest known polypeptide), twitchin (750 kDa), and projectin (>500 kDa) (11). The giant myosin-associated protein kinases are important cytoskeletal elements in muscle, where they may act as rulers for thick-filament assembly (11). These kinases contain catalytic protein kinase domains similar to those of the MLCKs near their respective COOH-termini, with the remainder of the enzyme made up predominantly of fibronectin type III (Fn-III) and immunoglobulin-like (Ig) repeats (12) (Fig. 1). Smooth-muscle MLCK has three Ig-like and one Fn-III repeat arranged around the catalytic domain, in the same juxtaposition as that found in twitchin and titin (Fig. 1).

Studies of the regulation of the MLCK family have played a major role in the development of the pseudosubstrate autoregulatory hypothesis (reviewed in [13]). It was recognized (14) that the basic residues present in the calmodulin-binding domain (15) closely resembled the number and juxtaposition of the basic residues identified as specificity determinants in MLCK (16) (see Table 4).

This led to the hypothesis that the sequence containing the calmodulin-binding site of MLCK may be positioned in the active site and render the MLCK inactive. Support for this concept was initially obtained with synthetic peptides corresponding to the pseudosubstrate sequence; these were found to inhibit smooth-muscle MLCK (14) and skeletal muscle MLCK (17). Detailed studies done with peptides and partial proteolysis have shown that the pseudosubstrate inhibitor and calmodulin-binding functions overlap but are not identical (13). While providing compelling support for the concept that the autoregulatory sequence bound to the active site of the enzyme, none of the studies done with MLCK provided direct evidence of this. This evidence

TABLE 4. *Pseudosubstrate Regulatory Hypothesis for MLCK*

substrate smMLC(1 to 21)	-**S**SK**R**A**K**AKTT**KKR**PQ**R** ATSN**V**^{21}- Substrate
smMLCK(787 to 807)	-**S**KD**R**M**K**KYMA**RRK**WQ**K**TGHA**V**^{807}- Pseudosubstrate

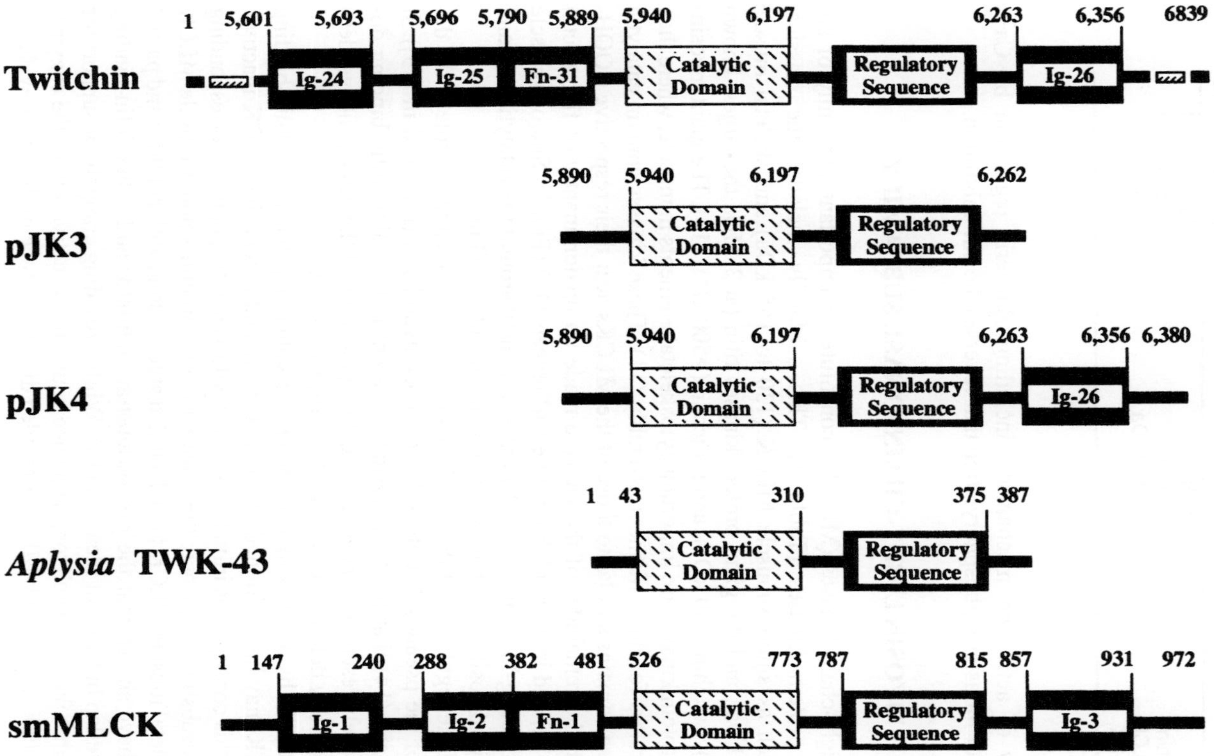

FIG. 1. Schematic diagrams of domain arrangements in protein kinases of the MLCK subfamily. Domain arrangements in the nematode twitchin (12), the crystallized fragments of nematode (pJK3 and pJK4) and *Aplysia* (TWK-43) twitchins, and chicken gizzard MLCK (26) (smMLCK) are shown.

came from the crystal structures of the fragments of twitchin kinase, another member of the MLCK subfamily (10,18).

STRUCTURE OF AUTOINHIBITED TWITCHIN KINASE

The giant protein kinase twitchin, encoded by the *Caenorhabditis elegans unc*-22 gene, contains 6,839 residues (19), with a catalytic core between residues 5,950 and 6,199. The catalytic core of twitchin kinase is 52% identical to the smooth-muscle MLCK sequence, without insertions or deletions. A similar protein is found in the marine mollusc *Aplysia* (20). Fragments of twitchin kinase containing the catalytic core have been successfully expressed in *Escherichia coli* (20,21). When the carboxyl-terminal sequence outside the catalytic core is removed, the enzyme fragment is constitutively active on peptide substrates (20). Three fragments of twitchin kinase containing the catalytic core and the carboxyl-terminal (C-terminal) regulatory domains were crystallized (18,22) and their structures solved (10,18) (Fig. 1); the structures of the nematode fragment pJK3, the nematode fragment pJK4, and the *Aplysia* fragment TWK-43 have been refined at 2.8-Å, 3.3-Å, and 2.3-Å resolution, respectively.

Protein Kinase Fold

The catalytic cores of twitchin kinases contain a canonical bilobate protein kinase structure (Fig. 2), first observed in the crystal structure of cAPK (23). The smaller N-terminal lobe is dominated by an antiparallel β-sheet, and the larger C-terminal lobe is built mainly of α-helices. These two lobes exhibit different relative rotations with respect to each other in different protein kinase structures, and twitchin kinases exhibit one of the most open structures observed, with the small lobe rotated approximately 30° relative to the large lobe when compared to the closed structures of active protein kinases. The rotation of the small lobe in twitchin kinase is readily achieved by a hinge motion around the adjacent Gly residues between the two lobes; hinging at the corresponding position has been determined in solution for cAPK (24).

The C-terminal autoinhibitory sequence is positioned in the cleft between the two lobes. It is built predominantly of irregular structure, but also contains a few elements of secondary structure, including two α-helices (H11 and H12), a few 3_{10} helices, and a short β-strand (S13) bound in an antiparallel manner to the strand S12 in the catalytic core.

The interface between the autoinhibitory region and the catalytic core of twitchin kinase is extensive. Between the autoinhibitory region and the catalytic domain are about 40 hydrogen bonds and 400 van der Waals contacts in the twitchin structures. The total accessible surface area buried from a 1.4-Å probe in the interface between the autoinhibitory sequence and the catalytic core amounts to about 6,500 Å^2, and approximately 20% of the surface of the catalytic core and 50% of the surface of the

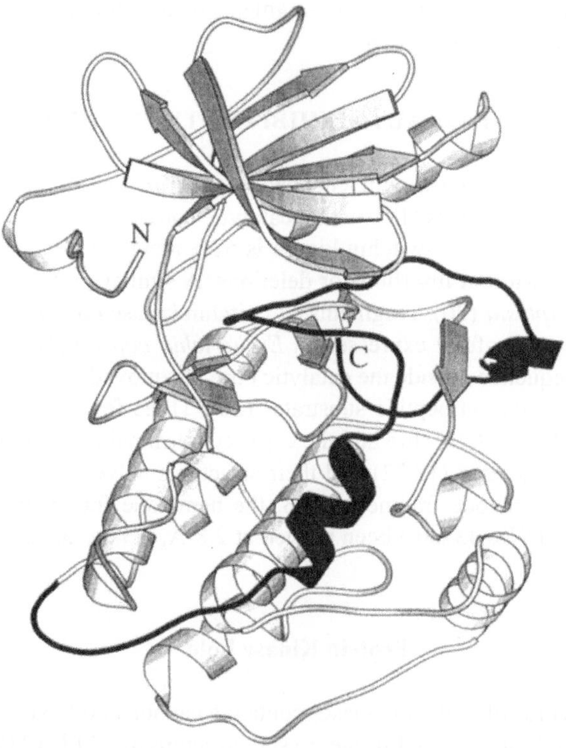

FIG. 2. A ribbon diagram of the structure of the *Aplysia* twitchin kinase fragment TWK-43 (drawn with the program MOLSCRIPT [34]). The autoregulatory sequence is shown dark.

inhibitory sequence get buried. The buried accessible surface area would correspond to a decrease in the calculated free energy of folding (25) of about 20 kcal/mol^{-1} for both the catalytic and the inhibitory regions. The degree of shape complementarity between the interacting surfaces of the catalytic and inhibitory regions is comparable to that of protein–protein interfaces.

The twitchin kinase fragments pJK3 from *C. elegans* and TWK-43 from *Aplysia* share 60% identity and have very similar structures (the root-mean-square deviation of the Cα atoms after superposition is 0.88 Å). The orientation of the two lobes is similar; the autoinhibitory sequences bind to the active site in a similar manner and form a similar number of interactions. The higher resolution of the structure of TWK-43 additionally reveals a number of water-mediated contacts between the inhibitory sequence and the catalytic core (18 water-mediated interactions with distances of less than 3.5 Å between the oxygen atom of water and the two hydrogen-bonding partners). At the N-terminus, TWK-43 is slightly shorter than the crystallized fragment of nematode twitchin pJK3, and does not form the two β-strands S1 and S2 observed in nematode twitchin. The sequence of TWK-43 only

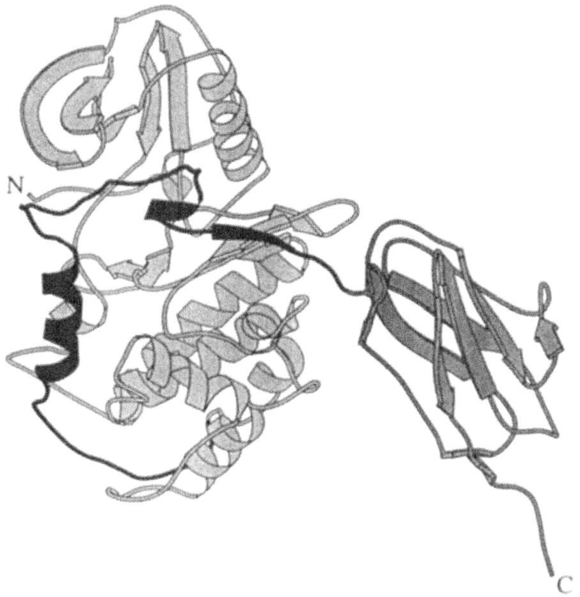

FIG. 3. A ribbon diagram of the structure of the *C. elegans* twitchin kinase fragment pJK4 (drawn using the program MOLSCRIPT [34]). The protein kinase catalytic core is the lightest area in the diagram, the Ig-26 domain is darker, and the autoregulatory sequence is darkest.

starts at the beginning of what is strand S2 in pJK3; the absence of the sequence corresponding to S1, which stabilizes the pair of strands in nematode twitchin, results in the disorder in TWK-43; the electron density is only apparent for residues C-terminal to residue 23. At the C-terminus, TWK-43 is slightly longer than pJK3; however, as compared to pJK3, only a few more C-terminal residues in TWK-43 are ordered.

Associated Noncatalytic Domains

In addition to the protein kinase domain, proteins of the smooth-muscle MLCK superfamily contain Fn-III-like and Ig-like domains (26) (Fig. 1). In some smooth muscles the C-terminal Ig-like domain of the MLCK is independently transcribed from the MLCK gene and expressed as the protein telokin. The crystal structure of telokin (27) shows a β-barrel fold related to the Ig-like domain of immunoglobulins. The Ig-like domain of titin kinase located on the amino terminal side of the protein kinase core, also contains β-structure (28).

The crystals of pJK4 allow us to observe the juxtaposition of the protein kinase catalytic core and the C-terminal Ig-26 domain (Fig. 3). The Ig-26 domain contacts the kinase domain on the opposite side of the active site. The domain is an elongated, antiparallel β-barrel with approximate dimensions of 45 × 25 × 30 Å; the β-strands run along its longest dimension. These strands project at an angle of 45° from the surface of the kinase domain, resulting in the longest dimension of Ig-26 projecting away from the surface of the kinase domain (Fig. 4). The Ig-26 domain interacts with the kinase domain through loops clustered near its N-terminus, and appears to be relatively rigidly associated with it.

Intrasteric Inhibition: Profile of Contacts

The mechanism of autoinhibition is very similar in *Aplysia* and *C. elegans* twitchin kinases. The autoinhibitory region inhibits the kinase activity by binding to residues of the catalytic core implicated in substrate binding, ATP binding, and catalysis. We will use the structure of *Aplysia* twitchin as the highest-resolution structure available to analyze the different contacts and compare them with the structure of cAPK bound to the PKI pseudosubstrate peptide.

Substrate-like Contacts

The inhibitory sequence of twitchin kinase traverses the entire substrate-binding groove. Substrate-specificity determinants in the local phosphorylation-site sequence are usually denoted P^{+1}, P^{+2} or P^{-1}, P^{-2}, and so forth, depending on their distance from the phosphate acceptor site, P0. Not much is known about the substrate specificity of twitchin kinase, except that the *Aplysia* enzyme can phosphorylate the regulatory myosin light chains (rMLC) in *Aplysia* (20). These contain arginines at positions P^{-3} and P^{-6}, but an Ala at position P^{-2}.

Helix H11 is parallel to the helical part of the inhibitory peptide PKI(5-24) bound to cAPK (29); the two helices are, however, about 10 Å away from each other. C-terminal to H11, residues 330 to 331 are in positions very similar to the PKI residues 16 and 17. Glutamic acids 203, 127, and 170 of cAPK, which are involved in electrostatic interactions with the basic substrate residues at positions P^{-6}, P^{-3}, and P^{-2}, respectively (29), are conserved in *Aplysia* TWK-43; the corresponding glutamic acids 215, 134, and 178 form close polar interactions with the basic residues Lys^{330}, Arg^{347}, and Lys^{334}, respectively (Fig. 4), of the inhibitory region. In the nematode twitchin kinase, Lys^{330} is replaced by a Ser and the charge of the glutamic acid is not neutralized. Helix H11 has been identified as the calmodulin-binding site of *Aplysia* twitchin kinase; however, calmodulin does not activate the kinase (30).

ATP-Binding-Site Contacts

The twitchin kinase structure does not contain a bound nucleotide, nor can nucleotides be soaked into the structure. The reason for this is that the autoinhibitory sequence sterically blocks the ATP-binding site. Essentially all the residues that form hydrogen bonds with ATP in cAPK (31) form close contacts with the inhibitory region in TWK-43. Along the same lines, some pseudosubstrate inhibitory peptides, including those of MLCK and the CaMK-II, act as both substrate and ATP antagonists.

```
                                                        β1
                                                    <-------
                                         34    40         50
cAPK                                   ...SQNTAQL-DQFDRIKTLG
twitchin                               ...EVKQGSVYDYYDILEELG
                                         43     50        60
                                                    <-------
                                                         S3

                 β2          β3      αB           αC
          >   <------->  <--------><----->    <----------->
                  60        70        80        90       100
cAPK      TGSFGRVMLVKHKESGNHYAMKILDKQKVVKLKQIEHTLNEKRILQAVNF
twitchin  SGAFGVVHRCVEKATGRVFVAKFINTP---YPLDKYTVKNEISIMNQLHH
                  70        80           90        100
          ->  <------->    <------->     <-------------->
                 S4           S5                H3

                 β4        β5            αD                αE
               <----->  <------>     <------->     <-----------
                 110       120       130        140       150
cAPK      PFLVKLEFSFKDNSNLYMVMEYVAGGEMFSHLRRI-GRFSEPHARFYAAQI
twitchin  PKLINLHDAFEDKYEMVLILEFLSGGELFDRIAAEDYKMSEAEVINYMRQA
          110       120       130       140       150
               <----->  <------>     <-->         <-----------
                 S6        S7          H4                  H5

                     β6         β7        β8      β9
          -------->  <>        <-->      <-->     <>
                 160       170         180       190        200
cAPK      VLTFEYLHSLDLIYRDLKPENLLIDQ--QGYIQVTDFGFAKRVKG-RTWTLC-G
twitchin  CEGLKHMHEHSIVHLDIKPENIMCETKKASSVKIIDFGLATKLNPDEIVKVTTA
         160       170       180       190       200       210
          -------->  <>        <-->      <-->     <>      <-->
                     S8         S9       S10      S11      S12

                  α                αF                     αG
                <-->       <-------------->        <--------
                 210       220       230       240       250
cAPK      TPEYLAPEIILSKGYNKAVDWWALGVLIYEMAAGYPPFFADQPIQIYEKI
twitchin  TAEFAAPEIVDREPVGFYTDMWAIGVLGYVLLSGLSPFAGEDDLETLQNV
               220       230       240       250       260
           <->  <-->      <---------------->        <--------
            H6   H7               H8                     H9

                              αH                     αI
          ->              <-------->               <--->
                    260       270       280       290      299
cAPK      VSGKVRFP----SHFSSDLKDLLRNLLQVDLTKRFGNLKNGVNDIKNHKWFAT...
twitchin  KRCDWEFDEDAFSSVSPEAKDFIKNLLQKEPRKRLT-----VHDALEHPWLKG...
               270       280       290           300       310
          ->              <------>                <---->
                             H10                    H11
```

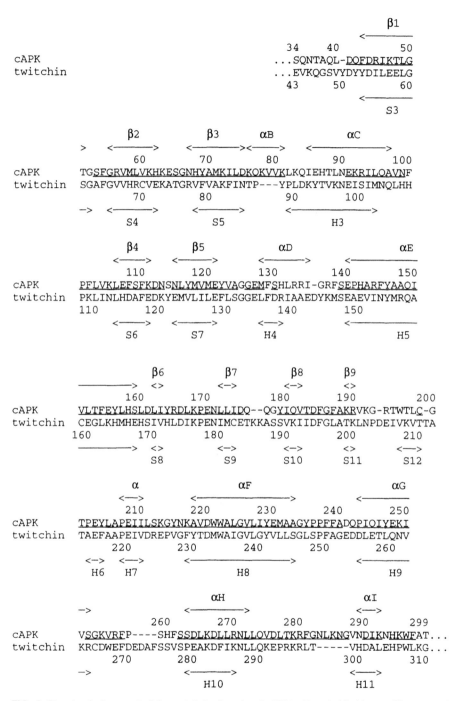

FIG. 4. Structural alignment of the catalytic domain of cAPK with twitchin kinase. Elements of secondary structure are indicated above the sequence of cAPK (α: α-helices; β: β-strands) and below the sequence of twitchin (H: α-helices; S: β-strands). The regions in which the root-mean-square deviations of the Cα atoms are less than 2 Å after separate superposition of small and large domains of the two proteins are underlined.

Catalytic Residue Contacts

Most of the twitchin kinase residues implicated in catalysis (31) contact the inhibitory region. The glycine-rich loop has been implicated in anchoring the β-phosphate of ATP; the corresponding sequence Gly^{60}–Thr-Gly-Ser-Phe-Gly^{65} in TWK-43 contacts many residues between residues 338 and 362 of the inhibitory region. Lys^{82}, equivalent to Lys^{72} of cAPK, which neutralizes the charges of α- and β-phosphates of ATP, contacts inhibitory residues 348 and 351; the side-chain nitrogen atom forms hydrogen bonds with the carbonyls of these residues. Glu^{98} (Glu^{91} of cAPK, involved in correctly positioning Lys^{72}) hydrogen-bonds Ser^{353} of the inhibitory region. Asp^{194} (Asp^{184} in cAPK, involved in Mg^{2+} binding) contacts Phe^{351}. The catalytic base Asp^{174} (Asp^{166} of cAPK) forms a salt link with Arg^{355}. These two aspartates are conserved in all protein kinases because of their crucial role in catalysis. Lys^{176} (Lys^{168} of cAPK suggested as binding the phosphate before and after transfer) is also close to Arg^{355}.

Implications for Other Autoregulated Protein Kinases

The rich diversity of contacts within the active site seen with the structure of twitchin kinase well exemplifies the features of intrasteric regulation. We can expect many of these features to also be found in other autoregulated protein kinases, especially the other members of the MLCK subfamily and the calmodulin-dependent and plant calcium-dependent protein kinases. There is clearly considerable room for diversity in terms of the detailed interactions within the active site and mechanisms of activation.

Inspection of the known autoregulatory sequences of the twitchin kinase family shows that it is exceptional in having a Pro-rich segment inserted in the autoinhibitory sequence. This sequence permits the sharp turning out of the substrate-binding groove occupied by the H11 helix, and penetrates deep into the active site to position two short 3_{10} helices to contact the key catalytic residues. In skeletal-muscle MLCK, for example, the autoinhibitory sequence does not appear to obstruct ATP binding as assessed with 5′-*p*-fluorosulfonylbenzoyl adenosine labeling (32), and is therefore less likely to have these contacts. Because the inactive form of CaMK-II does not bind ATP there may in this instance be some contacts that mimic those of twitchin kinase to block ATP binding (33).

CONCLUSION

Intrasteric regulation represents an important regulatory mechanism of protein kinases, which are involved in the control of almost all known physiologic processes. The crystal structures of autoregulated twitchin kinases provide essential insights into their mechanisms of autoinhibition and activation. However, because these protein kinases are so comprehensively turned off in their autoinhibited states, it is difficult to understand how they can be activated. Further structural and biochemical studies are required to understand the choreography of the entire regulatory process.

Intrasteric regulation as observed in twitchin kinase is only one example of the complex interplay of regulatory mechanisms governing the activity of the protein kinases. Even individual protein kinases of the MLCK subfamily, and more distant proteins belonging to the subfamilies of CaMK and the plant calcium-regulated protein kinases, exhibit high diversity of their regulatory sequences and mechanisms of activation. In addition, many protein kinases use complex combinations of different regulatory mechanisms to achieve particular regulatory properties; these include intrasteric autoinhibition, allosteric activation, and different types of phosphorylation events. With genes for more than a thousand different protein kinases estimated to be present in the mammalian genome, and every one of these enzymes playing a specific role in a specific biologic process, it is perhaps not surprising that such a sophisticated mechanism is required to keep the cell under control.

REFERENCES

1. Fischer EH, Krebs EG. Commentary on "The phosphorylase b to a converting enzyme of rabbit skeletal muscle." *Biochim Biophys Acta* 1989; 1000: 297–301.
2. Kemp BE, Faux MC, Means AR, House C, Tiganis T, Hu S-H, Mitchelhill KI. Structural aspects: pseudosubstrate and substrate interactions. In: Woodgett JR, ed., *Protein kinases*. Oxford: IRL Press; 1994; 30–67.
3. Corbin JD, Keely SL, Park CR. The distribution and dissociation of cyclic adenosine 3′:5′-monophosphate-dependent protein kinases in adipose, cardiac, and other tissues. *J Biol Chem* 1975; 250: 218–225.
4. Rosen OM, Erlichman J, Rubin CS. Molecular characterization of cyclic AMP-dependent protein kinases derived from bovine heart and human erythrocytes. In: Fischer EH, Krebs EG, Neurath H, Stadtman ER, eds., *International symposium of metabolic interconversion of enzymes*. Berlin: Springer-Verlag; 1973; 143–154.
5. Witt JJ, Roskoski R, Jr. Bovine brain adenosine 3′,5′-monophosphate dependent protein kinase. Mechanism of regulatory subunit inhibition of the catalytic subunit. *Biochemistry* 1975; 14: 4503–4507.
6. Corbin JD, Sugden PH, West L, Flockhart DA, Lincoln TM, McCarthy D. Studies on the properties and mode of action of the purified regulatory subunit of bovine heart adenosine 3′-5′-monophosphate-dependent protein kinase. *J Biol Chem* 1978; 253: 3997–4003.
7. Demaille JG, Peters KA, Fischer EH. Isolation and properties of the rabbit skeletal muscle protein inhibitor of adenosine 3′,5′-monophosphate dependent protein kinases. *Biochemistry* 1977; 16: 3080–3086.
8. Kemp BE, Barden JA, Kobe B, House C, Parker MW. Intrasteric regulation of calmodulin-dependent protein kinases. *Adv Pharmacol* 1996; 36: 221–249.
9. Taylor SS, Knighton DR, Zheng J, Sowadski JM, Gibbs CS, Zoller MJ. A template for the protein kinase family. *Trends Biochem Sci* 1993; 18: 84–89.
10. Hu S-H, Parker MW, Lei JY, Wilce MCJ, Benian GM, Kemp BE. Insights into autoregulation from the crystal structure of twitchin kinase. *Nature* 1994; 369: 581–584.
11. Trinick J. Titin and nebulin: protein rules in muscle. *Trends Biochem Sci* 1994; 19: 405–409.
12. Benian GM, Kiff JE, Neckleman N, Moerman DG, Waterston RH. Sequence of an unusually large protein implicated in regulation of myosin activity in *C. elegans*. *Nature* 1989; 342: 45–50.
13. Kemp BE, Pearson RB. Intrasteric regulation of protein kinases and phosphatases. *Biochim Biophys Acta* 1991; 1094: 67–76.
14. Kemp BE, Pearson RB, Guerriero V, Bagchi IC, Means AR. The calmodulin binding domain of chicken smooth muscle myosin light chain kinase contains a pseudosubstrate sequence. *J Biol Chem* 1987; 262: 2542–2548.

15. Lukas TJ, Burgess WH, Prendergast FG, Lau W, Watterson DM. Calmodulin binding domains: characterization of a phosphorylation and calmodulin binding site from myosin light chain kinase. *Biochemistry* 1986; 25: 1458–1464.
16. Kemp BE, Pearson RB. Spatial requirements for location of basic residues in peptide substrates for smooth muscle myosin light chain kinase. *J Biol Chem* 1985; 260: 3355–3359.
17. Kennely PJ, Edelman AM, Blumenthal DK, Krebs EG. Rabbit skeletal muscle myosin light chain kinase. The calmodulin binding domain as a potential active site-directed inhibitory domain. *J Biol Chem* 1987; 262: 11958–11963.
18. Kobe B, Heierhorst J, Feil SC, Parker MW, Benian GM, Weiss KR, Kemp BE. Giant protein kinases: domain interactions and structural basis of autoregulation. *EMBO J* 1996; 15: 6810–6821.
19. Benian GM, L'Hernault SW, Morris ME. Additional sequence complexity in the muscle gene, *unc-22*, and its encoded protein, twitchin, of *Caenorhabditis elegans*. *Genetics* 1993; 134: 1097–1104.
20. Heierhorst J, Probst WC, Kohanski RA, Buku A, Weiss KR. Phosphorylation of myosin regulatory light chains by the molluscan twitchin kinase. *Eur J Biochem* 1995; 233: 426–431.
21. Lei J, Tang X, Chamber TC, Pohl J, Benian GM. The protein kinase domain of twitchin has protein kinase activity and an autoinhibitory region. *J Biol Chem* 1994; 269: 21078–21085.
22. Hu S-H, Lei JY, Wilce MCJ, Valenzuela MRL, Benian GM, Parker MW, Kemp BE. Crystallization and preliminary X-ray analysis of the auto-inhibited twitchin kinase. *J Mol Biol* 1994; 236: 1259–1261.
23. Knighton DR, Bell SM, Zheng J, Ten Eyck LF, Xuong N-H, Taylor SS, Sowadski JM. 2.0 Å refined crystal structure of the catalytic subunit of cAMP-dependent protein kinase complexed with a peptide inhibitor and detergent. *Acta Crystallogr* 1993; D49: 357–361.
24. Olah GA, Mitchell RD, Sosnick TR, Walsh DA, Trewhella J. Solution structure of the cAMP-dependent protein kinase catalytic subunit and its contraction upon binding the protein kinase inhibitor peptide. *Biochemistry* 1993; 32: 3649–3657.
25. Eisenberg D, McLachlan AD. Solvation energy in protein folding and binding. *Nature* 1986; 319: 199–203.
26. Olson NJ, Pearson RB, Needleman DS, Hurwitz MY, Kemp BE, Means AR. Regulatory and structural motifs of chicken gizzard myosin light chain kinase. *Proc Natl Acad Sci USA* 1990; 87: 2284–2288.
27. Holden HM, Ito M, Hartshorne DJ, Rayment I. X-ray structure determination of telokin, the C-terminal domain of myosin light chain kinase at 2.8 Å resolution. *J Mol Biol* 1992; 227: 840–851.
28. Politou AS, Gautel M, Pfuhl M, Labeit S, Pastore A. Immunoglobulin-type domains of titin: same fold, different stability? *Biochemistry* 1994; 33: 4730–4737.
29. Knighton DR, Zheng J, Ten Eyck LF, Xuong N-H, Taylor SS, Sowadski JM. Structure of a peptide inhibitor bound to the catalytic subunit of cyclic adenosine monophosphate-dependent protein kinase. *Science* 1991; 253: 414–420.
30. Heierhorst J, Probst W, Vilim FS, Buku A, Weiss KR. Autophosphorylation of molluscan twitchin and interaction of its kinase domain with calcium/calmodulin. *J Biol Chem* 1994; 269: 21086–21093.
31. Bossemeyer D, Engh RA, Kinzel V, Ponstingl H, Huber R. Phosphotransferase and substrate binding mechanism of the cAMP-dependent protein kinase catalytic subunit from porcine heart as deduced from the 2.0 Å structure of the complex with Mn^{2+} adenyl imidodiphosphate and inhibitor peptide PKI(5-24). *EMBO J* 1993; 12: 849–859.
32. Kennely PJ, Colburn JC, Lorenzen J, Edelman AM, Stull JT, Krebs EG. Activation mechanism of rabbit skeletal muscle myosin light chain kinase. *FEBS Lett* 1991; 286: 217–220.
33. Shulman H. The multiufunctional Ca2+/calmodulin-dependent protein kinases. *Curr Opin Cell Biol* 1993; 5: 247–253.
34. Kraulis P. MOLSCRIPT: a program to produce both detailed and schematic plots of protein structures. *J Appl Cryst* 1991; 24: 946–950.

Signal Transduction in Health and Disease,
Advances in Second Messenger and Phosphoprotein Research, Vol. 31, edited by J. Corbin and S. Francis.
Lippincott–Raven Publishers, Philadelphia © 1997.

4

Specificity in Protein-Tyrosine Kinase Signaling

Lewis C. Cantley and Zhou Songyang

Department of Cell Biology, Harvard Medical School, and Division of Signal Transduction, Beth Israel Deaconess Medical Center, Boston, MA 02115; and Department of Biology, Massachusetts Institute of Technology, Cambridge, Massachusetts 02139

INTRODUCTION

In higher eukaryotic cells, protein-tyrosine kinases (PTKs) have evolved to provide specific responses to a host of different cellular activators including growth factors, differentiation factors, and hormones (1). On the basis of structure, these kinases fall into two major classes. The receptor PTKs (e.g., platelet-derived growth factor [PDGF], epidermal growth factor [EGF], fibroblast growth factor [FGF], nerve growth factor [NGF], and insulin receptors) span the lipid bilayer and have catalytic domains in the cytosol that are regulated by binding ligands to the extracellular domain. In contrast, the cytosolic PTKs (e.g., $pp60^{c\text{-}src}$, Janus kinases (JAKs), zeta-associated protein (ZAP70), and focal adhesion kinase (FAK) do not span the bilayer and are indirectly activated by extracellular ligands that bind to transmembrane receptors that are distinct from the receptor PTKs (e.g., cytokine receptors or integrins).

In mammalian organisms, genes for hundreds of PTKs have been cloned, and the functions of most of these enzymes are not understood. In a typical human cell, it is likely that more than 50 PTKs are expressed and play roles in growth, differentiation, or hormone responses. The functions of some of these kinases may be redundant. However, it is clear that cells have the capacity to respond differently to different growth factors and hormones, and there is clear evidence that this is a consequence of the ability of different PTKs (activated by distinct cellular activators) to stimulate different cellular responses.

Multiple factors could contribute to specificity in cellular responses to the activation of individual PTKs. For example, different PTKs might activate the same target(s) but do so at different subcellular locations and thereby provide distinct cellular responses. Alternatively, different PTKs might activate the same target(s) at the same location, but for different periods, thereby providing distinct cellular responses. There is evidence that variations in the location and duration of signals do contribute to specificity in signaling.

There is also ample evidence that distinct PTKs associate with and/or phosphorylate distinct cellular proteins. The structural basis for this specificity is the focus of this chapter. At least three factors contribute to the specificity of PTK-mediated signaling: (i) The location of the PTK in the cell restricts the number of possible targets. (ii) The specificity of the catalytic site of the PTK for Tyr residues in a specific sequence or structural context limits the number of productive targets. (iii) Domains outside the catalytic site interact with specific domains of target proteins to provide specific recruitment. Examples of this last case include src homology 2 (SH2) domains, src homology 3 (SH3) domains, and phosphoTyr-binding (PTB)/ phosphoTyr-interacting domain (PID). It is now clear that the catalytic site and these other interaction domains work in concert to provide specificity in signaling.

SPECIFICITIES OF CATALYTIC SITES OF PROTEIN-TYROSINE KINASES

The catalytic-site specificities of only a few protein kinases have been determined. Historically, this has been accomplished by mapping the sites that are phosphorylated by a given protein kinase *in vivo* or *in vitro*, synthesizing peptides based on the sequences surrounding the phosphorylation sites for use as substrates, and then determining the effect of single amino-acid changes in the peptide substrate on K_m and V_{max}. This approach is laborious and does not necessarily yield the optimal sequence for a peptide substrate, since it is not practical to make the billions of peptide substrates needed to test all possible sequence variations of an octapeptide substrate.

Recently, we designed an oriented peptide library technique for determining the optimal peptide substrate for protein kinases (2,3). In this approach, every peptide in the library has the same length and each has a Tyr residue at the same position (e.g., residue 7). The amino acids at positions N-terminal and C-terminal to the Tyr are variable (all amino acids except Tyr, Ser, Thr, Cys, and Trp). The initial library used for these studies had four positions of degeneracy N-terminal to the Tyr and four positions of degeneracy C-terminal to the Tyr for a total degeneracy of $15^8 > 2.5$ billion peptides. The peptides in this library all contained Met-Ala at the first two positions and Ala-Lys-Lys-Lys at the C-terminus, for a total length of 15 residues. This mixture of partly degenerate peptides is presented to the kinase of interest in the presence of adenosine triphosphate (ATP) and Mn^{2+} to allow phosphorylation of the subgroup of most favorable substrates. The mixture of phosphopeptide products is then separated from the nonphosphorylated peptides using a ferric-imido-diacetate (IDA) column, which specifically binds to the phosphate moiety. The mixture of phosphopeptides is then sequenced in bulk and this sequence is compared to the sequence of the starting mixture. The results provide a tremendous amount of information about the specificity of the kinase. For example, if the kinase has a strong preference for substrates with Ile located one residue to the N-terminal side of the phosphorylated Tyr (at residue 7), then Ile will be found as the most abundant amino acid at residue 6 (the P^{-1} position). The relative abundance of the 15 amino acids at residue 6 in the peptide products gives a prediction of which residues are preferred at this position and which are selected against.

An assumption implicit in this sort of analysis is that the selectivity for amino acids at each position N-terminal or C-terminal to the Tyr residue is independent of the sequence at the other positions. While this is a good starting point, it is unlikely to be strictly true for all sequences and all kinases. For example, if the kinase prefers peptides with a specific secondary structure that depends on the interaction of two different amino acids (e.g., Glu at P^{-2} forming a salt bond with Lys at P^{+2}), then sequences that contain only one of these two residues will not be selected. Thus, although there will still be a selection for the subgroup of peptides that contain both amino acids at the correct locations, these peptides will be in low abundance in the library (1/225 of the peptides have Glu at P^{-2} and Lys at P^{+2}) compared with peptides that have Ile at P^{-1} (1/15 of the peptides). Clearly, if the kinase has an absolute requirement for a combination of amino acids, then only the subgroup of peptides with that combination will be phosphorylated and the motif will be apparent.

Table 1 presents the optimal nonapeptide substrate predicted for a group of cytosolic and receptor-type PTKs. The residues in boldface type are those that are most strongly selected. Each PTK examined had a unique optimal substrate. However, the cytosolic PTKs all preferred to phosphorylate peptides with Ile at the P^{-1} position and with a negatively-charged or short side chain at the P^{+1} position. In contrast, the receptor PTKs preferentially phosphorylated substrates with Glu at the P^{-1} position and with a large hydrophobic amino acid at P^{+1}. All of the PTKs preferentially phosphorylated peptides with hydrophobic amino acids at P^{+3}, and most preferred negatively charged amino acids at P^{-3} and/or P^{-2} positions.

We have compared the optimal substrates of PTKs to the sequences at the *in vivo* phosphorylation sites that have been attributed to these kinases (4). It should be pointed out that there is considerable uncertainty in determining which kinase is responsible for phosphorylating a given site *in vivo*. For example, one cannot necessarily assume that tyrosine residues phosphorylated *in vivo* in response to addition of PDGF are phosphorylated by the PDGF receptor PTK. The PDGF receptor is known

TABLE 1. *Optimal Substrate Sequences Recognized by Different Protein-Tyrosine Kinases*

	Position								
	−4	−3	−2	−1	0	+1	+2	+3	+4
Tyrosine Kinases									
c-fps/fes	**E**	**E**	**E**	**I**	**Y**	**E**	**E**	**I**	**E**
MT/c-src	D	**E**	**E**	**I**	**Y**	**G/E**	E	**F**	F
v-src	E	**E**	**E**	**I**	**Y**	**G/E**	**E**	**F**	**D**
lck	X	E	X	**I**	**Y**	**G**	V	**L**	F
c-abl	A*	X	V	**I**	**Y**	**A**	A	**P**	F
EGF receptor	E	**E**	**E**	**E**	**Y**	**F**	E	L	V
PDGF receptor	E	**E**	**E**	**E**	**Y**	**V**	F	I	X
FGF receptor	A*	E	E	E	**Y**	F	**F**	L	F
Insulin receptor	X	E	E	E	**Y**	**M**	**M**	**M**	**M**

*Partly due to lag from previous sequencing cycle.

to bind and activate members of the src family of PTKs, which can in turn phosphorylate other proteins on Tyr. Similarly, it is clear that in cells transformed by the v-*src* oncogene, the FAK PTK is activated. The assignment of a particular PTK as responsible for the phosphorylation of a given site on a protein is strengthened if one can demonstrate that this same site is preferentially phosphorylated by the same kinase *in vitro*.

In any event, most of the sites phosphorylated on the PDGF receptor *in vivo* in response to PDGF are in good agreement with the optimal peptide substrate motifs found for this kinase. These sites are summarized in Table 2 and are compared with the optimal sequence predicted with the peptide library. Most of the sites have hydrophobic amino acids at P^{+1} and P^{+3} and negatively charged residues N-terminal to the Tyr.

It is likely that PTK autophosphorylation sites (or transphosphorylation sites of stable dimers) will have less evolutionary pressure to maintain optimal motifs for phosphorylation than sites on substrates that are not tightly bound. This is because the Tyr residue is never far from the catalytic site and thus does not require a low K_m in order to be phosphorylated. It is therefore not surprising that the various autophosphorylation sites on the PDGF receptor are not strictly conserved to the optimal motif. In fact, as pointed out subsequently, variations in the motifs at these sites provide binding sites for distinct SH2-containing proteins.

The expectation is that loosely bound substrates will have evolved phosphorylation sites that more closely resemble the optimal motif for the PTK catalytic site, since the local concentrations of these substrates will not be as high. In good agreement with this prediction, the phosphorylation sites on insulin receptor substrate-1 (IRS-1) (the major substrate of the insulin receptor) closely resemble the optimal motif of the insulin receptor PTK (Table 2).

The best evidence to date that the catalytic-site specificities of PTKs are critical for *in vivo* responses comes from studies of the RET receptor-type PTK. A single point mutation in the catalytic site of RET results in a dominantly inherited disease called multiple endocrine neoplasia type 2B (MEN2B). The point mutation occurs at a residue that is predicted to be involved in binding peptide substrates. Indeed, we showed that this mutation switches the peptide substrate specificity of the RET PTK (3). The mutant RET PTK preferentially phosphorylates substrates similar to those normally phosphorylated by cytosolic PTKs such as src and abl. These results indicate that inappropriate specificity of the catalytic site of a PTK in an otherwise normal protein can result in human cancers.

SH2 DOMAINS AND PTB/PID DOMAINS PROVIDE SEQUENCE-SPECIFIC AND PHOSPHORYLATION-DEPENDENT PROTEIN INTERACTIONS

Although catalytic-site specificities of PTKs are important, it is clear that other domains on these proteins play critical roles in specifying targets *in vivo*. In particu-

TABLE 2. *Comparison of Known Substrates of SH2-Containing PTKs and Receptor PTKs*

Substrates	Phosphorylation Sites	PTKs
Gastrin	E E A **Y** G W M D	v-src *in vitro*
Enolase	T G I **Y** E A L E	v-src *in vitro*
p34^{cdc2} peptide@	E G T **Y** G V V Y	v-src *in vitro*
Mouse Polyoma middle t	E E E **Y** M P M E	c-src *in vivo*
	N P T **Y** S V M	c-src *in vivo*
Hamster Polyoma middle t	E N E **Y** M P M A	c-src *in vivo*
Hamster Polyoma middle t	E P Q **Y** E E I P	c-src *in vivo*
PtdIns 3-kinase p85 subunit*	E D Q **Y** S L V E	c-src *in vivo*
v-src Autophosphorylation	D N E **Y** T A R Q	v-src *in vivo*
v-src Autophosphorylation?	L Y D **Y** E S W I	v-src *in vivo*
	L **Y** D Y E S	v-src *in vivo*
p125FAK	T D D **Y** A E I I	c-src?
"	E D T **Y** T M P S?	c-src? FAK?
"	D K V **Y** E N V T	c-src *in vivo*
T-cell receptor ζ+	E G L **Y** N E L Q	lck/fyn *in vivo*
T-cell receptor ζ+	A E A **Y** S E I G	lck/fyn *in vivo*
Annexin II heavy chain	P S A **Y** G S V K	v-src *in vivo*
ras-GAP	K E I **Y** N T I R	v-src *in vivo*
p130 in v-crk X-formed cells	X D I **Y** D V P P	v-src, abl? *in vivo*
(15 potential sites)	E V Q	
paxillin	N H T **Y** Q E I A?	v-src, abl *in vivo*
"	E T P **Y** S Y P T?	
"	E H V **Y** S F P N?	
chicken cortactin	S P V **Y** Q D A S?	v-src *in vivo*
"	E A E **Y** E P E T?	
"	E T V **Y** E V A G?	
"	E N T **Y** D E Y E?	
HS1	E P V **Y** E A E P?	
"	E N D **Y** E D V E?	
"	E G D **Y** E E V L?	
CRK II	P G P **Y** A Q P S	c-abl, *in vivo*
IRS-1	D D G **Y** M P M S	INR, *in vivo*
	I E E **Y** T E M M	
	S G D **Y** M P M S	
	P N G **Y** M M M S	
	T G D **Y** M N M S	
	T E E **Y** M K M D	
	R G D **Y** M T M Q	
	P V S **Y** A D M R	
	L S N **Y** I C M G	
EGF receptor	V P E **Y** I N Q S	EGFR, *in vivo*
"	N P V **Y** H N Q P	
"	N P E **Y** L N T V	
"	N P D **Y** Q Q D F	
"	N A E **Y** L R V A	
"	A D E **Y** L I P Q	
Lipocortin I	E Q E **Y** V Q T V	EGFR, *in vivo*
Myosin light chain	D E M **Y** R E A P	EGFR, *in vivo*
"	N F N **Y** V E F T	
Phospholipase C-γ	E P D **Y** G A L Y	EGFR & PDGFR, *in vivo*
"	P G F **Y** V E A N	
"	E A R **Y** Q Q P F	
ezrin	K S G **Y** L S S E	EGF receptor, *in vivo*
"	L Q D **Y** E E K T	

TABLE 2. *Continued*

Substrates	Phosphorylation Sites	PTKs
Human SHC	D P S **Y** V N V Q	EGFR, PDGFR, INR, *in vivo*
Hum PDGFR β	G H E **Y** I Y V D	PDGFR, *in vivo*
"	E Y I **Y** V D P M	
"	D G G **Y** M D M S	
"	S V D **Y** V P M L	
"	S S N **Y** M A P Y	
"	S V L **Y** T A V Q	
"	D N D **Y** I I P L	
ras-GAP	K E I **Y** N T I R	PDGFR,EGFR,*in vitro*
SHPTP2	A R V **Y** E N V G?	PDGFR, c-src?
"	G H E **Y** T N I K?	

lar, the autophosphorylation sites on receptor PTKs allow interaction with specific SH2-containing proteins. Since much of the binding energy of the SH2 domain comes from interaction with the phosphate residue, these interactions are triggered by phosphorylation and eliminated by dephosphorylation. Individual SH2 domains have evolved the ability to bind to Tyr-phosphate (Tyr-P) moieties in specific sequence contexts. Thus, when a PTK undergoes autophosphorylation, the sequence surrounding the Tyr-P moiety dictates which SH2-containing protein (or proteins) will subsequently bind.

Many cytosolic PTKs have endogenous SH2 domains. In some cases, these domains play a role in autoregulation, but in other cases they bring the PTK to a previously phosphorylated substrate (see below).

We designed an oriented Tyr-P peptide library for use in determining SH2-domain specificity (5,6). The approach was analogous to that discussed above for determining PTK specificity except that the subgroup of Tyr-P-containing peptides that bind with highest affinity to a given SH2 domain was selected, using an SH2-domain affinity column. These results, along with subsequent crystal structures and solution structures of SH2-domain/Tyr-P peptide complexes, have provided considerable insight into the structural basis for how selectivity is maintained (7–10).

A comparison of the optimal motifs of SH2 domains to optimal motifs of PTK catalytic sites reveals that these two unrelated structures have converged to recognize similar primary sequences. For example, SH2 domains found in cytosolic PTKs (i.e., group I SH2 domains containing Tyr or Phe at the βD5 position [5]) preferentially bind peptides with the motif Tyr-P-hydrophilic-hydrophilic-hydrophobic. This motif is in agreement with the optimal peptide substrate motif for the catalytic site of cytosolic PTKs. This observation has suggested a model in which the cytosolic PTKs preferentially phosphorylate sites that will bind to their own SH2 domains, allowing further phosphorylation of additional (less optimal) sites on the same protein. Indeed, there is evidence that the abl PTK progressively phosphorylates the $p130^{CAS}$ protein (11) and the tail of RNA polymerase II (12). Alternatively, phosphorylation of a substrate at an optimal site for one PTK catalytic site may create an optimal binding site for another SH2-containing PTK, which then phosphorylates additional sites on the same substrate.

In contrast to the group I SH2 domains, group III SH2 domains (those with aliphatic residues at the βD5 position) preferentially bind sequences with the motif Tyr-P-hydrophobic-X-hydrophobic. This motif is in good agreement with the optimal substrate motif of receptor PTKs (Table 1), and suggests that these kinases have evolved to preferentially interact with signaling molecules containing group III SH2 domains (e.g., phosphoinositide-3-kinase, phospholipase-C-γ, and SHPTP2).

Finally, another domain was recently shown to bind to Tyr-P moieties, the PTB domain or PID (13,14). This domain does not resemble SH2 domains and thus has independently evolved the ability to recognize Tyr-P. On the basis of results with two members of this family, phosphopeptide selectivity is based on residues N-terminal to the Tyr-P moiety (15,16). The PTB/PID domain of SHC prefers the motif Asn-Pro-Xxx-Tyr-P (15,16,17). Since the major selectivities of the PTK catalytic sites that have been investigated to date are based on residues at the P^{-1} position or at positions C-terminal to Tyr, and since the SHC PTB/PID domain is not selective at these sites, it is likely that the SHC PTB/PID domain will bind to sites phosphorylated by a broad group of both cytosolic and receptor PTKs.

In summary, by determining the optimal peptide-sequence specificity of the catalytic sites of PTKs, and the optimal Tyr-P peptide motifs of SH2 and PTB/PID domains, it has become apparent how these various domains have coevolved to ensure fidelity in PTK signaling. As with other biologic systems in which specificity is important (e.g., transcription factor/deoxyribonucleic acid [DNA] complexes), an ultimately high selectivity is provided by multiple interactions with moderate selectivity.

REFERENCES

1. Hunter T, Cooper JA. Protein-tyrosine kinases. *Ann Rev Biochem* 1985; 54: 897–930.
2. Songyang Z, Blechner S, Hoagland N, Hoekstra MF, Piwnica WH, Cantley LC. Use of an oriented peptide library to determine the optimal substrates of protein kinases. *Curr Biol* 1994; 4: 973–82.
3. Songyang Z, Carraway KL III, Eck MJ, et al. Catalytic specificity of protein-tyrosine kinases is critical for selective signaling. *Nature* 1995; 373: 536–9.
4. Songyang Z, Cantley LC. Recognition and specificity in protein tyrosine kinase-mediated signaling. *TIBS* 1995; 20: 470–5.
5. Songyang Z, Shoelson SE, Chaudhuri M, et al. SH2 domains recognize specific phosphopeptide sequences. *Cell* 1993; 72: 767–78.
6. Songyang Z, Shoelson SE, McGlade J, et al. Specific motifs recognized by the SH2 domains of Csk, 3BP2, fps/fes, GRB-2, HCP, SHC, Syk and Vav. *Mol Cell Biol* 1994; 14: 2777–85.
7. Waksman G, Shoelson SE, Pant N, Cowburn D, Kuriyan J. Binding of a high affinity phosphotyrosyl peptide to the Src SH2 domain: crystal structures of the complexed and peptide-free forms. *Cell* 1993; 72: 779–90.
8. Eck MJ, Shoelson SE, Harrison SC. Recognition of a high-affinity phosphotyrosyl peptide by the Src homology-2 domain of p56lck. *Nature* 1993; 362: 87–91.
9. Pascal SM, Singer AU, Gish G, et al. Nuclear magnetic resonance structure of an SH2 domain of phospholipase C-gamma 1 complexed with a high affinity binding peptide. *Cell* 1994; 77:461–72.

10. Lee CH, Kominos D, Jacques S, et al. Crystal structures of peptide complexes of the amino-terminal SH2 domain of the Syp tyrosine phosphatase. *Structure* 1994; 2: 423–38.
11. Mayer BJ, Hirai H, Sakai R. SH2 domains and processive phosphorylation. *Curr Biol* 1995; 5: 296–305.
12. Duyster J, Baskaran R, Wang JY. Src homology 2 domain as a specificity determinant in the c-Abl-mediated tyrosine phosphorylation of the RNA polymerase II carboxyl-terminal repeated domain. *Proc Natl Acad Sci USA* 1995; 92: 1555–9.
13. Kavanaugh WM, Williams L. An alternative to SH2 domains for binding tyrosine-phosphorylated proteins. *Science* 1994; 266: 1862–5.
14. Blaikie P, Immanuel D, Wu J, Li N, Yajnik V, Margolis B. A region in Shc distinct from the SH2 domain can bind tyrosine-phosphorylated growth factor receptors. *J Biol Chem* 1994; 269: 32031–4.
15. Gustafson TA, He W, Craparo A, Schaub CD, O'Neill TJ. Phosphotyrosine-dependent interaction of SHC and insulin receptor substrate 1 with the NPEY motif of the insulin receptor via a novel non-SH2 domain. *Mol Cell Biol* 1995; 15: 2500–8.
16. Kavanaugh WM, Turck CW, Williams LT. PTB domain binding to signaling proteins through a sequence motif containing phosphotyrosine. *Science* 1995; 268: 1177–9.
17. Songyang Z, Margolis B, Chaudhuri M, Shoelson SE, Cantley LC. The phosphotyrosine interaction domain of SHC recognizes tyrosine-phosphorylated NPXY motif. *J Biol Chem* 1995; 270: 14863–6.

Signal Transduction in Health and Disease, Advances in Second Messenger and Phosphoprotein Research, Vol. 31, edited by J. Corbin and S. Francis.
Lippincott–Raven Publishers, Philadelphia © 1997.

5

Historical Perspectives and New Insights Involving the MAP Kinase Cascades

*Lee M. Graves, †Karin E. Bornfeldt, and ‡Edwin G. Krebs

**Department of Pharmacology, University of North Carolina at Chapel Hill, The School of Medicine, Chapel Hill, North Carolina 27599-7365; †Department of Pathology, University of Washington, Seattle, WA 98195; and ‡Department of Pharmacology, Howard Hughes Medical Institute, School of Medicine, University of Washington Medical Center, Seattle, Washington 98195*

DEDICATION

We would like to dedicate this chapter to the late Earl W. Sutherland, who, together with his colleagues T. W. Rall and Walter D. Wosilait, discovered cyclic adenosine monophosphate (cAMP), and in so doing were responsible for initiating the modern era of research on intracellular signaling (reviewed in [1]). This epochal event occurred about 40 years ago. In the ensuing decades, many remarkable advances in this field have been made, undoubtedly including some very recent ones that will be reported at this meeting. A number of these advances can also be traced to a second major development for which Sutherland and his group were responsible: the ability of investigators to demonstrate the action of hormones in broken cell systems (i.e., in homogenates). A third major finding by these investigators was their codiscovery of reversible protein phosphorylation. Independently of the work that was done on interconversion of the two forms of glycogen phosphorylase in skeletal muscle (2), Sutherland and his colleagues determined that interconversion of inactive and active liver phosphorylase also involves protein phosphorylation–dephosphorylation (3).

HISTORICAL COMMENTS

Continuing in a historical vein, we would like to go back almost 70 years ago and briefly discuss the work of Carl and Gerty Cori, the mentors of Sutherland and one of us (E. G. K.). Based on research done at Washington University in the late 1920s (4), the Coris formulated what became known as the Cori cycle. From *in vivo* experiments, they determined that for animals in the postabsorptive state (between meals), blood sugar derived from liver glycogen is utilized in muscle. The part of lactic acid formed from glucose that escapes oxidation in muscle is carried to the liver, where it is reconverted to glycogen. In their studies the Coris also showed that the hormone epinephrine enhances blood sugar levels; this occurs as a result of glucose production exceeding glucose utilization. On the other hand, insulin increases the utilization of blood sugar and causes the formation of glycogen. The Coris were intrigued by the action of these hormones, and during the 1930s they did work that set the stage for an understanding of these phenomena. Their work was especially important with respect to the action of epinephrine. In their research they established the immediate metabolic process involved in the breakdown of glycogen, and they discovered the enzyme glycogen phosphorylase, which catalyzes glycogen phosphorolysis and the formation of glucose 1-phosphate.

In the early 1940s, at a time when Earl Sutherland was a medical student at Washington University, the Coris purified phosphorylase to homogeneity and demonstrated that it exists in two interconvertible forms, which they referred to as phosphorylase b, a relatively inactive form, and phosphorylase a, an active form. The chemical nature of the interconversion of the two enzyme forms was not determined, but in the late 1940s and early 1950s, Sutherland, who after serving in the Second World War had become a member of the Coris' department, showed that epinephrine causes conversion of phosphorylase b to phosphorylase a in intact cells (5). Later, while working at Western Reserve University in his own laboratory, he was able to demonstrate that this action of the hormone involved production of the "second messenger," cAMP (reviewed in [1]).

Formulation of the first known protein kinase cascade also grew out of early work on glycogen metabolism. On the occasion of the Second Symposium on Catecholamines, held in Milan in 1965, the scheme shown in Fig. 1 was presented (6). At that time, solid evidence for the existence of the enzyme pictured as catalyzing the conversion of nonactivated phosphorylase kinase to its activated form had not yet been obtained; this enzyme was later identified as the cAMP-dependent protein kinase (7). Also at the Milan symposium, Sutherland and Robison reviewed the status of the second-messenger hypothesis, which by this time had become a well-established concept applicable to the many different actions of epinephrine and to other hormones that cause changes in cellular cAMP levels.

Although the mechanism of action of epinephrine was thus reasonably well established more than 30 years ago, the way in which insulin acts remained obscure. It proved to be much more difficult to establish its mechanisms of action than had been true for epinephrine. Even today we cannot say exactly how insulin acts in some of its primary functions, such as promoting cellular glucose uptake and increasing glycogen synthesis in tissues. Nonetheless, during the past decade remarkable progress has been made in understanding insulin action, as well as that of the other hormones and growth factors whose receptors are protein tyrosine kinases (PTKs). Stimulation of cells by these growth factors leads to activation of the mitogen-activated protein (MAP) kinase cascade (reviewed in [8]) (Fig. 2). This branched multistep signaling pathway impinges on many different cellular functions, and, in turn is regulated by, or itself regulates, other signaling pathways. In this chapter we will examine some of the properties of the MAP kinase pathway, with particular reference to its interactions with other regulatory systems.

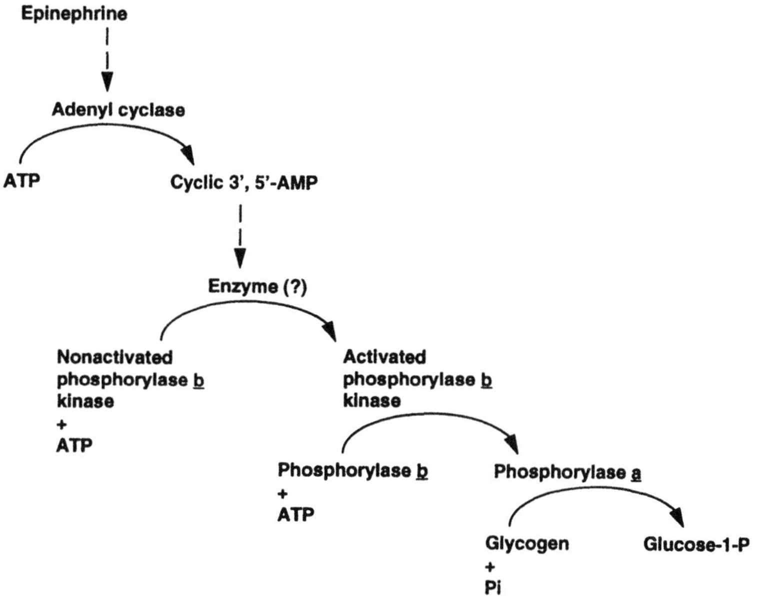

FIG. 1. Mechanism of action of epinephrine on glycogenolysis in muscle as presented at the Second Symposium on Catecholamines in Milan, 1965. (From the American Society for Pharmacology and Experimental Therapeutics; *Pharmacol Rev*, 1966;18:163–71 [6], with permission).

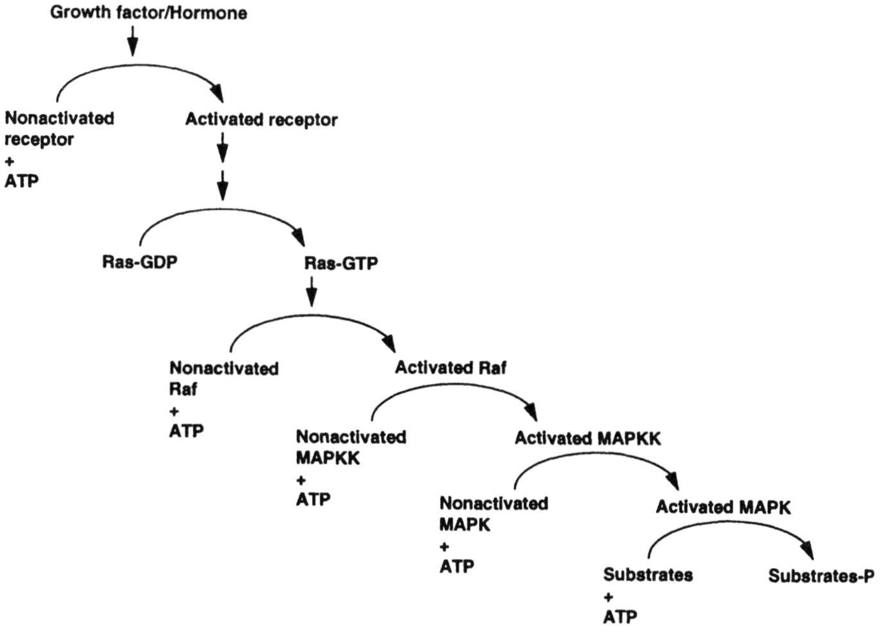

FIG. 2. The MAP kinase cascade as known today.

GENERAL ASPECTS OF PROTEIN-KINASE CASCADE SYSTEMS AND THEIR APPLICATION TO THE MAP KINASE CASCADE

With elucidation of the MAP kinase signaling pathway, it became apparent that the protein-kinase portion of this cascade contained many more successive cycles of phosphorylation–dephosphorylation than had previously been known to exist in other cascades. It is of interest to speculate about why such an elaborate pathway evolved. One possible answer is that a multistep pathway provides the means for signal amplification. For example, Stadtman and Chock (9) calculated the potential for amplification that exists in multicyclic protein kinase–protein phosphatase systems, and noted that under certain circumstances (appropriate K_m and V_{max} values for the enzymes involved) this potential can be exceedingly great. In fact, the extent of signal amplification could reach 100-fold with only two kinase steps. In the MAP kinase cascade, in which four consecutive kinase steps are present (i.e., from Raf to Rsk), the degree of amplification could be enormous indeed. It would therefore seem unlikely that amplification alone would constitute the *raison d'être* for the multiplicity of steps in the MAP kinase cascade.

A second possible reason for the existence of multiple steps in a protein-kinase cascade is that it provides numerous branch points that serve different sets of targets as the signal is relayed along the pathway. In the MAP kinase cascade, the degree of branching, as gauged by the number of substrates that have been identified for a given kinase, changes throughout the cascade. Raf and MAP kinase kinase (MAPKK), especially the latter, appear to be highly specific kinases (10). However, MAP kinase and Rsk (a MAP kinase substrate) have numerous substrates. MAP kinase transmits signals to many targets, both nuclear and nonnuclear, including, in addition to Rsk (11), several microtubule-associated proteins such as tau (12), cytosolic phospholipase A_2 ($cPLA_2$) (13), and transcription factors such as Elk-1 (reviewed in [14]). MAP kinase rapidly translocates to the nucleus upon activation (15), and recent findings have demonstrated that a substantial proportion of active MAP kinase is associated with the cytoskeleton (16). Consequently, MAP kinase signaling is effectively disseminated throughout the cell, a factor that is critical with respect to its activation by MAPKK and its ability to phosphorylate numerous substrates at diverse locations within the cell. In addition, Rsk, like MAP kinase, also phosphorylates nuclear as well as nonnuclear substrates, which will not be enumerated here.

A third benefit that derives from a multistep cascade system is that such a scheme provides an opportunity for the exertion of independent control over some (but not necessarily all) functions regulated by the cascade. How widely disseminated and how great an impact a given controlling signal will have depends on its site of action in the cascade. For example, an agent that acts on an early step in the cascade would be expected to affect multiple functions, whereas agents acting at the end of the cascade would affect only a few functions. The effect of cAMP and other interacting components on the MAP kinase cascade (see following sections) illustrates this point.

A multistep protein kinase cascade also lends itself to interesting patterns of feedback control, which will not be discussed in this chapter.

INTERACTIONS OF THE MAP KINASE PATHWAY WITH OTHER SIGNALING SYSTEMS

Signaling Pathways that Impinge on the MAP Kinase Pathway

Interaction with Cyclic Adenosine Monophosphate Signaling

Signaling through kinase cascades would be expected to be influenced by cross-talk with other protein kinase pathways. The result of this type of interaction is acute inhibition (or amplification) at specific points within the cascade, such as that described for regulation of the MAP kinase cascade by cAMP (reviewed in [17]). In many cell types, elevation of cAMP, through the action of the cAMP-dependent protein kinase A (PKA), inhibits the activation of MAP kinase by a wide array of stimuli. The list of such stimuli includes growth factors that activate receptor tyrosine kinases, G-protein-coupled signals, activators of PKC (e.g., phorbol esters), and inhibitors of protein serine phosphatases (e.g., okadaic acid) (18–22). Given the generality of this effect, a common upstream component is suggested, and the majority of evidence points to Raf as the site of regulation by PKA (20–22,23–28).

The mechanism by which cAMP regulates Raf is not completely resolved. Raf-1 (one of the major isoforms of Raf) is phosphorylated by PKA on Ser^{43}, and the result of this phosphorylation appears to be a decreased ability of Raf-1 to bind to Ras-guanosine triphosphate (GTP) (22,29); however, although lacking an analogous phosphorylation site at Ser^{43}, the other major isoform of Raf (B-Raf) is also inhibited in cells in which cAMP is elevated (30). In this instance, the binding of B-Raf to Ras-GTP is similarly inhibited, presumably through the phosphorylation of different sites catalyzed by PKA. For example, there are potential PKA phosphorylation sites at Ser^{429} and Ser^{446} in B-Raf (30), which might explain this antagonism. Hence, some evidence suggests that cAMP exerts an inhibitory effect on Raf signaling by preventing the association of this kinase with Ras-GTP and activation of the MAP kinase cascade.

Alternatively, it may be that active Raf itself is directly affected by PKA. Häfner et al. (31) reported that PKA directly inhibited the phosphotransferase activity of Raf through the phosphorylation of a site, or sites, within the catalytic domain of a truncated form of Raf (BXB-Raf, lacking the N-terminus). However, identification of these inhibitory site(s) remains to be established *in vivo*, and the inhibition of BXB-Raf by PKA has not been consistently found in all cell types (32). Instead, it seems likely that cAMP influences other aspects of Raf activation that have yet to be discovered.

To further complicate matters, in some cells cAMP neither inhibits the activation of MAP kinase nor correlates well with effects seen on Raf. This phenomenon seems evident in cells in which cAMP is generally found to play a stimulatory role on cell growth, such as pheochromocytoma (PC12) cells, thyroid cells, or Swiss 3T3 cells (26,33–35). In PC12 cells, cAMP stimulates neurite outgrowth, an effect that is causally associated with stimulation of the MAP kinase cascade (36,37). Surprisingly, in these cells, cAMP increases MAP kinase activity despite a simultaneous inhibition of Raf (30,33,37). The dissociation of Raf and MAP kinase activity in PC12 cells suggests that an alternative kinase(s) may substitute for Raf in a cAMP-dependent manner, an interesting possibility that remains to be established.

Heterotrimeric G-Protein-Coupled Signals

A number of stimuli activate MAP kinase through seven-transmembrane receptors that are coupled to the heterotrimeric G-proteins G_q and G_i (reviewed in [38]). Stimulation as such appears to be mediated by tyrosine phosphorylation and the small GTP-binding protein Ras. Although the mechanism of regulation is not completely understood, some studies have demonstrated a direct role for $G_{\beta\gamma}$ subunits in this signaling process (39,40,41). By contrast, stimulation of G_s-coupled receptors (i.e., β-adrenergic receptors) increases intracellular cAMP in many cell types (42), an effect that may be expected to inhibit the activation of MAP kinase (as described above). In support of this, Crespo and coworkers (40) found that in Cos-7 cells, MAP kinase signaling from β-adrenergic receptors was subject to dual regulation; MAP kinase was stimulated by the release of $G_{\beta\gamma}$ subunits and inhibited by the elevation of cAMP. Thus, the extent of MAP kinase activation is likely to depend on the relative activation of the competing stimulatory and inhibitory pathways present in a given cell type.

Interaction with Calcium-dependent Signaling Processes

Of further interest is the way in which additional second messengers such as calcium may also influence the MAP kinase pathway. Chao and coworkers (43) were the first to demonstrate that MAP kinase was activated in a calcium-dependent manner by treating cells with compounds such as thapsigargin or ionomycin, which increase intracellular calcium. The mechanism by which this was achieved was not elucidated, but one possible explanation included activation of PKC, which has been implicated in the activation of Raf (44). Recently, a novel calcium- and PKC-activated tyrosine kinase known as calcium-dependent tyrosine kinase (CADTK or PYK-2) was isolated, revealing new possibilities. CADTK was purified and identified as a major calcium-activated tyrosine kinase in rat liver epithelial cells (45). Simultaneously, the complementary deoxyribonucleic acid (cDNA) encoding a human calcium-dependent tyrosine kinase (PYK-2), was independently isolated by PCR cloning (46). Subsequent experiments have confirmed that CADTK and PYK-2 are identical kinases (Dr. H. S. Earp, *personal communication*). Treatment of PC12 cells with agents that increased intracellular calcium or activated PKC (phorbol esters) stimulated PYK-2 activity. Furthermore, overexpression of PYK-2 increased MAP kinase activity in these cells (46). Thus, an important connection linking calcium, PKC, tyrosine phosphorylation, and MAP kinase may be present in some cells. Of future interest will be to determine which additional calcium-stimulated kinase pathways are coupled to PYK-2 activation. Moreover, since there are numerous examples of antagonism and/or synergism between calcium and cAMP signaling, it may be rewarding to explore the influence of cAMP on this novel pathway.

Signaling Pathways Regulated by MAP Kinase

Rsk and Glycogen Synthase Kinase-3

MAP kinase has been shown to regulate transcription through the phosphorylation of nuclear substrates such as tertiary complex factor (TCF)/Elk-1 or serum response factor (SRF) transcription factors (for review see [14]). In addition, the serine/threonine kinase Rsk, which was one of the first substrates identified for MAP kinase, also contributes to transcriptional regulation, as well as targeting one or more cytosolic proteins (15,47). Another protein kinase that may be regulated through the MAP kinase cascade and which is also implicated in nuclear as well as cytosolic events (48,49) is the glycogen synthase kinase 3 (Gsk-3). Rsk and the p70 S6 kinase have been shown to phosphorylate and inhibit Gsk-3 activity *in vitro* (50,51), and Gsk-3 activity is reduced in cells incubated with growth factors or insulin, which activate both Rsk and p70 S6 kinase (52). In NIH 3T3 cells expressing a dominant negative form of MAPKK, the growth factor-induced inhibition of Gsk-3 activity is prevented, an effect that correlates with an inability to increase Rsk activity in these cells (53a). Thus, in these studies, a role for Rsk in regulating Gsk-3 *in vivo* was suggested. In similar studies, activation of the MAP kinase cascade was prevented by the use of the phosphotidyl inositol (PI)-3 kinase inhibitor wortmannin, an effect that also correlated with a loss of Gsk-3 inhibition (52,53b). However, recent studies using inhibitors of the MAP kinase cascade (cAMP or PD098059, an MAPKK inhibitor) (54), or of the p70 S6 kinase pathway (rapamycin [53b]) have suggested that an additional growth factor-activated kinase is involved. A candidate for this enzyme is the protein kinase Akt/RAC (55), also known as PKB, which may also be responsible for the phosphorylation and inhibition of Gsk-3 in response to growth factors (56a).

Protein Phosphatases

It is known that insulin and other growth factors, acting through the MAP kinase cascade, can regulate the serine/threonine phosphatase protein phosphatase-1 (PP-1) by mechanisms that involve modification of its regulatory subunits. For example, Cohen and his collaborators have proposed that in skeletal muscle, phosphorylation of the glycogen-binding subunit of PP-1 (i.e., the "G-subunit") by activated Rsk increases the affinity of the catalytic subunit of PP-1 for the G-subunit, thereby increasing the effective phosphatase activity at the glycogen particle in skeletal muscle (56b). This event is antagonized by phosphorylation of the G-subunit by PKA (57,58); however, given the inhibitory effects of cAMP on the MAP kinase cascade (described above), it seems plausible that cAMP could also influence PP-1 activity by interfering with the activation of Rsk. In support of this idea, it has been reported that cAMP agonists (Sp-cAMP) prevent the activation of PP-1 by insulin in adipocytes (59).

A second mechanism of PP-1 regulation involves the form of the enzyme in which the PP-1 catalytic subunit is bound to the regulatory subunit known as inhibitor 2 (I-2). Phosphorylation of I-2 by ATP-Mg results in activation of the enzyme. Early studies (60) suggested that this phosphorylation (involving Thr^{72}) was catalyzed by Gsk-3, but insofar as Gsk-3 is known to be inactivated through growth factor stimulation, this would result in decreased PP-1 activity. However, Depaoli-Roach and coworkers (61) recently demonstrated stoichiometric phosphorylation of I-2 by MAP kinase, thereby providing a mechanism by which MAP kinase may directly activate this phosphatase in response to growth factors. Thus, stimulation of the MAP kinase cascade may regulate protein phosphatase activity, an event that is predicted to be influenced by increasing cAMP in cells.

Phospholipase A2

Another well-characterized substrate for MAP kinase is the cytosolic phospholipase A_2 ($cPLA_2$) (13). Phosphorylation and activation of this enzyme provide a connection between MAP kinase and arachidonic acid metabolism; however, the physiologic consequence(s) of this signaling is not fully understood. Recently, we obtained evidence for a link between activation of the MAP kinase cascade and the regulation of cAMP synthesis by growth factors (62). In our studies, platelet-derived growth factor (PDGF) (in the absence of phosphodiesterase inhibitors) potently stimulated the formation of cAMP and activated the cAMP-dependent protein kinase (PKA) in newborn human arterial smooth-muscle cells (hSMC) (Fig. 3A,B). The stimulation of cAMP synthesis and PKA activity was rapid and paralleled the activation of MAP kinase in these cells. Furthermore, insulin-like growth factor-I (IGF-I), which did not increase MAP kinase activity (63), failed to increase cAMP synthesis in hSMC. Hence, the connection between growth factors and cAMP synthesis appeared to require MAP kinase activity.

Using indomethacin to inhibit prostaglandin formation, we determined that PDGF stimulated cAMP accumulation through a prostaglandin-dependent pathway, consistent with the findings in earlier studies by Rozengurt and coworkers (64). In accord with this, we found that PDGF potently stimulated arachidonic acid release and prostaglandin E_2 synthesis, which resulted in concomitant cAMP synthesis in these cells (18,62). We found that phosphorylation of the rate-limiting enzyme in arachi-

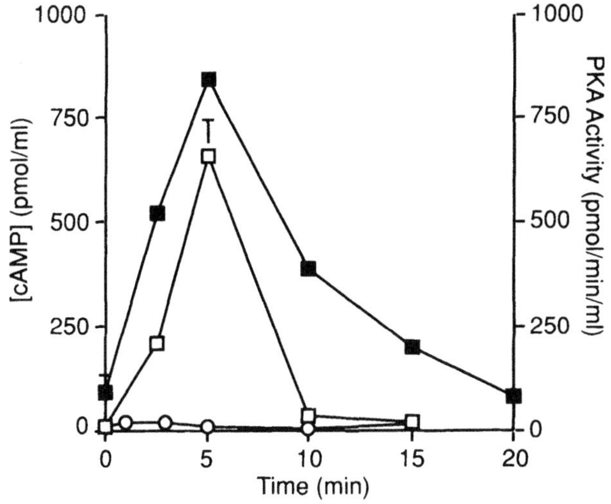

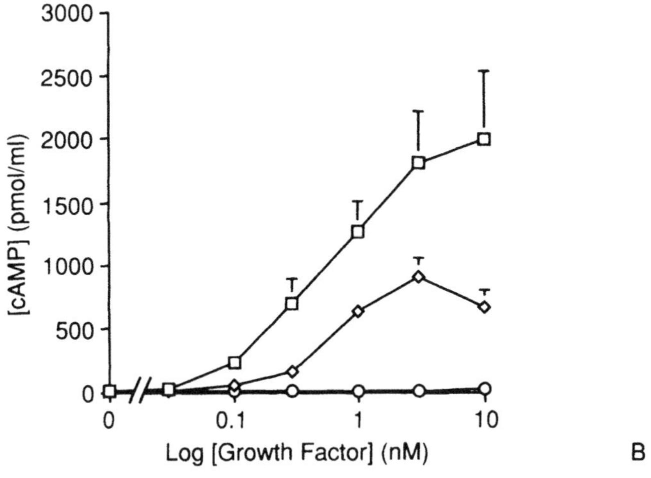

FIG. 3. PDGF increases cAMP and activates PKA. Human newborn arterial smooth-muscle cells (hSMC) were obtained from the thoracic aorta of infants following death due to congenital heart defects or sudden infant death syndrome. The hSMC (3×10^6 cells/sample) were incubated in the presence of 1% human plasma-derived serum for 48 h. **A:** Time course of cAMP formation by 1 nM PDGF-BB (□) and 1nM IGF-I (○), and PKA activation by 1 nM PDGF-BB (■). Samples were rapidly harvested and the cAMP concentration measured in triplicate by radioimmunoassay, whereas PKA activity was measured as phosphorylation of Kemptide (67) in the presence or absence of the PKA inhibitor PKI. **B:** Shows dose–response curves for effect of PDGF-BB (□), PDGF-AA (◊), and IGF-I (○) on cAMP formation. (From [62] *J Biol Chem* 1996; 271:505–11; with permission.)

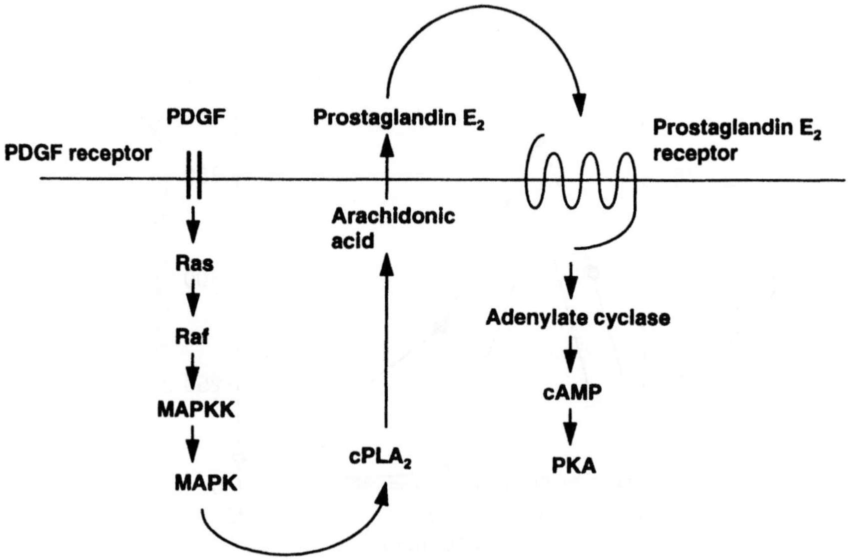

FIG. 4. Activation of the MAP kinase cascade results in activation of PKA in newborn human arterial smooth-muscle cells.

donic acid release, $cPLA_2$, correlated with the activation of MAP kinase in these cells, in agreement with previous reports that $cPLA_2$ is regulated by MAP kinase (13). In further support of a role for MAP kinase in the activation of $cPLA_2$, the MAPKK inhibitor PD098059 (54), abolished PDGF-stimulated PKA activity (K. E. Bornfeldt, *unpublished results*). Thus, our results demonstrate that a growth-factor signal (i.e., PDGF) can be converted into a PKA response through the MAP kinase-dependent activation of $cPLA_2$ and arachidonic acid metabolism in these cells of newborn origin (Fig. 4).

What are the physiologic consequences of PDGF-stimulated cAMP-dependent protein kinase activity? In light of our previous results showing that cAMP inhibits the MAP kinase cascade in hSMC (18), we examined whether PDGF-stimulated PKA activity could influence MAP kinase signaling in a "feedback" manner. Recently, Pyne and coworkers (65) reported that bradykinin stimulated an increase in cAMP accumulation, an effect that counteracted the activation of MAP kinase (Erk-2) in airway smooth-muscle cells. We therefore investigated whether the PDGF-stimulated increase in cAMP affected either the magnitude or the time course of MAP kinase activation in hSMC. Using indomethacin to block the PDGF-stimulated increase in cAMP, we found that activation of MAP kinase was not substantially changed by this treatment. One possible explanation for these findings is that the formation of cAMP occurs too transiently to sufficiently impede the activation of MAP kinase. Alternatively, these two signaling pathways may be physically compartmentalized to prevent "feedback" antagonism. It is anticipated that PDGF-stimulated PKA activity may also be utilized to influence other aspects of cellular signaling, such as cytoskeletal remodeling and/or transcriptional regulation. How the balance of these two signaling events ultimately influences cellular proliferation remains to be established. In other cell types, depending on the type of arachidonyl metabolites generated (thromboxanes, leukotrienes, etc.), $cPLA_2$ activation by MAP kinase may have alternative biologic effects.

CONCLUSION

Based on the examples discussed here, it is apparent that various signaling pathways are intertwined in ways that we are only now beginning to appreciate. With the recent finding of several MAP kinase-like pathways in animal cells, as well as in yeast, the interconnections between different forms of signaling will become even more complicated than they are now. Research in this field is now at a point in which much of it necessarily results in "descriptive" information. Uniform principles that will make intracellular signaling more comprehensible, as in the case of the discovery of cAMP will, however, inevitably emerge.

ACKNOWLEDGMENTS

Supported by grants from the National Institutes of Health DK42528 to Edwin G. Krebs and HL18645 and HL47151 to Russell Ross and Elaine W. Raines and a grant from the Washington State affiliate of the American Heart Association to Lee M. Graves.

REFERENCES

1. Rall TW, Sutherland EW. The regulatory role of adenosine-3′,5′-phosphate. *Cold Spring Harbor Symp Quant Biol* 1961;26:347–54.
2. Fischer EH, Krebs EG. Conversion of phosphorylase *b* to phosphorylase *a* in muscle extracts. *J Biol Chem* 1955;216:121–32.
3. Wosilait WD, Sutherland EW. The relationship of epinephrine and glucagon to liver phosphorylase. Enzymatic inactivation of liver phosphorylase. *J Biol Chem* 1956;218:469–81.
4. Cori C, Cori GT. The mechanism of epinephrine action. *J Biol Chem* 1928;79:309–55.
5. Sutherland EW. Influence of hormones on phosphate metabolism. The effect of epinephrine and the hyperglycemic factor on liver and muscle metabolism in vitro. In: McElroy WD, Glass B, eds. *A symposium on phosphorus metabolism.* Baltimore: The Johns Hopkins Press;1952:577–96.
6. Krebs EG, DeLange RJ, Kemp RG, Riley WD. Activation of skeletal muscle phosphorylase. *Pharmacol Rev* 1966;18:163–71.
7. Walsh DA, Perkins JP, Krebs EG. An adenosine 3′,5′-monophosphate-dependent protein kinase from rabbit skeletal muscle. *J Biol Chem* 1968;243:3765–75.
8. Ahn NG. The MAP kinase cascade. Discovery of a new signal transduction pathway. *Mol Cell Biochem* 1993;127/128:201–9.
9. Stadtman ER, Chock PB. Interconvertible enzyme cascades in metabolic regulation. *Curr Top Cell Regul* 1971;13:53–95.
10. Seger R, Ahn NG, Posada J, et al. Purification and characterization of mitogen-activated protein kinase activator(s) from epidermal growth factor-stimulated A431 cells. *J Biol Chem* 1992;267:14373–81.
11. Sturgill TW, Ray LB, Erickson E, Maller JL. Insulin-stimulated MAP-2 kinase phosphorylates and activates ribosomal protein S6 kinase II. *Nature* 1988;334:715–8.

12. Drewes G, Lichtenberg-kraag B, Doring F, et al. Mitogen-activated protein kinase (MAP) kinase transforms tau protein into an Alzheimer like state. *EMBO J* 1992;11:2131–8.
13. Lin LL, Wartmann M, Lin AY, Knopf JL, Seth A, Davis RJ. $cPLA_2$ is phosphorylated and activated by MAP kinase. *Cell* 1993;72:269–78.
14. Davis RJ. The mitogen-activated protein kinase signal transduction pathway. *J Biol Chem* 1993;268: 14553–6.
15. Chen RH, Sarnecki C, Blenis J. Nuclear localization and regulation of erk- and rsk-encoded protein kinases. *Mol Cell Biol* 1992;12:915–27.
16. Reszka AA, Seger R, Diltz CD, Krebs EG, Fischer EH. Association of mitogen-activated protein kinase with the microtubule cytoskeleton. *Proc Natl Acad Sci USA* 1995;92:8881–5.
17. Graves LM, Lawrence JC. Insulin, growth factors, and cAMP: Antagonism in the signal transduction pathways. *Trends Endocrin* 1996;7:43–50.
18. Graves LM, Bornfeldt KE, Raines EW, et al. Protein kinase A antagonizes platelet-derived growth factor-induced signaling by mitogen-activated protein kinase in human arterial smooth muscle cells. *Proc Natl Acad Sci USA* 1993;90:10300–4.
19. Sevetson BR, Kong X, Lawrence JC. Increasing cAMP attenuates activation of mitogen-activated protein kinase. *Proc Natl Acad Sci USA* 1993;90:10305–9.
20. Burgering BMT, Pronk GJ, van Weeren PC, Chardin P, Bos JL. cAMP antagonizes $p21^{ras}$-directed activation of extracellular signal-regulated kinase 2 and phosphorylation of mSos nucleotide exchange factor. *EMBO J* 1993;12:4211–20.
21. Cook SJ, McCormick F. Inhibition by cAMP of Ras-dependent activation of Raf. *Science* 1993;262: 1069–72.
22. Wu J, Dent P, Jelinek T, Wolfman A, Weber MJ, Sturgill TW. Inhibition of the EGF-activated MAP kinase signaling pathway by adenosine 3′,5′-monophosphate. *Science* 1993;262:1065–9.
23. Schramm K, Niehof M, Radziwill G, Rommel C, Moelling K. Phosphorylation of c-Raf-1 by protein kinase A interferes with activation. *Biochem Biophys Res Commun* 1994;201:740–7.
24. Buhl AM, Avdi N, Worthen GS, Johnson GL. Mapping of the C5a receptor signal transduction network in human neutrophils. *Proc Natl Acad Sci USA* 1994;91:9190–4.
25. Worthen GS, Avdi N, Buhl AM, Suzuki N, Johnson GL. FMLP activates Ras and Raf in human neutrophils. Potential role in activation of MAP kinase. *J Clin Invest* 1994;94:815–23.
26. Vaillancourt RR, Gardner AM, Johnson GL. B-Raf-dependent regulation of the MEK-1/mitogen-activated protein kinase pathway in PC12 cells and regulation by cyclic AMP. *Mol Cell Biol* 1994;14: 6522–30.
27. al-Alawi N, Rose DW, Buckmaster C, et al. Thyrotropin-induced mitogenesis is Ras dependent but appears to bypass the Raf-dependent cytoplasmic kinase cascade. *Mol Cell Biol* 1995;15:1162–8.
29. Chuang E, Barnard D, Hettich L, Zhang XF, Avruch J, Marshall MS. Critical binding and regulatory interactions between Ras and Raf occur through a small, stable N-terminal domain of raf and specific ras effector residues. *Mol Cell Biol* 1994;14:5318–25.
30. Peraldi P, Frodin M, Barnier JV, et al. Regulation of the MAP kinase cascade in PC12 cells: B-Raf activates MEK-1 and is inhibited by cAMP. *FEBS Lett* 1995;357:290–6.
31. Häfner S, Adler HS, Mischak H, et al. Mechanism of inhibition of Raf-1 by protein kinase A. *Mol Cell Biol* 1994;14:6696–703.
32. Whitehurst CE, Owaki H, Bruder JT, Rapp UR, Geppert TD. The MEK kinase activity of the catalytic domain of RAF-1 is regulated independently of Ras binding in T cells. *J Biol Chem* 1995;270: 5594–9.
33. Frodin M, Peraldi P, Van Obberghen E. Cyclic AMP activates the mitogen-activated protein kinase cascade in PC12 cells. *J Biol Chem* 1994;269:6207–14.
34. Lamy F, Wilkin F, Baptist M, et al. Phosphorylation of mitogen-activated protein kinases is involved in the epidermal growth factor and phorbol ester, but not in the thyrotropin/cAMP, thyroid mitogenic pathway. *J Biol Chem* 1993;268: 8398–401.
35. Withers DJ, Bloom SR, Rozengurt E. Dissociation of cAMP-stimulated mitogenesis from activation of the mitogen-activated protein kinase cascade in Swiss 3T3 Cells. *J Biol Chem* 1995;270:21411–9.
36. Traverse S, Gomez N, Paterson H, Marshall C, Cohen P. Sustained activation of the mitogen-activated protein (MAP) kinase cascade may be required for differentiation of PC12 cells. *Biochem J* 1992;288:351–5.
37. Young SW, Dickens M, Tavare JM. Differentiation of PC12 cells in response to a cAMP analogue is accompanied by sustained activation of mitogen-activated protein kinase. Comparison with the effects of insulin, growth factors and phorbol esters. *FEBS Lett* 1994;338:212–6.

38. Johnson GL, Vaillancourt RR. Sequential protein kinase reactions controlling cell growth and differentiation. *Curr Opin in Cell Biology* 1994;6:230–8.
39. Faure M, Voyno-Yasenetskaya TA, Bourne HA. cAMP and $\beta\gamma$ subunits of heterotrimeric g-proteins stimulate the mitogen-activated protein kinase pathway in Cos-7 Cells. *J Biol Chem* 1994;269: 7851–4.
40. Crespo P, Cachero TG, Xu N, Gutkind JS. Dual effect of β-adrenergic receptors on mitogen-activated protein kinase. Evidence for a $\beta\gamma$-dependent activation and $G\alpha s$-cAMP-mediated inhibition. *J Biol Chem* 1995;270:25259–65.
41. van Biesen T, Hawes BE, Lutrell DK, et al. Receptor-tyrosine-kinase-and $G\beta\gamma$-mediated MAP kinase activation by a common signalling pathway. *Nature* 1995;376:781–4.
42. Lefkowitz RJ, Caron M. Regulation of adrenergic receptor function by phosphorylation. *Curr Top Cell Regul* 1986;28:209–31.
43. Chao TO, Byron KL, Lee KM, Villereal M, Rosner MR. Activation of MAP kinases by calcium-dependent and calcium-independent pathways. *J Biol Chem* 1992;267:19876–83.
44. Kolch W, Heldecker G, Kochs G, et al. Protein kinase $C\alpha$ activates RAF-1 by direct phosphorylation. *Nature* 1993;364:249–50.
45. Earp HS, Huckle WR, Dawson TL, Li X, Graves LM, Dy R. Angiotensin II activates at least two tyrosine kinases in rat liver epithelial cells. Separation of the major calcium-regulated tyrosine kinase from $p125^{FAK}$. *J Biol Chem* 1995;270:28440–7.
46. Lev S, Moreno H, Martinez R, et al. Protein tyrosine kinase PYK2 involved in Ca^{+2}-induced regulation of ion channel and MAP kinase functions. *Nature* 1995;376:737–45.
47. Chen RH, Tung R, Abate C, Blenis J. Cytoplasmic to nuclear signal transduction by mitogen-activated protein kinase and 90 kDa ribosomal S6 kinase. *Biochemical Society Trans* 1993;21:895–900.
48. Pierce SB, Kimelman D. Regulation of Speman organizer formation by the intracellular kinase Xgsk-3. *Development* 1995;121:755–65.
49. He X, Saint-Jeannet JP, Woodgett JR, Varmus HE, Dawid IB. Glycogen synthase kinase-3 and dorsal ventral patterning in *Xenopus* embryos. *Nature* 1995;374:617–22.
50. Sutherland C, Cohen P. The α-isoform of glycogen synthase kinase-3 from rabbit skeletal muscle is inactivated by p70 S6 kinase or MAP kinase-activated protein kinase-1 *in vitro*. *FEBS Lett* 1994; 338:37–42.
51. Sutherland C, Leighton IA, Cohen P. Inactivation of glycogen synthase kinase-3 beta by phosphorylation: new kinase connections in insulin and growth-factor signalling. *Biochem J* 1993;296:15–9.
52. Welsh GI, Foulstone EJ, Young SW, Tavare JM, Proud CG. Wortmannin inhibits the effects of insulin and serum on the activities of glycogen synthase kinase-3 and mitogen-activated protein kinase. *Biochem J* 1994;303:15–20.
53a. Eldar-Finkelman H, Seger R, Vandenheede JR, Krebs EG. Inactivation of glycogen synthase kinase-3 by epidermal growth factor is mediated by mitogen-activated protein kinase /p90 ribosomal protein S6 kinase signaling pathway in NIH/3T3 cells. *J Biol Chem* 1995;270:987–90.
53b. Cross DAE, Alessi DR, Vandenheede JR, et al. The inhibition of glycogen synthase kinase-3 by insulin or insulin-like growth factor 1 in rat skeletal muscle cell line L6 is blocked by wortmannin, but not by rapamycin: evidence that wortmannin blocks activation of the mitogen-activated protein kinase pathway in L6 cells between Ras and Raf. *Biochem J* 1994;303:21–6.
54. Alessi DR, Cuenda A, Cohen P, Dudley DT, Saltiel AR. PD098059 is a specific inhibitor of the activation of mitogen-activated protein kinase *in vitro* and *in vivo*. *J Biol Chem* 1995;270:27489–94.
55. Jones PF, Jakubowicz T, Pitossi FJ, Maurer F, Hemmings BA. Molecular cloning and identification of a serine/threonine protein kinase of the second-messenger subfamily. *Proc Natl Acad Sci USA* 1991;88:4171–5.
56a. Cross DAE, Alessi DR, Cohen P, Andjelkovich M, Hemmings BA. Inhibition of glycogen synthase kinase-3 by insulin mediated by protein kinase B. *Nature* 1995;378:785–9.
56b. Dent P, Lavoinne A, Nakielny S, Caudwell FB, Watt P, Cohen P. The molecular mechanism by which insulin stimulates glycogen synthesis in mammalian skeletal muscle. *Nature* 1990;348:302–8.
57. Hubbard MJ, Cohen P. Regulation of protein phosphatase-1_G from rabbit skeletal muscle. *Eur J Biochem* 1989;186:701–9.
58. Hubbard MJ, Cohen P. On target with a new mechanism for the regulation of protein phosphorylation. *Trends Biochem Sci* 1993;18:172–7.
59. Srinivasan M, Begum N. Regulation of protein phosphatase 1 and 2A activities by insulin during myogenesis in rat skeletal muscle cells in culture. *J Biol Chem* 1994;269:12514–20.

60. Hemmings BA, Resink TJ, Cohen P. Reconstitution of a Mg-ATP-dependent protein phosphatase and its activation through a phosphorylation mechanism. *FEBS Lett* 1982;150:319–24.
61. Wang QM, Guan KL, Roach PJ, Depaoli-Roach AA. Phosphorylation and activation of the ATP-Mg-dependent protein phosphatase by the mitogen-activated protein kinase. *J Biol Chem* 1995;270: 18352–8.
62. Graves LM, Bornfeldt KE, Sidhu JS, et al. Platelet-derived growth factor stimulates protein kinase A through a mitogen-activated protein kinase-dependent pathway in human arterial smooth muscle cells. *J Biol Chem* 1996;271:505–11.
63. Bornfeldt KE, Raines EW, Nakano T, Graves LM, Krebs EG, Ross R. Insulin-like growth factor-I and platelet-derived growth-BB induce directed migration of human arterial smooth muscle cells via signaling pathways that are distinct from those of proliferation. *J Clin Invest* 1994;93:1266–74.
64. Rozengurt E, Stroobant P, Waterfield MD, Deuel TF, Keehan M. Platelet-derived growth factor elicits cyclic AMP accumulation in Swiss 3T3 cells: role of prostaglandin production. *Cell* 1983;34:265–72.
65. Pyne NJ, Moughal N, Stevens PA, Tolan D, Pyne S. Protein kinase C-dependent cyclic AMP formation in airway smooth muscle: the role of type II adenylate cyclase and the blockade of extracellular-signal-regulated kinase-2 (ERK-2) activation. *Biochem J* 1994;304, 611–16.
66. Crespo P, N Xu, WF Simonds, Gutkind JS. Ras-dependent activation of MAP kinase pathway mediated by G-protein βγ subunits. *Nature* 1994;369:418–20.
67. Kemp BE, Pearson RB. Design and use of peptide substrates for protein kinases. In: T Hunter, BM Setton, eds. *Methods in enzymology.* 1991;200:121–34.

Signal Transduction in Health and Disease, Advances in Second Messenger and Phosphoprotein Research, Vol. 31, edited by J. Corbin and S. Francis.

6

Coupling Transcription to Signaling Pathways

cAMP and Nuclear Factor cAMP-Responsive Element Modulator

Lucia Monaco, Monica Lamas, Katherine Tamai, Enzo Lalli, Emmanuel Zazopoulos, Lucia Penna, François Nantel, Nicholas S. Foulkes, Cristina Mazzucchelli, and Paolo Sassone-Corsi

Institut de Génétique et de Biologie Moléculaire et Cellulaire, Strasbourg, France

INTRODUCTION

The structural organization of most transcription factors is intrinsically modular, in most cases including a deoxyribonucleic acid (DNA)-binding domain and an activation domain. It has been shown that these domains can be interchanged between different factors and still retain their functional properties. This modularity suggests that during evolution, increasing complexity of gene expression may have resulted not only from the duplication and divergence of existing genes, but also from a domain-shuffling process to generate factors with novel properties (1).

An important step forward in the study of transcription factors has been the discovery that many constitute the final targets of specific signal-transduction pathways. The two major signal-transduction systems are those including cyclic adenosine monophosphate (cAMP) and diacylglycerol (DAG) as secondary messengers (2). Each pathway is also characterized by specific protein kinases (protein kinase A [PKA] and protein kinase C [PKC], respectively) and its ultimate target DNA-control element (cAMP-responsive element [CRE] and TPA-responsive element [TRE], respectively). Although these two pathways were initially characterized as distinct systems, accumulating evidence points toward extensive cross-talk between them (3–5).

Intracellular levels of cAMP are regulated primarily by adenylate cyclase. This enzyme is in turn modulated by various extracellular stimuli mediated by receptors and their interaction with G proteins (6). cAMP binds cooperatively to two sites on the regulatory subunits of PKA, releasing the active catalytic subunits of the enzyme (7,8). These are translocated from cytoplasmic and Golgi-complex anchoring sites, and phosphorylate a number of cytoplasmic and nuclear proteins on serines in the context X-Arg-Arg-X-*Ser*-X (7). In the nucleus, PKA-mediated phosphorylation ultimately influences the transcriptional regulation of various genes through distinct, cAMP-inducible promoter-responsive sites (9,10).

INDUCTION OF GENE EXPRESSION BY ACTIVATION OF THE cAMP SIGNALING PATHWAY

The consensus cAMP-responsive element (CRE) is constituted of an 8-bp palindromic sequence (TGACGTCA) with greater conservation in the 5′ half of the palindrome than in the 3′ sequence. Several genes that are regulated by a variety of endocrine stimuli contain similar sequences in their promoter regions, although at different positions (9,11).

The first CRE-binding factor to be characterized was CRE-binding protein (CREB) (12), but subsequently, at least 10 additional CREB-factor complementary DNAs (cDNAs) have been cloned. They were obtained by screening a variety of cDNA expression libraries, with CRE and activator transcription factor (ATF) sites (13,14). These proteins belong to the bZip transcription factor class.

The different factors can heterodimerize with each other, but only in certain combinations. A "dimerization code" exists that seems to be a property of the "leucine zipper" structure of each factor. Some ATF/CREB factors can heterodimerize with Fos and Jun, and this may change the specific affinity of binding to a CRE with respect to a Fos–Jun binding site (15). This property resides in the similarity between the CRE (TGACGTCA) and TRE (TGACTCA) sequences (5,16), and demonstrates the versatility of the transcriptional response to signal transduction.

There are both activators and repressors of cAMP-responsive transcription. Some alternatively spliced CRE modulator (CREM) isoforms act as antagonists of cAMP-induced transcription. The cAMP-inducible ICER product deserves special mention, since it is generated from an alternative promoter of the cAMP-responsive element modulator (CREM) gene and is responsible for its early-response inducibility, which is unique among CRE-binding factors (17,18).

ACTIVATION BY PHOSPHORYLATION

The characterization of the transcriptional activators CREB and CREM (19,20) has helped to elucidate the molecular mechanisms involved in transcriptional activation. These factors contain a transcriptional-activation domain that is divided into two independent regions (7). The first, known as the phosphorylation box (P-box), contains several consensus phosphorylation sites for various kinases, such as PKA, PKC, $p34^{cdc2}$, glycogen synthase kinase-3 (Gsk-3), and casein kinases (CK) I and II

(19–23). The second region flanks the P-box, and is constituted of domains rich in glutamine residues (7).

Upon activation of the adenylyl cyclase pathway, serine residues at position 133 of CREB and at position 117 of CREM are phosphorylated by PKA (19,22). The major effect of this phosphorylation is to convert CREB and CREM into powerful transcriptional activators. Within the P-box, serine 133/117 is located in a region of about 50 amino acids containing an abundance of phosphorylatable serines and acidic residues, which was shown to be essential for transactivation by CREB and CREM (21,22).

Interestingly, in PC12 cells, increases in the levels of intracellular Ca^{2+} caused by membrane depolarization have been shown to induce the phosphorylation of Ser^{133} in CREB and a concomitant activation of c-*fos* gene expression mediated by a CRE in the promoter (24,25). Although Ca^{2+}-dependent CAMK was shown to be able to phosphorylate Ser^{133} *in vitro* (26), the *in vivo* significance of this remains unclear, since PKA also seems to be necessary for c-*fos* induction mediated by Ca^{2+} influx into PC12 cells (27).

An important finding that reveals the complexity of the transcriptional response elicited by these factors concerns the mitogen-induced p70 S6 kinase, which phosphorylates and activates CREM (28). This finding also implicates $p70^{s6k}$, a kinase generally considered cytoplasmic, in the mitogenic response at the nuclear level. Interestingly, since CREM and other factors of the CREB/ATF family represent the final targets of the cAMP-pathway, these results show that they may also act as effectors of converging signaling systems, and possibly as mediators of pathway cross-talk (28).

INTERACTION WITH COACTIVATOR CBP

The two domains flanking the P-box contain about three times more glutamine residues than does the remainder of the protein in both CREB and CREM. Glutamine-rich domains have been characterized in other factors, such as activator protein-2 (AP-2) and Sp1 (29,30), as transcriptional-activation domains. The current notion is that they constitute surfaces of the protein that can interact with other components of the transcriptional machinery. Indeed, further steps toward an understanding of the mechanism of action of the P-box have come with the identification of a 265K, 2,441-amino-acid protein, a CBP that can interact specifically with the phosphorylated CREB P-box domain (31). The CBP sequence reveals two zinc-finger domains, a glutamine-rich domain at its C-terminus, and a single consensus PKA recognition site. Phosphorylation of Ser^{133} in CREB promotes binding to CBP and consequently the interaction with transcription factor IIB (TFIIB), a general transcription factor involved in ribonucleic acid (RNA) polymerase II activity (32). Thus, CBP may act as a link between CREB and the transcription preinitiation complex. This interaction may need some RNA polymerase II cofactors, such as TAF110 (Fig. 1). Finally, the adenoviral E1A oncoprotein-associated p300, which is

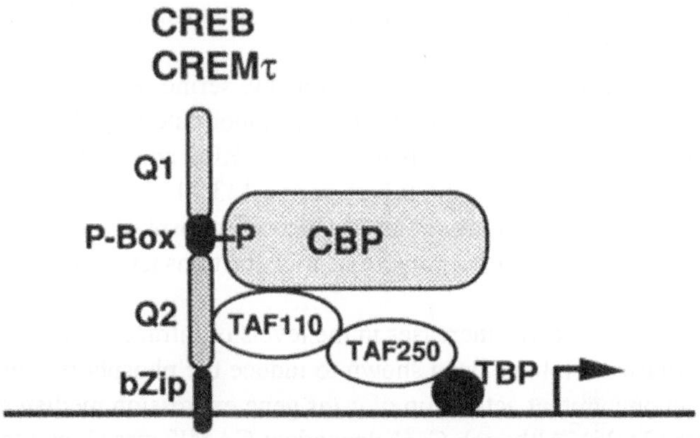

FIG. 1. Scheme of the possible interaction between the transcriptional activators CREB and CREM τ with the coactivator CBP (31). CREMτ is the activator isoform generated by the CREM gene. The equivalent protein is highly similar in structure to CREB and is particularly abundant in male germ cells (20). Interaction is phosphorylation-dependent and allows contact with other elements of the transcriptional machinery, such as TAF110, TAF250, and the TATA-binding protein TBP.

thought to play a role in preventing the cell cycle G0/G1 transition, is structurally very closely related to CBP (33). Both CBP and p300 appear to have intrinsic activating properties which are inhibited by the E1A protein (33). Thus, it is clear that studies of the transcriptional-activation domain of CRE-binding bZip factors continues to provide important insights into the function of transcription factors in general.

CREM IS AN EARLY-RESPONSE GENE

During studies of CREM expression within the neuroendocrine system, an unexpected new facet emerged: that transcription of the CREM gene is inducible by cAMP (17). Furthermore, the kinetics of this induction are those of an early-response gene (34). This important finding further reinforces the notion that CREM products play a fulcral role in the nuclear response to cAMP, since the expression of no other CREB factor has been shown to be inducible to date.

The demonstration that the CREM gene was cAMP inducible first came from the finding that adrenergic signals direct CREM transcription in the pineal gland (18). The inducibility phenomenon was then characterized in detail in the pituitary corticotroph cell line AtT20. In unstimulated cells the level of CREM transcript is undetectable. However, upon treatment with forskolin (or other cAMP analogues), there is within 30 min a rapid increase in CREM transcript levels, which peak after 2 h and then progressively decline to basal levels by 5 h. These characteristic kinetics classify CREM as an early-response gene and thus for the first time directly implicate the cAMP pathway in the cell's early response. CREM inducibility is specific for the cAMP pathway, since CREM is not inducible by TPA or dexamethasone treatment.

The inducible CREM transcript corresponds to a truncated product, termed the ICER (17,18).

The 5′ end of ICER clones corresponds to an alternative transcription start site. The start of transcription, which identifies the P2 promoter, is within the 10-kb intron that is C-terminal to the Q2 glutamine-rich domain exon. In contrast to the promoter generating all of the previously characterized CREM isoforms (P1), which is GC-rich and not inducible by cAMP (N. S. Foulkes, unpublished results), the P2 promoter has a normal A–T and G–C content and is strongly inducible by cAMP. It contains two pairs of closely spaced CRE elements organized in tandem, in which the separation between each pair is only three nucleotides (Fig. 2). These features make P2 unique among cAMP-regulated promoters, and suggest cooperative interactions among the factors binding to these sites.

The ICER open-reading frame is constituted by the C-terminal segment of CREM. The predicted open-reading frame encodes a small protein of 120 amino acids with a predicted molecular weight of 13.4 kD. This protein, compared with the previously described CREM isoforms, essentially consists only of the DNA binding domain, which is constituted by the leucine zipper and basic region. The structure of ICER is suggestive of its function and makes it one of the smallest transcription factors ever described (17,18).

The intact DNA-binding domain directs specific ICER binding to a consensus CRE element. Importantly, ICER is able to heterodimerize with the other CREM proteins and with CREB. ICER functions as a powerful repressor of cAMP-induced transcription in transfection assays using an extensive range of reporter plasmids carrying individual CRE elements or cAMP-inducible promoter fragments (17). Interestingly, ICER-mediated repression is obtained at substoichiometric concentrations, as with previously described CREM antagonists (35). ICER escapes from PKA-dependent phosphorylation and thus constitutes a new category of CRE-binding factor for which the principal determinant of activity is intracellular concentration and not the degree of phosphorylation. Recent data implicate dynamic ICER expression as a more general feature of neuroendocrine systems (36,37).

ATTENUATION AFTER INDUCTION

Dephosphorylation appears to represent a key mechanism in the negative regulation of CREB activation function. It has been proposed that a mechanism explaining the attenuation of CREB activity following induction by forskolin is dephosphorylation by specific phosphatases (38; see Fig. 3). After the initial burst of phosphorylation in response to cAMP, CREB is dephosphorylated *in vivo* by protein phosphatase-1 (PP-1). However, the situation is more complex, since it has been shown that both PP-1 and PP-2A can dephosphorylate CREB *in vitro* (39), resulting in an apparent decreased binding to low-affinity CRE sites *in vitro*. Therefore, the precise roles of PP-1 and PP-2A in the dephosphorylation of CREB remain to be determined.

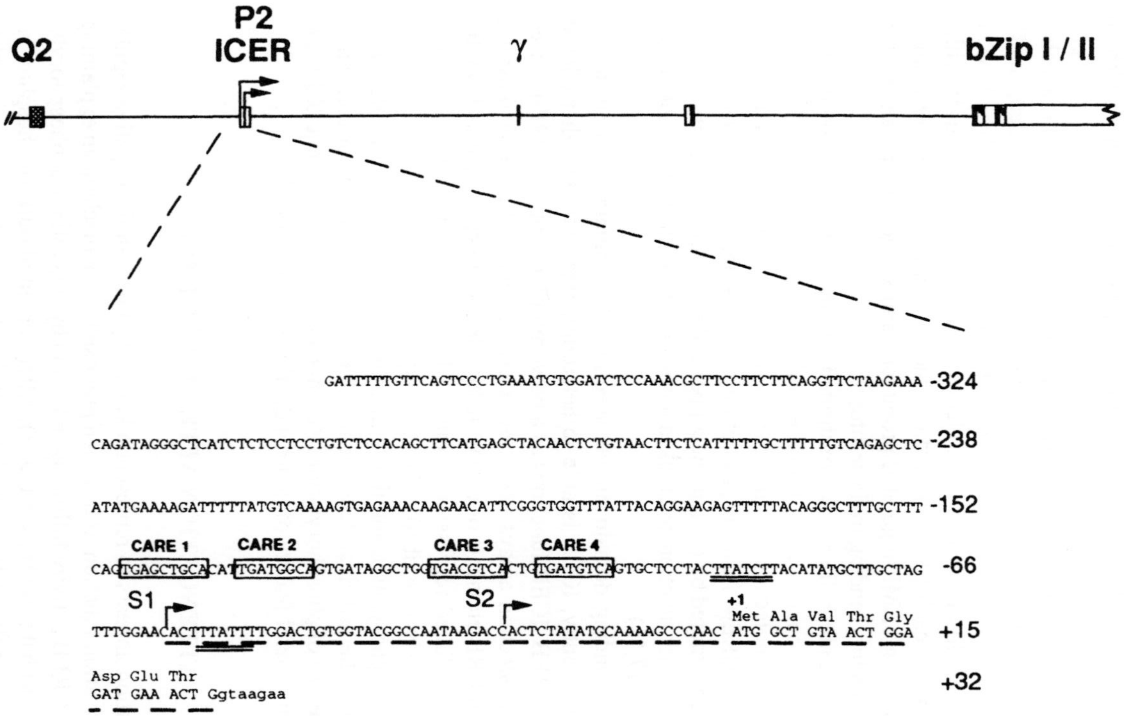

FIG. 2. The ICER promoter. Schematic representation of the ICER 5′ flanking region. The positions of the two starts of transcription (S1 and S2) and the Kozak ATG codon are indicated. A 400-bp genomic sequence including the ICER 5′ exon is shown. *Dashed underlining* delineates the ICER 5′ exon. Lower-case sequence represents the beginning of the first intron of the ICER transcript. Putative TATA elements are indicated by double underlining, while the four CRE-like elements (CAREs) are boxed and labeled. The position +1 corresponds to the A of the Kozak ATG initiation codon (17).

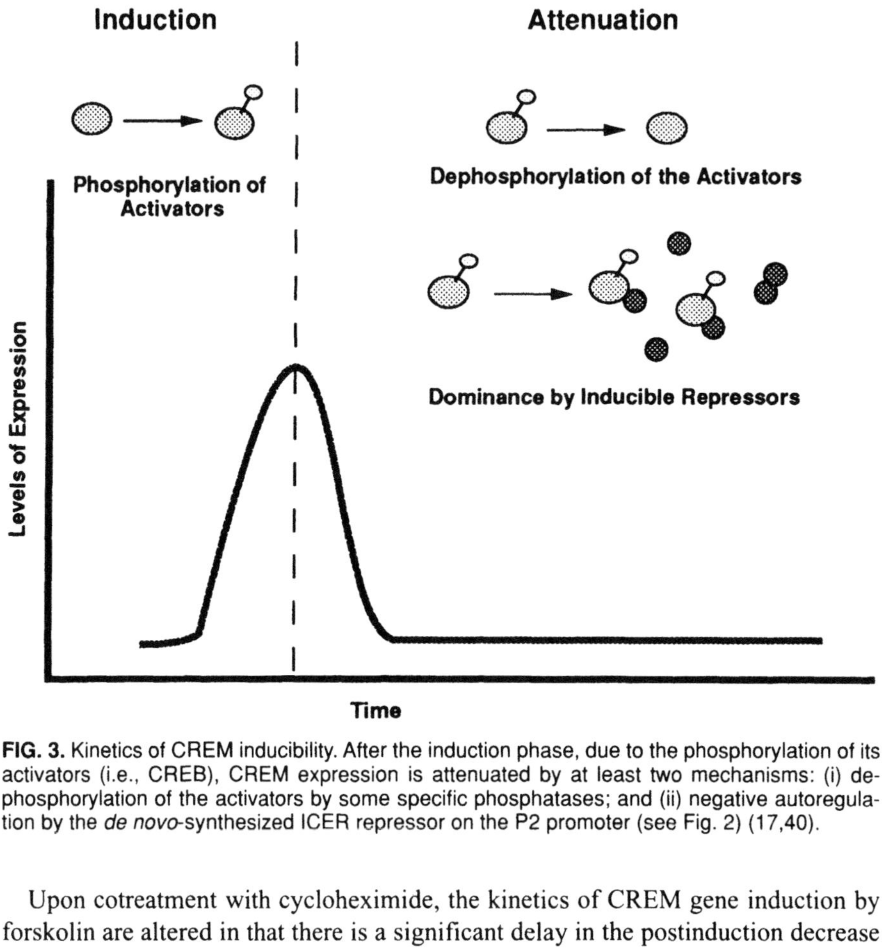

FIG. 3. Kinetics of CREM inducibility. After the induction phase, due to the phosphorylation of its activators (i.e., CREB), CREM expression is attenuated by at least two mechanisms: (i) dephosphorylation of the activators by some specific phosphatases; and (ii) negative autoregulation by the *de novo*-synthesized ICER repressor on the P2 promoter (see Fig. 2) (17,40).

Upon cotreatment with cycloheximide, the kinetics of CREM gene induction by forskolin are altered in that there is a significant delay in the postinduction decrease in the transcript; increased levels of the transcript persist for as long as 12 h. This implicates a *de novo* synthesized factor that might downregulate CREM transcription (17). This observation, combined with the presence of CRE elements in the P2 promoter, suggested that the transient nature of the inducibility could be due to ICER (Fig. 3). Consistently, the CRE elements in the P2 promoter have been shown to bind to the ICER proteins. Detailed studies have shown that the ICER promoter is indeed a target for ICER negative regulation (17). Thus, there exists a negative autoregulatory mechanism controlling ICER expression. The CREM feedback loop predicts the presence of a refractory inducibility period in the gene's transcription (40).

CREM AND SPERMATOGENESIS

CREM is a highly abundant transcript in adult testis, while in prepubertal animals it is expressed at very low levels. Thus, in testis, expression of CREM is the subject of a developmental switch (20). Further characterization revealed that the abundant

CREM transcript encodes exclusively the activator form, while in prepubertal testis only the repressor forms were detected at low levels. Thus, the developmental switching of CREM expression also constitutes a reversal of function (41).

Spermatogenesis is a process occurring in a precise and coordinated manner within the seminiferous tubules (42). During this entire developmental process the germ cells are maintained in intimate contact with the somatic Sertoli cells. As the spermatogonia mature, they move from the periphery toward the lumen of the tubule until the mature spermatozoa are conducted from the lumen to the collecting ducts.

A remarkable aspect of the CREM developmental switch in germ cells is its exquisite hormonal regulation. The spermatogenic differentiation program is under the tight control of the hypothalamic–pituitary axis (42). The regulation of CREM function in testis seems to be intricately linked to follicle-stimulating hormone (FSH) both at the level of the control of transcript processing and at the level of protein activity (Fig. 4). For example, surgical removal of the pituitary gland leads to the loss of CREM expression in the rat adult testis (43). Furthermore, hypophysectomy in prepubertal animals prevents the switch in CREM expression at the pachytene spermatocyte stage, thus implicating the pituitary directly in the maintenance of as well as the switch to high levels of CREM expression. Injection of FSH leads to a rapid and significant induction of the CREM transcript. The hormonal induction of CREM by FSH is not transcriptional, which is consistent with the housekeeping nature of the P1 promoter. Instead, by a mechanism of alternative polyadenylation, AUUUA destabilizer elements present in the 3′ untranslated region of the gene are excluded, dramatically increasing the stability of the CREM message. CREM is the first example of a gene whose expression is modulated by a pituitary hormone during spermatogenesis (43). The implication of these findings is that hormones can regulate gene expression at the level of RNA processing and stability. Importantly, the effect of FSH cannot be direct, since germ cells do not have FSH receptors. Recent data suggest that another hormonal message originating from the Sertoli cells upon FSH stimulation mediates CREM activation in germ cells (L. Monaco, *unpublished results*).

CREM, A REGULATOR OF GENE EXPRESSION IN HAPLOID GERM CELLS

A first hint about the role of CREM during spermatogenesis was indicated by its protein expression pattern. In the seminiferous epithelium, CREM transcripts accumulate in spermatocytes and spermatids, but CREM protein is detected only in haploid spermatids (44). The absence of CREM protein in spermatocytes reflects a strict translational control and indicates multiple levels of regulation of gene expression in testis. It will be extremely important to further analyze the mechanism of the delay in CREM translation, and to define whether it is also hormonally dependent.

Phosphorylation by PKA activates CREM function, allowing the relay of the hormonal signal from the cytoplasm to the nucleus (7). The CREM activator is efficiently phosphorylated by cAMP-dependent PKA activity endogenous to the sper-

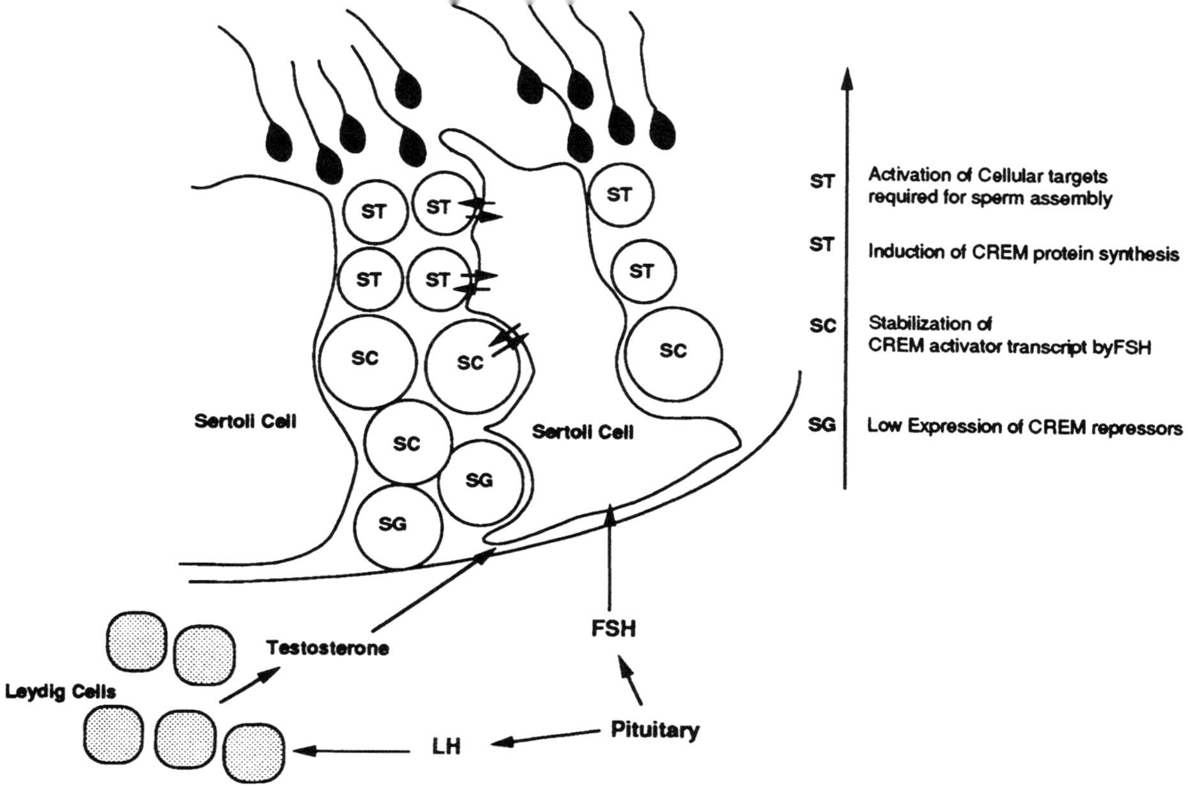

FIG. 4. Schematic representation of a section of a seminiferous tubule, in which the CREM expression pattern is indicated. CREM expression is regulated at multiple levels during spermatogenesis. Premeiotic germ cells (spermatogonia [SG]) express a low level of CREM repressor isoforms. During meiotic prophase, pituitary FSH is responsible for stabilizing CREM activator transcripts in spermatocytes (SC); CREM protein, on the other hand, is detected only after meiosis in haploid spermatids (ST).

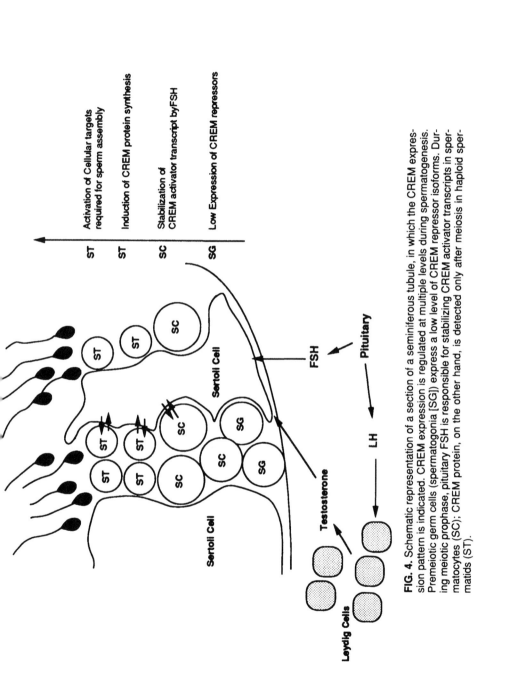

matids, indicating that the CREM protein is a nuclear target for the cAMP pathway in haploid spermatogenic cells (44).

The expression of CREM activator protein in spermatids coincides with the transcriptional activation of several genes containing a CRE motif in their promoter region. These genes encode mainly structural proteins required for spermatozoa assembly (transition protein, protamine, etc.), suggesting a role for CREM in the activation of genes required for the late phase of spermatid differentiation (Fig. 4). This observation implies that the transcription of some key structural genes is directly linked to hormonal control and consequently to the level of cAMP present in seminiferous epithelium. To date at least three genes—RT7 (44), transition protein-1 (45), and calspermin (46)—have been shown to be targets of CREM-mediated transactivation in germ cells.

The role of CREM in the expression of one of these genes, RT7, was demonstrated with *in vitro* transcription experiments. A CREM-specific antibody blocks RT7 transcription *in vitro* with nuclear extracts from seminiferous tubules but not with extracts from liver (44). In conclusion, CREM might participate in testis- and developmental-specific regulation of genes containing a CRE in their promoter region, through expression of its repressor isoforms before meiosis and high levels of the activator after meiosis.

REFERENCES

1. Harrison SC. A structural taxonomy of DNA-binding domains. *Nature* 1991; 353: 715–9.
2. Nishizuka Y. Studies and perspectives of protein kinase C. *Science* 1986; 233: 305–12.
3. Cambier JC, Newell NK, Justement JB, McGuire JC, Leach KL, Chen KK. Iα binding ligands and cAMP stimulate nuclear translocation of PKC in β lymphocytes. *Nature* 1987; 327: 629–32.
4. Yoshimasa T, Sibley DR, Bouvier M, Lefkowitz RJ, Caron MG. Cross-talk between cellular signalling pathways suggested by phorbol ester adenylate cyclase phosphorylation. *Nature* 1987; 327: 67–70.
5. Masquilier D, Sassone-Corsi P. Transcriptional cross-talk: nuclear factors CREM and CREB bind to AP-1 sites and inhibit activation by Jun. *J Biol Chem* 1992; 267: 22460–6.
6. McKnight SG, Clegg CH, Uhler SR, Chrivia JC, Cadd GG, Correll LL. Analysis of the cAMP-dependent protein kinase system using molecular genetic approaches. *Rec Progr Horm Res* 1988; 44: 307–35.
7. Lalli E, Sassone-Corsi P. Signal transduction and gene regulation: the nuclear response to cAMP. *J Biol Chem* 1994; 269: 17359–62.
8. Roesler WJ, Vanderbark GR, Hanson RW. Cyclic AMP and the induction of eukaryotic gene expression. *J Biol Chem* 1988; 263: 9063–6.
9. Borrelli E, Montmayeur JP, Foulkes NS, Sassone-Corsi P. Signal transduction and gene control: the cAMP pathway. *Crit Rev Oncogen* 1992; 2: 321–38.
10. Ziff EB. Transcription factors: a new family gathers at the cAMP response site. *Trends Genet* 1990; 6: 69–72.
11. Sassone-Corsi P. Cyclic AMP induction of early adenovirus promoters involves sequences required for E1A-transactivation. *Proc Natl Acad Sci USA* 1988; 85: 7192–6.
12. Hoeffler JP, Meyer TE, Yun Y, Jameson JL, Habener JF. Cyclic AMP-responsive DNA-binding protein: structure based on a cloned placental cDNA. *Science* 1988; 242: 1430–3.
13. Hai TY, Liu F, Coukos WJ, Green MR. Transcription factor ATF cDNA clones: an extensive family of leucine zipper proteins able to selectively form DNA binding heterodimers. *Genes Dev* 1989; 3: 2083–90.

14. Foulkes NS, Borrelli E, Sassone-Corsi P. CREM gene: use of alternative DNA binding domains generates multiple antagonists of cAMP-induced transcription. *Cell* 1991; 64: 739–49.
15. Hai TY, Curran T. Cross-family dimerization of transcription factors Fos:Jun and ATF/CREB alters DNA binding specificity. *Proc Natl Acad Sci USA* 1991; 88: 3720–4.
16. Sassone-Corsi P, Ransone LJ, Verma IM. Cross-talk in signal transduction: TPA-inducible factor Jun/AP-1 activates cAMP responsive enhancer elements. *Oncogene* 1990; 5: 427–31.
17. Molina CA, Foulkes NS, Lalli E, Sassone-Corsi P. Inducibility and negative autoregulation of CREM: an alternative promoter directs the expression of ICER, an early response repressor. *Cell* 1993; 75: 875–86.
18. Stehle JH, Foulkes NS, Molina CA, Simonneaux V, Pévet P, Sassone-Corsi P. Adrenergic signals direct rhythmic expression of transcriptional repressor CREM in the pineal gland. *Nature* 1993; 365: 314–20.
19. Gonzalez GA, Montminy MR. Cyclic AMP stimulates somatostatin gene transcription by phosphorylation of CREB at Ser 133. *Cell* 1989; 59: 675–80.
20. Foulkes NS, Mellström B, Benusiglio E, Sassone-Corsi P. Developmental switch of CREM function during spermatogenesis: from antagonist to transcriptional activator. *Nature* 1992; 355: 80–4.
21. Lee CQ, Yun Y, Hoeffler JP, Habener JF. Cyclic-AMP-responsive transcriptional activation involves interdependent phosphorylated subdomains. *EMBO J* 1990; 9: 4455–65.
22. de Groot RP, den Hertog J, Vandenheede JR, Goris J, Sassone-Corsi P. Multiple and cooperative phosphorylation events regulate the CREM activator function. *EMBO J* 1993; 12: 3903–11.
23. de Groot RP, Derua R, Goris J, Sassone-Corsi P. Phosphorylation and negative regulation of the transcriptional activator CREM by p34cdc2. *Mol Endocrinol* 1993; 7: 1495–1501.
24. Sassone-Corsi P, Visvader J, Ferland L, Mellon PL, Verma IM. Induction of proto-oncogene *fos* transcription through the adenylate cyclase pathway: characterization of a cAMP-responsive element. *Genes Dev* 1988; 2: 1529–38.
25. Sheng M, McFadden G, Greenberg ME. Membrane depolarization and calcium induce c-*fos* transcription via phosphorylation of transcription factor CREB. *Neuron* 1990; 4: 571–82.
26. Dash PK, Karl KA, Colicos MA, Prywes R, Kandel ER. cAMP response element-binding protein is activated by Ca^{2+}/calmodulin- as well as cAMP-dependent protein kinase. *Proc Natl Acad Sci USA* 1991; 88: 5061–5.
27. Ginty DD, Glowacka D, Bader DS, Hidaka H, Wagner JA. Induction of immediate early genes by Ca^{2+} influx requires cAMP-dependent protein kinase in PC12 cells. *J Biol Chem* 1991; 266: 17454–8.
28. de Groot RP, Ballou LM, Sassone-Corsi P. Positive regulation of the cAMP-responsive activator CREM by the p70 S6 kinase: an alternative route to mitogen-induced gene expression. *Cell* 1994; 79: 81–91.
29. Courey AJ, Tjian R. Analysis of Sp1 *in vivo* reveals multiple transcriptional domains, including a novel glutamine activation motif. *Cell* 1989; 55: 887–98.
30. Williams T, Admon A, Luscher B, Tjian R. Cloning and expression of AP-2, a cell-type-specific transcription factor that activates inducible enhancer elements. *Genes Dev* 1988; 2: 1557–69.
31. Chrivia JC, Kwok RPS, Lamb N, Haniwawa M, Montminy M, Goodman RH. Phosphorylated CREB binds specifically to the nuclear protein CBP. *Nature* 1993; 365: 855–9.
32. Kwok RP, Lundblad JR, Chrivia JC, et al. Nuclear protein CBP is a coactivator for the transcription factor CREB. *Nature* 1994; 370: 223–6.
33. Arany Z, Newsome D, Oldread E, Livingston DM, Eckner R. A family of transcriptional adaptor proteins targeted by the E1A oncoprotein. *Nature* 1995; 374: 81–4.
34. Verma IM, Sassone-Corsi P. Proto-oncogene *fos*: complex but versatile regulation. *Cell* 1987; 51: 513–4.
35. Laoide BM, Foulkes NS, Schlotter F, Sassone-Corsi P. The functional versatility of CREM is determined by its modular structure. *EMBO J* 1993; 13: 1179–91.
36. Lalli E, Sassone-Corsi P. Long-term desensitization of the TSH receptor involves TSH-directed induction of CREM in the thyroid gland. *Proc Natl Acad Sci USA* 1995; 92: 9633–7.
37. Monaco L, Foulkes NS, Sassone-Corsi P. Pituitary follicle-stimulating hormone (FSH) induces CREM gene expression in Sertoli cells: Involvement in long-term desensitization of the FSH receptor. *Proc Natl Acad Sci USA* 1995; 92: 10673–7.
38. Hagiwara M, Alberts A, Brindle P, et al. Transcriptional attenuation following cAMP induction requires PP-1–mediated dephosphorylation of CREB. *Cell* 1992; 70: 105–13.

39. Nichols M, Weih F, Schmid W, et al. Phosphorylation of CREB affects its binding to high and low affinity sites: implications for cAMP induced gene transcription. *EMBO J* 1992; 11: 3337–46.
40. Sassone-Corsi P. Rhythmic transcription and autoregulatory loops: winding up the biological clock. *Cell* 1994; 78: 361–4.
41. Foulkes NS, Sassone-Corsi P. More is better: activators and repressors from the same gene. *Cell* 1992; 68: 411–4.
42. Jégou B. The Sertoli-germ cell communication network in mammals. *Int Rev Cytol* 1993; 147: 25–96.
43. Foulkes NS, Schlotter F, Pévet P, Sassone-Corsi P. Pituitary hormone FSH directs the CREM functional switch during spermatogenesis. *Nature* 1993; 362: 264–7.
44. Delmas V, van der Hoorn F, Mellström B, Jégou B, Sassone-Corsi P. Induction of CREM activator proteins in spermatids: downstream targets and implications for haploid germ cell differentiation. *Mol Endocrinol* 1993; 7: 1502–14.
45. Kistler M, Sassone-Corsi P, Kistler SW. Identification of a functional cAMP response element in the 5′-flanking region of the gene for transition protein 1 (TP1), a basic chromosomal protein of mammalian spermatids. *Biol Reprod* 1994; 51: 1322–9.
46. Sun Z, Sassone-Corsi P, Means A. Calspermin gene transcription is regulated by two cyclic AMP response elements contained in an alternative promoter in the calmodulin kinase IV gene. *Mol Cell Biol* 1995; 15: 561–71.

Signal Transduction in Health and Disease, Advances in Second Messenger and Phosphoprotein Research, Vol. 31, edited by J. Corbin and S. Francis. Lippincott–Raven Publishers, Philadelphia © 1997.

7

Cyclic Nucleotide-Gated Channels and Calcium

An Intimate Relation

Stephan Frings

Institut für Biologische Informationsverarbeitung, Forschungszentrum Jülich GmbH, D-52425 Jülich, Germany

INTRODUCTION

Cyclic nucleotides regulate many diverse cellular functions. Both cyclic guanosine monophosphate (cGMP) and cyclic adenosine monophosphate (cAMP) were identified as intracellular messengers in phototransduction and olfaction in vertebrates. In photoreceptor cells and olfactory sensory neurons (OSNs), the ion permeability of the plasma membrane strongly depends on cation channels that are directly gated by cAMP or cGMP. The first cyclic nucleotide-gated (CNG) channel subunit was cloned from vertebrate rod photoreceptor cells. Additional members of this channel family were subsequently identified in olfactory sensory neurons, in cone photoreceptor cells, and in vertebrate spermatozoa (1). All CNG channels known today are cation-selective and show interesting structural similiarities to voltage-gated K^+ channels, as well as functional similarities to voltage-gated Ca^{2+} channels. Like Ca^{2+} channels, CNG channels conduct monovalent cations in the absence of Ca^{2+}. Micromolar concentrations of Ca^{2+}, however, block this monovalent current. Both types of channels mediate a substantial Ca^{2+} influx under physiologic conditions, and achieve this through a set of glutamate residues within the channel pore (2). This chapter will focus mainly on the physiologic role of CNG channels in sensory cells. In vertebrate photoreceptor cells and olfactory sensory neurons, both excitatory and adaptive processes are controlled by Ca^{2+} influx through CNG channels. Recent studies suggest that CNG channels are also involved in sperm chemotaxis, probably by generating a Ca^{2+} signal in response to chemoattractive stimuli.

CYCLIC NUCLEOTIDE-GATED CHANNELS IN VERTEBRATE PHOTOTRANSDUCTION

The native CNG channel in vertebrate rod photoreceptor cells is composed of α-subunits (63-kDa protein) (3) and β-subunits (240-kDa protein) (4). The channels are localized in the plasma membrane of the photoreceptor outer segments at a density of about 500 μm^{-2}. CNG channels are probably the only Ca^{2+}-permeable channels in the outer-segment membrane. In the dark, an elevated level of cGMP keeps the channels open, giving rise to a steady inward current, the so-called *dark current* (Fig. 1). Approximately 10% of the dark current is carried by Ca^{2+}, and this Ca^{2+} influx is balanced by an Na/Ca,K-exchanger that extrudes Ca^{2+} from the cytosol (5,6). The combined activity of CNG channels and Na/Ca,K-exchangers maintains a stable free Ca^{2+} concentration of about 500 nM in the dark (7). Upon illumination, the photopigment rhodopsin is converted into its active form (Rh*), which leads to activation of cGMP-specific phosphodiesterase (PDE*) through the guanosine triphosphate (GTP)-binding protein transducin (T*). PDE* rapidly hydrolyzes cGMP to guanosine monophosphate (GMP). As a result of the decrease in cGMP concentration, CNG channels close, leading to hyperpolarization of the photoreceptor membrane. The immediate consequence of CNG channel closure in light is a decrease in the cytosolic Ca^{2+} concentration due to continued activity of the Na/Ca,K-exchanger. Several Ca^{2+}-binding proteins utilize this Ca^{2+} signal to initiate recovery of the photoresponse by terminating excitatory processes and reopening CNG channels. One mechanism affects the lifetime of PDE* and hence the rate of cGMP hydrolysis: activation of PDE by Rh* is suppressed when Rh* is phosphorylated by rhodopsin kinase (RhK) (8). The enzymatic activity of RhK is in turn controlled by the Ca^{2+}-binding protein recoverin (Rec). At low Ca^{2+} concentrations, Rec releases RhK from inhibition. Rh* is phosphorylated and thereby uncoupled from the signaling cascade, leading to inhibition of PDE (9). The low Ca^{2+} concentration also accelerates cGMP synthesis through activation of a retinal guanylate cyclase (retGC), which is controlled by another Ca^{2+}-binding protein, the guanylyl cyclase-activating protein (GCAP) (10). GCAP is inhibited at high Ca^{2+} concentrations in the dark, but becomes activated upon illumination of the photoreceptor cell, causing increased retGC activity. Thus, Rec and GCAP mediate an increase in cGMP concentration at low Ca^{2+} by inhibiting the hydrolysis and promoting the synthesis of this second messenger. In addition to the rising cGMP level, a third Ca^{2+}-dependent mechanism contributes to the opening of CNG channels. The ligand sensitivity of the channels is controlled by calmodulin (CaM). At high Ca^{2+} levels (in the dark-adapted cell), Ca^{2+}/calmodulin binds to the β-subunit of the native channel and causes a decrease in its apparent cGMP affinity (11). When the Ca^{2+} concentration decreases upon illumination, CaM dissociates from the channels, restoring the high cGMP affinity, which leads to opening of the channels and recovery of the photoresponse.

Interestingly, the speed of the Ca^{2+} recycling process that controls the dynamics of phototransduction appears to differ in rod and cone photoreceptors. More Ca^{2+} enters cone outer segments in the dark, since Ca^{2+} contributes about 20% to the dark

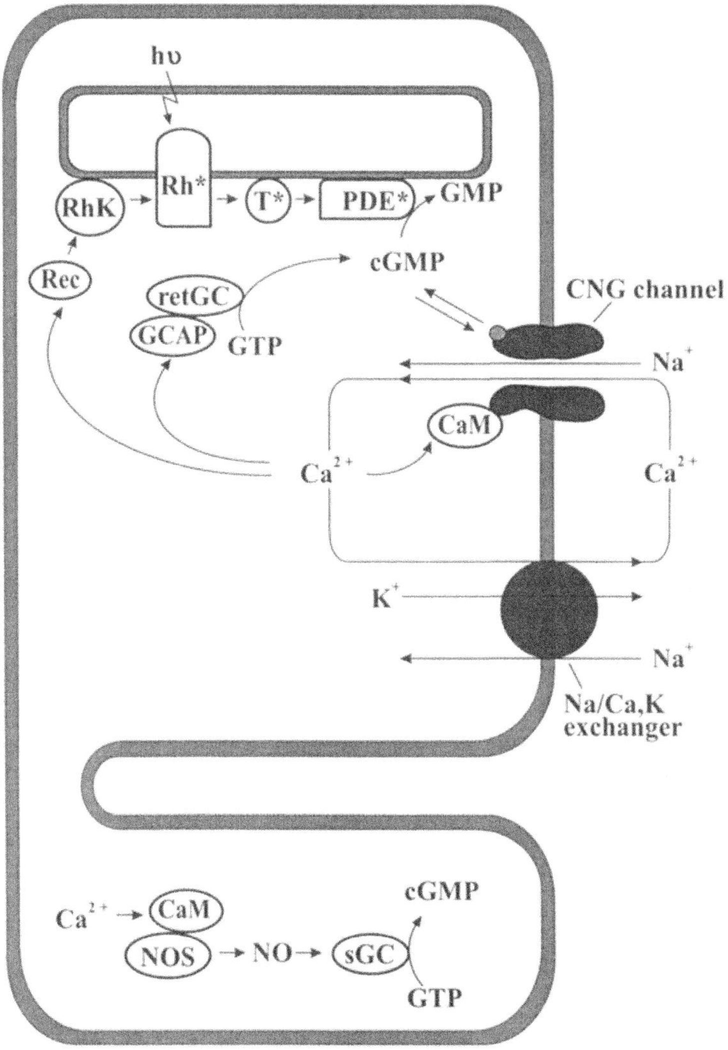

FIG. 1. Signal transduction in vertebrate photoreceptor cells. The transmitter molecule cGMP is synthesized by a membrane-bound guanylyl cyclase (retGC) and hydrolyzed by a phosphodiesterase (PDE). Hydrolysis is triggered by absorption of a photon (hν). The Ca^{2+} cycle is controlled by Ca^{2+} influx through CNG channels and Ca^{2+} efflux via Na/Ca,K-exchangers. The cytosolic Ca^{2+} concentration is sensed by various Ca^{2+}-binding proteins such as calmodulin (CaM), recoverin (Rec), and guanylyl cyclase-activating protein (GCAP). Ca^{2+} modulates phototransduction through the indicated feedback loops. A second pathway of cGMP synthesis, involving a Ca^{2+}/calmodulin-dependent nitric oxide synthase (NOS) and a soluble NO-sensitive guanylyl cyclase (sGC), was recently described in the inner segment (8).

current (12). Cone photoreceptor cells express different although homologous CNG channel subunits (62% identity to the rod α-subunits) (13), which display a remarkably high Ca^{2+} permeability (14,15). The high rate of Ca^{2+} influx mediated by these channels is balanced by a more efficient Na/Ca,K-exchanger that extrudes Ca^{2+} at a correspondingly higher rate. Consequently, light-induced closure of cone CNG channels results in a faster decline of Ca^{2+} concentration and a faster onset of signal recovery. It appears that differences in the Ca^{2+} conductance between rod and cone CNG channels play a critical role in shaping the amplitude and time course of the photoresponse in these cells (16).

CYCLIC NUCLEOTIDE-GATED CHANNELS IN VERTEBRATE OLFACTORY TRANSDUCTION

The chemosensory membrane of vertebrate OSNs consists of a set of thin, elongated cilia that are embedded in a mucus layer exposed to air within the nasal cavity. These cilia contain all components involved in the chemoelectrical signal transduction that is triggered by the detection of an odor (Fig. 2). Odorants can activate two alternative pathways, each of which represents a highly amplifying biochemical cascade that leads to depolarization and electrical excitation of the neuron. Many odorants bind to receptor proteins (R_1) in the ciliary membrane (17) and activate adenylyl cyclase (AC) through the stimulatory GTP-binding protein G_{olf} (18,19). The consequent increase in ciliary cAMP concentration activates olfactory neuron-specific CNG channels (57% identity to rod α-subunits [20]) that are expressed at a density of 1,000 to 1,500 μm^{-2} in the ciliary membrane. A large fraction of the current conducted by these channels is carried by Ca^{2+} (estimated: 40% to 80%). The exact contribution of Ca^{2+} to the cAMP-induced current depends on the free Ca^{2+} concentration in the olfactory mucus (15). This value is difficult to determine and still has to be established. However, considering the high channel density, the high Ca^{2+} conductance of the channels, and the small volume of the ciliary lumen (diameter: 0.1 μm to 0.2 μm), a rapid increase can be expected in the ciliary Ca^{2+} concentration following activation of CNG channels. This odor-induced Ca^{2+} signal triggers both excitatory and adaptive processes in the sensory neuron. Ca^{2+}-activated Cl^- channels open in response to the increasing Ca^{2+} concentration and amplify the receptor current by conducting a depolarizing Cl^- efflux (21–23). Furthermore, Ca^{2+} can potentiate cAMP synthesis through a positive feedback mechanism affecting AC, which is activated by CaM (24). Two adaptive processes are also mediated by CaM. A Ca^{2+}/CaM-sensitive phosphodiesterase hydrolyzes cAMP and leads to a decrease in ciliary cAMP concentration (25). In addition, Ca^{2+}/CaM also binds to CNG channels and dramatically decreases their cAMP sensitivity. This promotes closure of the channels and termination of the neuronal response to odors (26,27). The relative contributions of these Ca^{2+}-dependent processes to the shaping of the olfactory receptor current are currently being investigated with considerable effort.

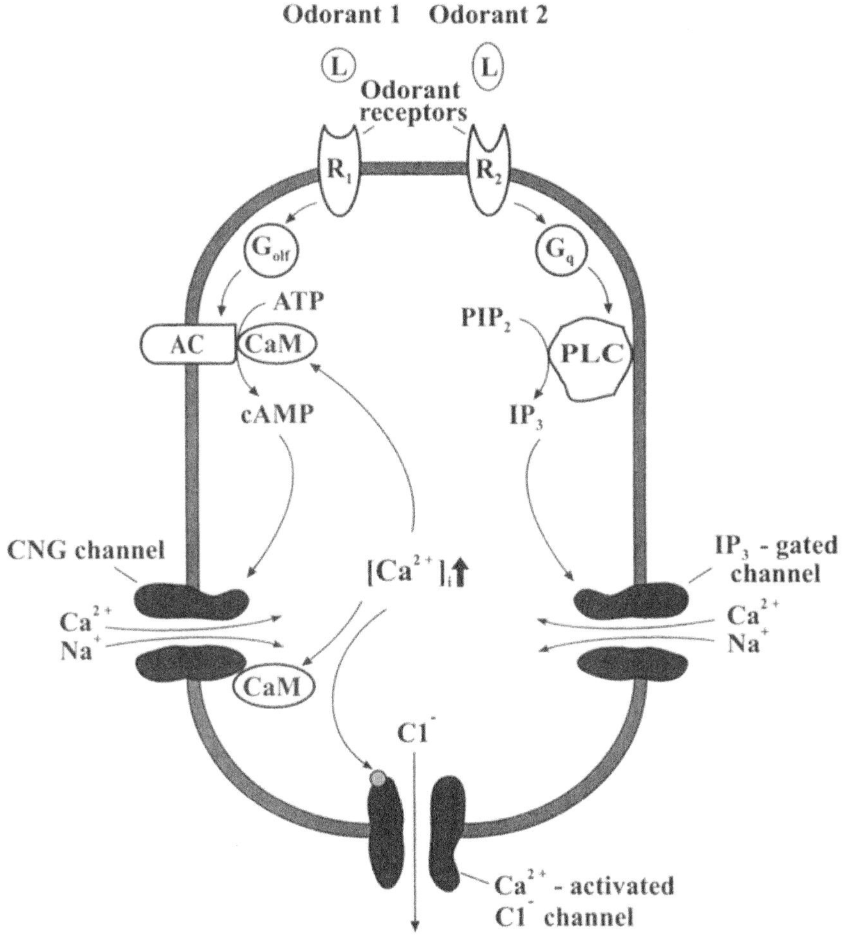

FIG. 2. Signal transduction in olfactory sensory neurons. Two molecular pathways mediate olfactory signal transduction. Many odorants bind to receptor proteins coupled to adenylyl cyclase (AC). This signaling pathway includes cAMP, CNG channels, and Ca^{2+}-activated Cl^- channels. Other odorants use an alternative pathway employing phospholipase C (PLC), inositol-1,4,5-trisphosphate (IP_3), and IP_3-gated cation channels. The cytosolic Ca^{2+} concentration ($[Ca^{2+}]_i$) controls Cl^- channels and, through activation of calmodulin (CaM), also CNG channels and AC activity.

CYCLIC NUCLEOTIDE-GATED CHANNELS IN SPERM CELLS

It has been known for some time that sperm cells are guided toward egg cells by chemical stimuli. For example, sea urchin sperm cells are attracted by the peptide speract, which is released by sea urchin eggs (28). Stimulation by speract causes an increase in cGMP and Ca^{2+} concentrations in the sperm cell, indicating involvement of these messengers in processing the chemoresponse (28,29). Egg-derived or egg-associated agents stimulate a variety of biologic responses in sperm cells, including chemotaxis, species-specific adhesion to the egg, acrosome reaction, and fusion with the egg plasma membrane. For vertebrates, there is now convincing evidence that a small fraction of sperm cells (capacitated cells) can follow chemoattractive factors in the follicular fluid (30). Interestingly, vertebrate sperm cells express proteins that were previously considered to be specific components of the olfactory signal-transduction chain (Fig. 3). In particular, odorant receptor proteins (31), adenylyl cyclase

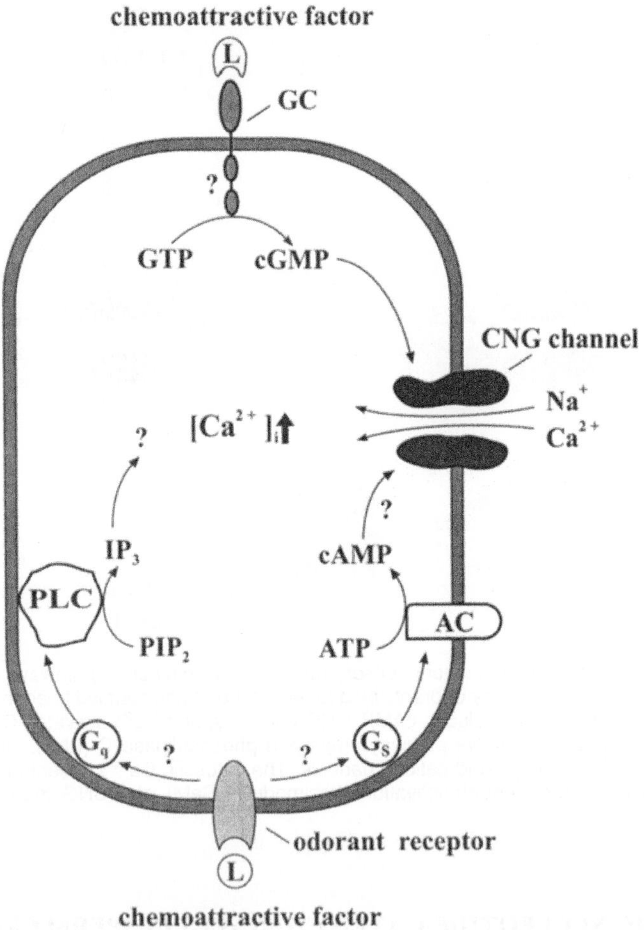

FIG. 3. Hypothetical signaling pathways in vertebrate sperm chemoattraction. Binding of chemoattractive factors to membrane receptors causes a transient increase in $[Ca^{2+}]_i$. The change in $[Ca^{2+}]_i$ could lead to alteration of sperm swimming behavior. Membrane receptors for chemoattractants may belong to the family of membrane-bound guanylyl cyclases (GC), as shown for invertebrate sperm, or to the family of G-protein-coupled odorant receptors. CNG channels may provide an entry pathway for Ca^{2+}.

(AC), and GTP-binding proteins (G_s) (32), as well as CNG channels (13), have been identified in vertebrate sperm. The CNG channel α-subunit expressed in bovine sperm is identical to the α-subunit of bovine cone photoreceptor cells. Its high Ca^{2+} conductance makes it well suited to generate cyclic nucleotide-dependent Ca^{2+} signals in response to chemical stimuli recognized by the sperm cell. While the context of biochemical interactions that control chemotaxis in sperm cells is not well understood, the striking similarities to olfactory transduction components offer a promising approach to its investigation.

CONCLUSION

CNG channels have now been identified in a number of different vertebrate tissues, including retina, olfactory epithelium, testis, heart, kidney, and brain. Although various invertebrate organisms were also found to express CNG channels, only one CNG channel subunit has so far been cloned and characterized in detail (33). The *Drosophila* channel is expressed in the visual and olfactory system of the fly, and like other CNG channels also shows a very high Ca^{2+} conductance. The most important property of CNG channels, which appears to be common to all CNG channels, is their ability to efficiently mediate Ca^{2+} entry into cells. These channels therefore serve two biologic functions: (i) their activation depolarizes cells to some degree; (ii) they generate a Ca^{2+} signal that links cyclic nucleotide metabolism to Ca^{2+}-regulated mechanisms of the cell.

REFERENCES

1. Finn JT, Grunwald ME, Yau K-Y. Cyclic nucleotide-gated ion channels: An extended family with diverse functions. *Annu Rev Physiol* 1996;58:395–426.
2. Kaupp UB. Family of cyclic nucleotide gated ion channels. *Curr Opin Neurobiol* 1995;5:434–42.
3. Kaupp UB, Niidome T, Tanabe T, et al. Primary structure and functional expression from complementary DNA of the rod photoreceptor cyclic GMP-gated channel. *Nature* 1989;342:762–6.
4. Körschen H, Illing G, Seifert R, et al. A 240 kDa protein represents the complete β subunit of the cyclic nucleotide-gated channel from rod photoreceptor. *Neuron* 1995;15:627–36.
5. Nakatani K, Yau K-Y. Calcium and magnesium fluxes across the plasma membrane of the toad rod outer segment. *J Physiol* 1988;395:695–729.
6. Schnetkamp PPM. How does the retinal rod Na/Ca⁺K exchanger regulate free cytosolic Ca^{2+}? *J Biol Chem* 1995;270:13231–9.
7. Gray-Keller MP, Detwiler PB. The calcium feedback signal in the phototransduction cascade of vertebrate rods. *Neuron* 1994;13:849–61.
8. Kawamura S. Rhodopsin phosphorylation as a mechanism of cGMP phosphodiesterase regulation by S-modulin. *Nature* 1993;362:855–7.
9. Koch K-W. Control of photoreceptor proteins by Ca^{2+}. *Cell Calcium* 1995;18:314–21.
10. Gorczyca WA, Gray-Keller MP, Detwiler PB, Palczewski K. Purification and physiological evaluation of a guanylate cyclase activating protein from retinal rods. *Proc Natl Acad Sci USA* 1994;91:4014–8.
11. Chen TY, Illing M, Molday LL, Hsu YT, Yau K-W, Molday RS. Subunit 2 (or β) of retinal rod cGMP-gated cation channel is a component of the 240–kDa channel-associated protein and mediates Ca^{2+}-calmodulin modulation. *Proc Natl Acad Sci USA* 1994:91:11757–61.
12. Perry RJ, McNaughton PA. Response properties of cones from the retina of the tiger salamander. *J Physiol* 1991;455:111–42.

13. Weyand I, Godde M, Frings S, et al. Cloning and functional expression of a cyclic nucleotide-gated channel from mammalian sperm. *Nature* 1994; 368:859–63.
14. Picones A, Korenbrot JI. Permeability and interaction of Ca^{2+} with cGMP-gated channels differ in retinal rod and cone photoreceptors. *Biophys J* 1995;69:120–7.
15. Frings S, Seifert R, Godde M, Kaupp UB. Profoundly different calcium permeation and blockage determine the specific function of distinct cyclic nucleotide-gated channels. *Neuron* 1995;15:169–79.
16. Korenbrot JI. Ca^{2+} flux in retinal rod and cone outer segments: differences in Ca^{2+} selectivity of the cGMP-gated ion channels and Ca^{2+} clearance rates. *Cell Calcium* 1995;18:285–300.
17. Buck L, Axel R. A novel multigene family may encode odorant receptors: a molecular basis for odor recognition. *Cell* 1991;65:175–87.
18. Bakalyar HA, Reed RR. Identification of a specialized adenylyl cyclase that may mediate odorant detection. *Science* 1990;250: 1403–6.
19. Jones DT, Reed RR. G_{olf}: an olfactory neuron-specific G protein involved in odorant signal transduction. *Science* 1989;244: 790–5.
20. Dhallan RS, Yau K-Y, Schrader KA, Reed RR. Primary structure and functional expression of a cyclic nucleotide-gated channel from olfactory neurons. *Nature* 1990;347: 184–7.
21. Kleene SJ, Gesteland RC. Calcium-activated chloride conductance in frog olfactory cilia. *J Neurosci* 1991;11:3624–9.
22. Kurahashi T, Yau K-Y. Co-existence of a cationic and chloride component in odorant-induced current of vertebrate olfactory receptor cells. *Nature* 1993; 363: 71–4.
23. Lowe G, Gold GH. Non-linear amplification by calcium-dependent chloride channels in olfactory receptor cells. *Nature* 1993; 366: 283–6.
24. Anholt RRH, Rivers AM. Olfactory transduction: cross-talk between second-messenger systems. *Biochemistry* 1990; 29: 788–95.
25. Borisy FF, Ronnett GV, Cunningham AM, Juilfs D, Beavo J, Snyder SH. Calcium/calmodulin-activated phosphodiesterase expressed in olfactory receptor neurons. *J Neurosci* 1992; 12: 915–23.
26. Chen T-Y, Yau K-Y. Direct modulation by Ca^{2+}-calmodulin of cyclic nucleotide-activated channel of rat olfactory receptor neurons. *Nature* 1994;368:545–8.
27. Liu M, Chen T-Y, Ahamed B, Li J, Yau K-Y. Calcium-calmodulin modulation of the olfactory cyclic nucleotide-gated cation channel. *Science* 1994; 266: 1348–54.
28. Hansbrough JR, Garbers DL. Speract—purification and characterization of a peptide associated with eggs that activates spermatozoa. *J Biol Chem* 1981;256:1447–52.
29. Schackmann RW, Chock PB. Alterations of intracellular [Ca^{2+}] in sea urchin sperm by the egg peptide speract. *J Biol Chem* 1986;261:8719–28.
30. Ralt D, Manor M, Cohen-Dayag A, et al. Chemotaxis and chemokinesis of human spermatozoa to follicular factors. *Biol Reprod* 1994; 50: 774–85.
31. Vanderhaeghen P, Schurmans S, Vassart G, Parmentier M. Olfactory receptors are displayed on dog mature sperm cells. *J Cell Biol* 1993; 123: 1441–52.
32. Ward CR, Kopf GS. Molecular events mediating sperm activation. *Dev Biol* 1993;158:9–34.
33. Baumann A, Frings S, Godde M, Seifert R, Kaupp UB. Primary structure and functional expression of a Drosophila cyclic nucleotide-gated channel present in eyes and antenna. *EMBO J* 1994; 13: 5040–50.

Signal Transduction in Health and Disease, Advances in Second Messenger and Phosphoprotein Research, Vol. 31, edited by J. Corbin and S. Francis.

8

Chemoattractant Receptor Signaling

G Protein-Dependent and -Independent Pathways

Jacqueline L. S. Milne, Ji-Yun Kim, and Peter N. Devreotes

Department of Biological Chemistry, The Johns Hopkins School of Medicine, Baltimore, MD 21205-2185

Many hormones, neurotransmitters, chemoattractants, and environmental stimuli influence cellular functions through seven helix receptors that couple to heterotrimeric G proteins. Several hundred of these receptors have been identified, as well as multiple G protein α, β, and γ subunits that regulate the activity of effectors such as adenylyl cyclases, phospholipases, phosphodiesterases, ion channels, and mitogen-activated protein (MAP) kinases (1–6). A wealth of evidence has defined receptor domains that are important for ligand binding, activation of G proteins, and desensitization (7–9). Many questions, however, remain. How does ligand binding induce or stabilize activated receptor states? How many conformations do receptors adopt during activation? How do these trigger the G protein cascade? How do the dissociated G protein subunits ultimately regulate downstream effectors? Identification of mutant receptors, G proteins, and effectors that are blocked in their respective activation cycles, together with the ability to obtain high-resolution structural information, will be essential to answering these questions.

We are using the biochemically and genetically amenable model of *Dictyostelium* to analyze G protein-mediated signal transduction (10). This simple protozoan utilizes homologues of mammalian receptors, G proteins, and effectors for numerous processes. The targeted disruption of many of the genes encoding these proteins has clarified their roles in signal transduction and has laid the foundation for detailed structure/function studies using random mutagenesis. It has also led to the surprising discovery that seven helix receptors activate a novel signal-transduction pathway that acts independently of G proteins. Moreover, the development of insertional mutagenesis has led to the identification of novel modulator proteins, which add a rich layer of complexity to G protein-dependent signaling.

DICTYOSTELIUM LIFE CYCLE

Dictyostelium live as single amoebae in decaying leaves, and respond chemotactically to folate and other compounds secreted by their bacterial food source (11). Upon depletion of nutrients, the amoebae become responsive to extracellular cyclic adenosine monophosphate (cAMP) and enter a developmental program. Within a few hours, $\sim 10^5$ cells aggregate into a multicellular structure that eventually forms a fruiting body composed of stalk and spore cells. During aggregation, centrally located cells periodically synthesize and secrete waves of cAMP that induce adjacent cells to move toward the center and to relay the cAMP signal to distally located cells. cAMP also regulates gene-expression events essential for the differentiation, morphogenesis, and pattern formation that underlie formation of fruiting bodies (12–14).

cAMP-RECEPTOR SUBTYPES

The physiologic responses of *Dictyostelium* elicited by cAMP depend upon the presence of cell-surface cAMP receptors. cAMP-receptor occupancy activates several effectors, including adenylyl (15,16) and guanylyl cyclases (17,18), phospholipase C (PLC) (19,20), and an MAP kinase (21). cAMP elevates intracellular Ca^{2+} (22,23) by inducing both release of Ca^{2+} from IP_3-sensitive stores (24,25) and a transient entry of extracellular Ca^{2+} (26–28). Extrusion of K^+ (29) and H^+ (30) also occurs, as do changes in cytoskeletal components required for chemotaxis (31). To date, four highly homologous cAMP receptors (cARs 1 to 4) have been cloned; these receptors have seven predicted membrane-spanning regions and a marked similarity to rhodopsin and the β-adrenergic receptor (32–34). Each of these receptors elicits many of the same biochemical responses (35,36) (J. Y. Kim and P. N. Devreotes, *unpublished data*; J. M. Louis and A. Kimmel, *personal communication*).

Each receptor is transiently expressed at distinct developmental stages. The highest-affinity receptor (37), cAR1, is expressed during aggregation (32,38). Cells lacking cAR1 do not aggregate under normal conditions, but can be induced to develop and show cAMP-mediated responses (36,38–40). This occurs because the cells express cAR3, a receptor of slightly lower affinity (41) whose expression overlaps that of cAR1 (42). Deletion of the genes for both receptors through homologous recombination effectively abolishes development of the organism (36). cAR2 and cAR4 have a much lower affinity for cAMP than do the other cARs, and are expressed in prestalk cells found in multicellular aggregates (33,34,43). Deletion of either of these receptors arrests development after the formation of multicellular structures. The different affinities and developmental expression of the four cARs enables extracellular cAMP to regulate, in a precise temporal and cell-specific manner, the diverse events required for aggregation, cell–cell signaling, and morphogenesis in *Dictyostelium*.

MULTIPLE Gα-SUBUNITS

In addition to the cARs, eight G protein α subunits and a single Gβ subunit have been cloned (44–48). These are remarkably homologous to their mammalian counterparts. The Gγ subunit has recently been identified (N. Zhang and P. W. Devreotes, *unpublished data*). Most of the Gα subunits contain the highly conserved domains required for guanosine triphosphatase (GTPase) function. Sequence analysis indicates that the Gα subunits share only 35% to 50% sequence identity and do not fall into classes or belong to a specific mammalian Gα subfamily (48). Many of the Gα subunits (1,5,7,8) can be deleted from the cells of the organism without blocking growth or development, although alterations in the timing of development or morphogenesis have been observed with $g\alpha 1^-$ and $g\alpha 5^-$ cells or in cells overexpressing Gα1 or Gα7 (49–51) (R. A. Firtel, *personal communication*). Owing to their low homology, it is unlikely that these Gα subunits are functionally redundant. It is more likely that their roles are too subtle to detect in the types of assays applied thus far.

In contrast, deletion of Gα2 or Gα4 has dramatic phenotypic effects on familiar developmental events. The $g\alpha 4^-$ cells do not respond to folic acid, but remain sensitive to cAMP (52). Gα4 is likely to couple to folate receptors (53), which have properties reminiscent of G-protein-coupled receptors (54), although they have not been cloned. Gα2 couples specifically to the cARs. The $g\alpha 2^-$ cells show wild-type responses to folic acid (55,56), but despite expressing cAR1, do not aggregate (57), move toward cAMP gradients, or show cAMP-mediated synthesis of cAMP, cyclic guanosine monophosphate (cGMP), or inositol 1,4,5-triphosphate (IP_3) (58,59). These responses are not restored by overexpressing cAR1 or cAR3, suggesting that other Gα subunits cannot replace Gα2 (35).

G PROTEIN-DEPENDENT SIGNAL TRANSDUCTION THROUGH cAR1

G2 is essential for many of the cAR1-mediated responses described above, including activation of PLC and guanylyl and adenylyl cyclases, and alterations in the cytoskeleton (Fig. 1). Activation of this aggregation-specific adenylyl cyclase (ACA) requires Gβγ (60) analogous to mammalian type II and IV adenylyl cyclases (4). Although Gα2 is necessary for cAR1-mediated synthesis of cAMP, *in vitro* studies show that guanosine 5′-[8-thio]triphosphate (GTPγS) activates ACA in lysates of $g\alpha 2^-$ cells. Presumably, GTPγS acts to free Gβγ subunits from other endogenous G proteins present in the $g\alpha 2^-$ cells. Construction of cells lacking the unique Gβ provides convincing evidence for the role of Gβ in ACA activation. In contrast to $g\alpha 2^-$ cells, $g\beta^-$ cells expressing cAR1, Gα2, and ACA do not exhibit ACA activity *in vivo* or *in vitro* (60). The $g\beta^-$ cells do not aggregate or move toward cAMP or any other chemoattractant (60). Together, these results suggest that the $g\beta^-$ cells lack functional G proteins.

The $g\beta^-$ cells are useful for assessing the role of G proteins in a variety of pro-

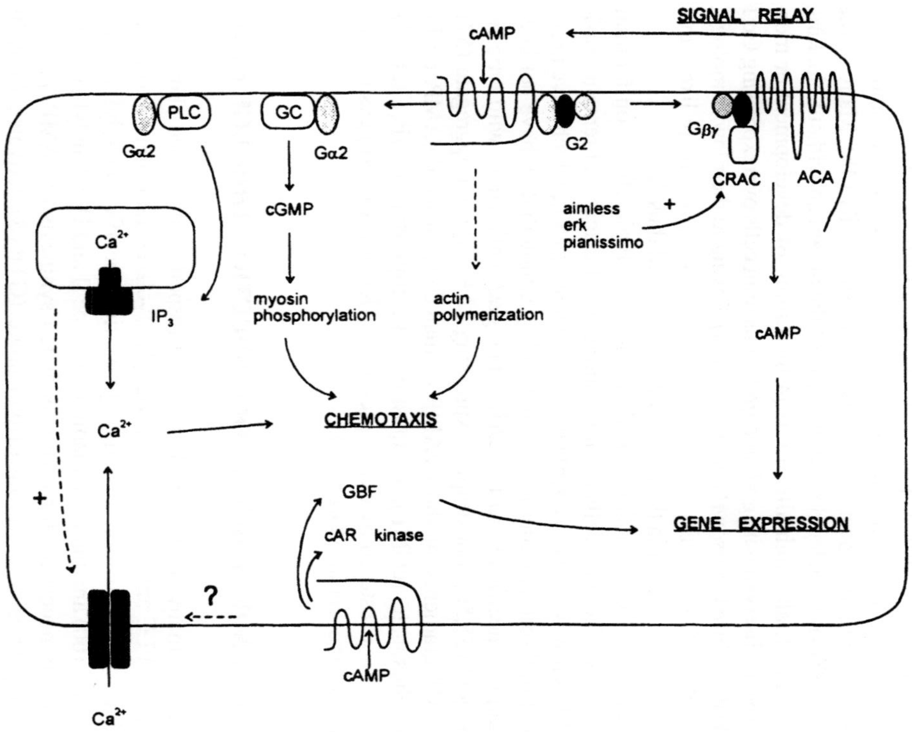

FIG. 1. Schematic representation of cAMP-mediated signal transduction through cAR1, leading to chemotaxis, gene expression, and signal relay. Activation of phospholipase C (PLC), guanylyl cyclase (GC), adenylyl cyclase (ACA), myosin phosphorylation, and actin polymerization require the G protein G2. cAR1 regulates Ca^{2+} entry, cAR kinase, and the transcription factor GBF through G protein-independent signaling pathways. Additional details are provided in the text.

cesses. Wild-type cells possess a single Gβ protein that is expressed at constant levels throughout growth and development. Disruption of the Gβ gene leads to loss of the corresponding messenger ribonucleic acid (mRNA) and the 36 kDa protein that immunoreacts with a Gβ-specific antiserum. Low-stringency Southern blot analysis indicates the absence of any closely related genes (47), although there are several proteins with regions of homology to Gβ, including coronin, an actin-binding protein (61). Neither coronin nor other proteins with limited homology to Gβ are likely to couple to cAR1 in the absence of Gβ. *gβ*⁻ cells lack G protein-dependent responses *in vivo*. Membranes from these cells show a single type of cAMP binding site with low affinity for cAMP (K_d = ~400 nM), and these sites are not influenced by GTPγS (60), a classical feature of G protein-coupled receptors (1). Moreover, a high degree of sequence conservation is evident among the Gβ subunits of organisms ranging from *Dictyostelium* to mammals (47). Highly conserved domains of Gβ are probably essential for interactions with receptors and the G protein α and γ subunits required for G protein-mediated signaling.

ADDITIONAL REGULATORS OF ADENYLYL CYCLASE IDENTIFIED BY INSERTIONAL MUTAGENESIS

The complexity of G protein-mediated signaling is further highlighted by the identification of novel regulators of ACA (Fig. 1). Many of these have been discovered through restriction enzyme-mediated insertion (REMI) (62). Linearized plasmid deoxyribonucleic acid (DNA) containing a selectable marker is electroporated into the cells with a restriction enzyme to enhance its insertion into restriction sites distributed throughout the genome. The tagged DNA adjacent to the insertion site is rescued and used to disrupt the same locus in wild-type cells in order to recapitulate the phenotype (63). A variety of developmental phenotypes arise from REMI mutagenesis; characterization of aggregation-deficient clones has yielded regulators of ACA (31).

One protein, the cytosolic regulator of adenylyl cyclase (CRAC), was originally identified by its ability to restore cAMP- or GTPγS-stimulated adenylyl cyclase activity to lysates of an unresponsive mutant (64). Coincidentally, the N-terminal sequence of partially purified CRAC was found to be identical to the sequence of a gene isolated by REMI (65,66). CRAC translocates to the membrane fraction when cells are exposed to cAMP or when lysates are treated with GTPγS (67). Mutant analysis indicates that CRAC translocation specifically requires Gβ, suggesting that Gβγ may be required to form a CRAC binding site (67). In support of this concept, CRAC contains a pleckstrin homology (PH) domain (65), and Gβ from other systems facilitates the membrane localization of proteins containing PH domains (68–70). Whether ACA interacts with CRAC alone or with the CRAC/Gβ complex is unclear. Additional regulators of ACA have been identified by REMI, as discussed in detail elsewhere (31). These include the MAP kinase ERK2 (extracellular signal-regulated kinase 2); aimless, a protein with homology to ras exchange factors such as son-of-sevenless (SOS); and a newly discovered protein, pianissimo, which has homology to recently identified open-reading frames in yeasts. Given the diversity of adenylyl cyclases in mammalian cells, it is plausible that mammalian homologues of these proteins exist and regulate certain of these cyclases.

G PROTEIN-INDEPENDENT SIGNAL TRANSDUCTION THROUGH THE cARS

The absence of apparent G protein-dependent signal transduction events in *gα2*⁻ or *gβ*⁻ cells has been instrumental for identification of the G protein-independent responses illustrated in Fig. 1. The first evidence for a separate signaling pathway was the discovery of cAMP-dependent Ca^{2+} entry in *frigid A* (28), an aggregation-minus mutant lacking functional Gα2 (71), and later in *gα2*⁻ cells overexpressing cAR1 or cAR3 (35). Although the other Gα subunits appeared unessential for Ca^{2+} entry, the possibility of functional redundancy among these proteins was eliminated by analysis of the *gβ*⁻ cells. As mentioned above, the receptors in these cells do not couple to G proteins. Yet *gβ*⁻ cells expressing cAR1 or cAR3 mediate a cAMP-dependent Ca^{2+} influx that has kinetics and agonist dependence like those of wild-type controls (72). The EC_{50} values for cAMP in the Ca^{2+} entry response closely match the dissociation constant of cAMP binding to the low-affinity forms of the cARs, suggesting that uncoupled forms of the cARs trigger this response. While G protein-independent signals provide the primary on/off switch for Ca^{2+} entry, this response is probably also modulated by G protein-dependent events, since *gα2*⁻ and *gβ*⁻ cells accumulate 50% to 75% less Ca^{2+} in response to agonist than do control cells (35,72) (Fig. 1). Depletion of intracellular IP_3-sensitive Ca^{2+} stores may produce a positive feedback signal that enhances Ca^{2+} entry, analogous to the Ca^{2+}-store-operated Ca^{2+} entry pathways of mammalian cells (73).

In addition to Ca^{2+} fluxes, other cAR-mediated responses occur in the absence of functional G proteins. The time course and agonist sensitivity of cAMP-mediated cAR1 phosphorylation is comparable in wild-type and *gβ*⁻ cells (72). Moreover, the MAP kinase ERK2 (74) is activated by cAMP in *gα2*⁻ and *gβ*⁻ cells (21). Interestingly, ERK2 activity may be regulated, at least in part, by Ca^{2+}. Addition of extracellular ethylene glycol-bis-(β-amino ethyl ether)-N,N,N′,N′-tetraacetic acid (EGTA) reduces the level of stimulated ERK2 activity; no activation is observed in cells treated for 15 min with EGTA and thapsigargin, a compound that induces efflux of Ca^{2+} from intracellular stores (M. Maeda and R. A. Firtel, *personal communication*). cAMP-induced Ca^{2+} influx occurs normally in *erk*⁻ cells (J. L. S. Milne, *unpublished data*), supporting the idea that ERK2 activation may be downstream of increases in cytosolic free Ca^{2+}.

Several gene-expression events require cARs, but appear to be G protein-independent. Two genes critical for the proper regulation of extracellular cAMP signals during early development, those for the extracellular phosphodiesterase and its inhibitor, are both expressed normally in *gα2*⁻ and *gβ*⁻ cells (75). The receptor-mediated activation of G-box binding factor (GBF), a transcriptional regulator that controls expression of RasD, loose aggregate C (LagC), and pst-cathespin (CP2) during multicellular development, also does not require these G protein subunits (76). Whether these gene-expression events require ERK2 activity or influx of extracellular Ca^{2+}, and whether they are influenced by G protein-dependent signals remains to be determined.

G protein-independent signal transduction through seven helix receptors is likely to be a general phenomenon in eukaryotic cells. In addition to the cARs, a number of G protein-coupled receptors are known to undergo phosphorylation and/or internalization without coupling to G proteins (77–81). These receptors probably undergo conformational changes that permit their interaction with receptor kinases in the absence of G proteins (82). The genetic evidence from *Dictyostelium* for G protein-independent Ca^{2+} influx and gene induction raises the possibility that seven helix receptors can also activate downstream effectors, either directly or through unidentified adaptors that recognize agonist-activated state(s) of the receptor. While genetic evidence for G protein-independent signaling is not available in systems other than *Dictyostelium*, recent biochemical evidence indicates that these signaling pathways are present in mammalian cells. For example, M3 muscarinic receptors regulate Ca^{2+} entry (83), and μ-opioid receptors modulate a Ca^{2+}-dependent K^+ channel (84) without G proteins.

MUTAGENESIS OF SIGNAL-TRANSDUCTION COMPONENTS

Many null mutants, such as *car1⁻*, *aca⁻*, *gβ⁻*, *gα2⁻*, and *crac⁻* are aggregation-deficient, but can be rescued by constitutive expression of the wild-type version of the missing protein. This phenotypic screen permits structure/function analysis of each component; cells can be efficiently transformed with segregating plasmids that contain selectable markers (85). This approach permits the use of random mutagenesis to assess thousands of amino-acid substitutions with no bias in the location or type of substitution. Following polymerase chain reaction (PCR) amplification, a library is created by cloning into an extrachromasomal plasmid under the control of a promoter that is active during the growth stage. Null cells transformed with a degenerate library are plated as single cells on a bacterial lawn. After ~1 wk, the dividing cells clear the bacteria, forming plaques of ~1 cm, and starve. Figure 2A illustrates the identification of loss-of-function mutations in ACA. Clones of *aca⁻* expressing functional ACA will undergo development, while clones expressing unstable or functionally impaired ACA remain as a flat layer of cells (86). Loss-of-function mutations have also been found in cAR1, using rescue of *car1⁻* cells (see below). Gain-of-function mutations in ACA (Fig. 2A) have also been identified using a variation of the screen. Random mutagenesis is a powerful extension of site-directed mutagenesis of conserved residues in other receptor-signaling systems (7), and of the identification of mutant receptors and G proteins by PCR of genetic loci corresponding to mammalian diseases (87,88).

Figure 2 illustrates several convenient biochemical screens used to analyze the

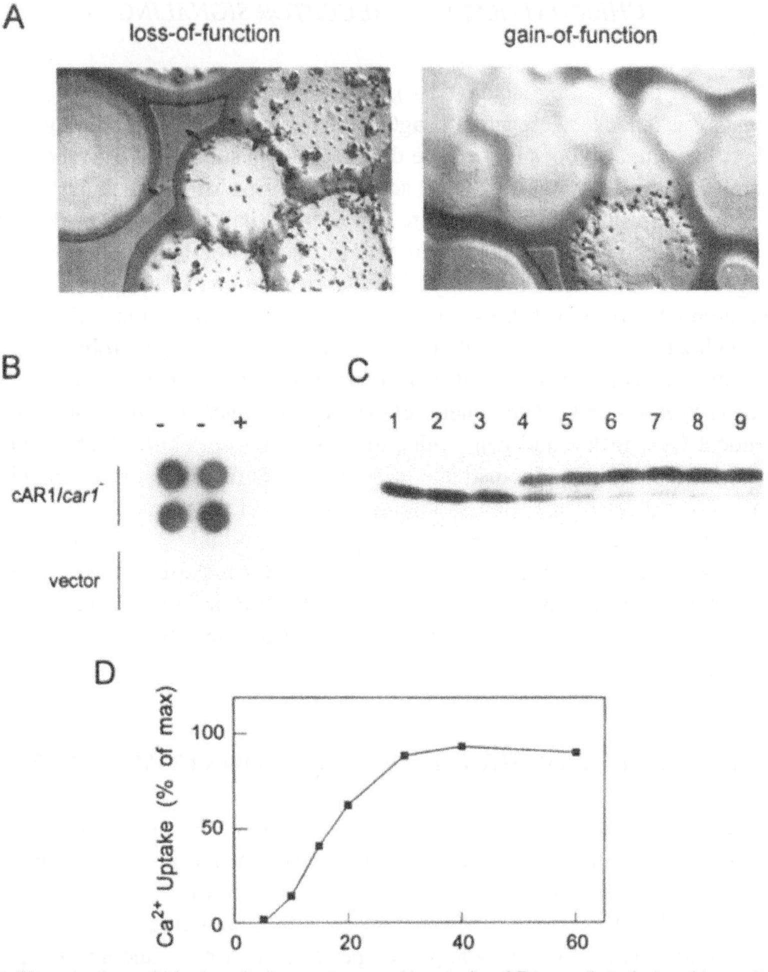

FIG. 2. Phenotypic and biochemical screens used to study cAR1-mediated signal transduction. **A:** Random mutagenesis of ACA and expression of the mutant cyclase library in *aca*$^-$ (left panel) or *crac*$^-$ (right panel) cells permits identification of loss-of-function and gain-of-function mutants, respectively. **B:** [^{32}P]cAMP binding to cAR1. *car1*$^-$ cells expressing cAR1 or empty expression vector were loaded into 96-well, filter-bottomed tissue-culture plates and incubated with 0.5 nM [^{32}P]cAMP in the absence (-) or presence (+) of 100 μM nonradioactive cAMP as described (89). The plate was washed, dried, and autoradiographed. **C:** cAMP-induced cAR1 phosphorylation. Amoebae were shaken for 15 min in 10 mM phosphate buffer containing 10 mM dithiothreitol to inhibit cellular phosphodiesterases and various concentrations of cAMP. Following solubilization, proteins were separated electrophoretically and immunoblotted with cAR1 antiserum as described (89). Phosphorylation of cAR1 induces a change in the mobility of the receptor from 40 kDa to 43 kDa. Lanes 1 to 9 represent the following concentrations of cAMP: 0, 1 nM, 5 nM, 10 nM, 50 nM, 100 nM, 1 μM, 10 μM, and 100 μM. A representative experiment is shown. **D:** cAR1-mediated Ca^{2+} entry. cAR1/*car1*$^-$ cells were added to a Ca^{2+} uptake medium containing $^{45}Ca^{2+}$ and 10 μM Ca^{2+} or in the same medium containing 100 μM cAMP. The reaction was terminated at the indicated times by adding 225 mM $CaCl_2$, the cells were washed, and cell-associated radioactivity was determined (35). cAMP-induced uptake (■) is equal to the amount of Ca^{2+} taken up by the cAMP-treated cells minus the amount of Ca^{2+} taken up by nonstimulated cells. Stimulated Ca^{2+} uptake is typically between 50 and 100 pmol/mg protein, or about ~1- to 4-fold greater than basal Ca^{2+} uptake. Data from a single representative experiment are shown. The loss-of-function data in A are from Parent and Devreotes, refs. 86 and 112, with permission.

functions of cAR1. First, *car1*⁻ cells contain low levels of surface cAMP binding sites, and these levels increase dramatically upon expression of cAR1. With a rapid 96-well plate assay, [^{32}P]cAMP binding to cAR1 can be measured in the presence of ammonium sulfate, which stabilizes cAMP binding to cAR1 (Fig. 2B). Second, cAR1-expressing clones can be readily analyzed for agonist-induced activation by following cAR1 phosphorylation. Phosphorylation of serines in the C-terminal domain of cAR1 changes the apparent molecular mass of the receptor from 40 kDa to 43 kDa on sodium dodecyl sulfate (SDS)-polyacrylamide gels (Fig. 2C). Through the use of several cAMP concentrations, mutants with impaired cAMP-binding affinities can be detected. Influx of extracellular Ca^{2+} can be readily determined (Fig. 2D). Third, development is a good initial screen to determine whether G protein-signaling pathways are functional. GTP-induced inhibition of cAMP binding to membranes (120), and ACA activation (Fig. 5 on page 97) provide accurate, but more labor-intensive, measures of G protein-dependent signal transduction.

SIGNAL TRANSDUCTION AND CYCLIC ADENOSINE MONOPHOSPHATE BINDING AFFINITY MUTANTS OF cAR1

We are mapping domains of cAR1 essential for ligand binding, signal transduction, and desensitization, using random and site-directed mutagenesis and chimeragenesis. In one study, mutations were targeted to the third intracellular loop of cAR1, since this loop is critical for interactions between other seven helix receptors and G proteins (89,121). In a second study, random substitutions were introduced through PCR mutagenesis into a region extending from the third transmembrane domain to the seventh transmembrane domain (122).

A striking observation in both studies was that cAR1 showed remarkable functional resilience. In a screen of about 2,000 receptor mutants generated by PCR mutagenesis, each with an estimated two amino-acid substitutions, 90% of the mutants restored aggregation to *car1*⁻ cells. This implies that the receptor folds properly, binds cAMP, and transduces signals. Only 3.3% of the expressed receptors were nonfunctional. Moreover, analysis of the third-intracellular-loop mutants indicated that at least one substitution at every position at which substitutions were observed (20 of 24 amino acids) permitted cAMP binding, signal transduction, and rescue of aggregation. This capacity for plasticity may explain the notable divergence of sequence among the seven helix receptors.

Clearly, not all changes are tolerated; analysis of both sets of mutants has led to the identification of affinity mutants, general activation mutants, and signal-transduction mutants. The most prevalent mutant requires higher concentrations of cAMP to induce maximal cAR1 phosphorylation as well as all other receptor-mediated responses. Since saturating concentrations of cAMP elicit full responses, these are affinity, rather than signaling, mutants. Figure 3 shows typical phosphorylation responses of these mutants, designated as class III. They fall into three subclasses;

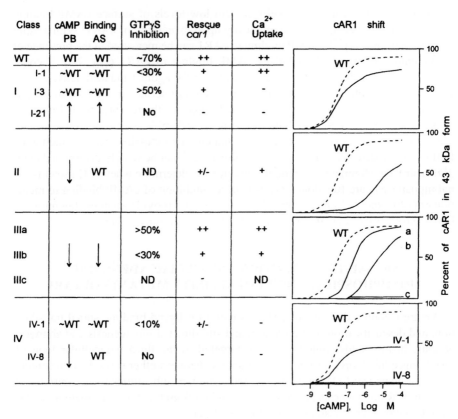

Class		cAMP Binding PB	cAMP Binding AS	GTPγS Inhibition	Rescue *car1*	Ca^{2+} Uptake
WT		WT	WT	~70%	++	++
I	I-1	~WT	~WT	<30%	+	++
	I-3	~WT	~WT	>50%	+	-
	I-21	↑	↑	No	-	-
II		↓	WT	ND	+/-	+
IIIa		↓	↓	>50%	++	++
IIIb				<30%	+	+
IIIc				ND	-	ND
IV	IV-1	~WT	~WT	<10%	+/-	-
	IV-8	↓	WT	No	-	-

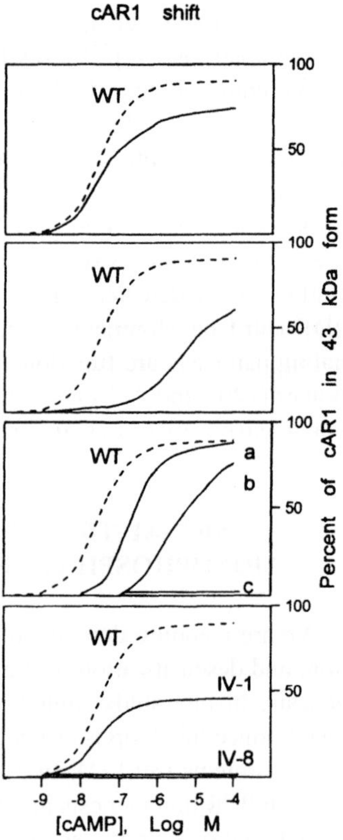

FIG. 3. Biochemical properties of wild-type cAR1 and affinity and activation mutants of cAR1 expressed in *car1⁻* cells. The affinity of cAMP binding to cAR1 was assessed in the presence or absence of 3 M ammonium sulfate. GTPγS inhibition of $[^3H]$cAMP binding to membranes was measured (120). Cells washed free of nutrients were plated on nonnutrient agar to initiate development. cAMP-stimulated Ca^{2+} uptake was followed for 30 s as detailed in the legend to Fig. 2. cAR1 shifting was measured as described in Fig. 2C.

each group shows a quantal shift in affinity. Classes IIIa and IIIb bind cAMP with 5-fold and 100-fold less efficiency than cAR1, while class IIIc does not bind cAMP. Importantly, these differences in affinity are maintained in the presence of ammonium sulfate, which induces all of the cARs to bind cAMP with high affinity. These affinity mutants probably have intrinsic changes in the receptor rather than an inability to couple with G proteins, since the low-affinity form of cAR1 seen in $g\beta^-$ cells under physiologic conditions shows wild-type affinity in the presence of ammonium sulfate (60).

The mutations underlying these defects in affinity map to a wide area on the receptor, and are present in the transmembrane domains and the extracellular and intracellular loops. The involvement of transmembrane domains is not unexpected: ligand interactions with the β-adrenergic receptor and rhodopsin are thought to occur within the plane of the membrane, where a considerable mass of the receptor resides (8). The involvement of extracellular and intracellular domains suggests that multiple domains can influence the agonist-binding pocket of the cARs, a feature that is shared by other seven helix receptors (9,90–92). In certain of these receptors, alterations in intracellular domains could reduce binding affinity by disrupting interactions with G proteins. Alternatively, mutations in the intracellular and extracellular domains could influence the arrangement of the helices within the membrane.

A second type of affinity mutant, designated class II, resembles cAR2 in that it binds cAMP poorly under physiologic conditions but displays high affinity in the presence of ammonium sulfate. Other cAMP-dependent responses are impaired or absent (Fig. 3). This mutant has a single amino-acid substitution, in the second extracellular loop, that removes a positive charge. Using a technique called random chimeragenesis, we have previously shown that sequences within the second extracellular loop account for most of the large difference in affinity between cAR1 and cAR2 (93). A chimeric receptor containing all cAR2 sequences except this domain, which came from cAR1, bound cAMP with an affinity only slightly lower than that of wild-type cAR1. This slight difference in affinity appears to reside in amino acids of the second intracellular loop and the fourth transmembrane domain. The presence of a major affinity determinant outside the transmembrane domains suggests a unique mode of regulation for binding affinity. Introduction of a net negative charge to amino acids in this region appears to act as a gate, controlling access of cAMP to its binding pocket. Examination of the highly conserved cAR sequences support this idea. cAR1, with the highest binding affinity, has no negative charges in this loop, cAR3 has lower affinity and one negative charge, while cAR2 has the lowest affinity and two negative charges.

We have also identified general activation mutants. These mutants, designated class IV, bind cAMP with high affinity but show submaximal levels of receptor phosphorylation at saturating concentrations of cAMP. They fail to rescue the aggregation-minus phenotype of *car1*⁻ cells and thus appear not to couple to G proteins (Fig. 3). All of the substitutions that induce this phenotype map to the third intracellular loop. One of these, IV-1, introduces a single Val→Asp change in the N-terminal side, while the same change in the adjacent Val has no effect (89). A particularly severe allele binds cAMP but lacks all cAMP-induced responses. This mutant, IV-8, introduces two substitutions, Ser→Pro and Thr→Ser, into the N-terminal end of the third intracellular loop.

Together, analysis of mutant classes II, III, and IV suggests that binding of agonists cause a series of coordinate transitions in the receptor (Fig. 4). Class II mutants most likely limit access of cAMP to its binding pocket. The incremental decreases in affinity seen in the class III mutants suggest that the mutations that produce them disrupt specific receptor/agonist contact sites within the binding pocket. Reduced affinity for cAMP impairs the subsequent generation of active receptor conformations: this can be overcome by saturating concentrations of cAMP, which completely restores downstream signaling events in most of these affinity mutants. Lig-

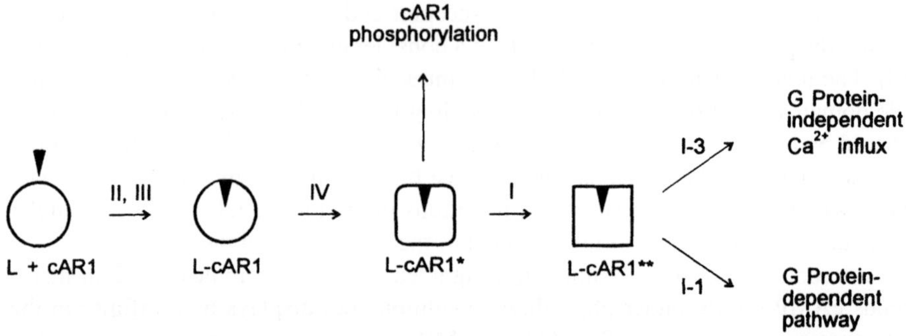

FIG. 4. Model depicting cAMP-induced activation of G protein-dependent and -independent signal transduction through cAR1. Details are presented in the text. L = cAMP. From Milne et al., 121, with permission.

and binding probably perturbs critical interactions within the helix bundle, leading to spatial reorganization of the intracellular domains and formation of activated receptor states. The class IV mutants suggest that disruption of domains in the third intracellular loop also block the appearance of active cAR intermediates, but not by influencing cAMP-binding affinity. The impairment of both G protein-dependent and -independent signals in these activation mutants suggests that regions in the third intracellular loop act as a hinge. Agonist binding relieves a constraint on the wild-type receptor that holds it in a resting conformation and controls the formation of receptor conformation(s) that interact with G proteins, factors for G protein-independent signaling, and the receptor kinases that mediate desensitization. The third intracellular loop will probably be essential for the activation of many, if not all, seven helix receptors, given the recent plethora of activation mutants and constitutively active mutants that map to this region (87,88,94–96).

In addition to its proposed role as a hinge region, the third intracellular loop of many of the seven helix receptors appears to be essential for the binding and activation of G proteins (8,97,98). Extensive analysis of mutants obtained from the third-loop mutagenesis of cAR1 has yielded surprising results. Mutant cARs defective in one response were defective in all others, whereas functional receptors elicited all responses. At least one substitution at each position in which amino acid substitutions were observed preserved both G protein-dependent and -independent responses. Many of these functional replacements involved the introduction or deletion of charged residues, suggesting that G proteins and the factors that mediate G protein-independent signal transduction may not interact with the third intracellular loop. Thus, certain receptors, including cAR1 and the N-formyl peptide receptor (99,100), may typify a subclass of receptors that use other domains to couple to G proteins.

An intriguing class of signal-transduction mutants, designated class I, suggests that there are substrates of receptor activation. One of these mutants, I-21, binds cAMP with higher affinity than does wild-type cAR1, and is phosphorylated in response to cAMP, but does not couple to G proteins, rescue aggregation, or show G protein-independent Ca^{2+} entry (Fig. 3) (122). This implies that association of agonist with wild-type receptors alters the receptor structure so that kinases can bind and access their phosphorylation sites, but that additional conformational intermediate(s) interact with the proteins that mediate G protein-dependent and -independent signaling. Interestingly, mutants of rhodopsin (97,98) and the angiotensin II receptor type I (101) with very similar properties have been reported.

Other interesting class I mutants have selective defects in G protein-dependent signaling or in G protein-independent Ca^{2+} influx (121). These receptors become phosphorylated in response to agonist and function sufficiently well to restore the development of *car1*⁻ cells. Certain mutants, typified by receptor I-1, have normal levels of cAMP-dependent Ca^{2+} influx but couple inefficiently to G proteins, as assessed by GTPγS inhibition of cAMP binding. These receptors may bind G proteins with reduced affinity or fail to activate them once they are bound. Other mutants, such as I-3, can activate G protein-dependent signaling but not the stimulated Ca^{2+} entry. These receptors may be unable to bind or activate the yet unidentified downstream effectors(s) required for this response; the receptor itself probably does not serve as a Ca^{2+} channel, since there are insufficient numbers of acidic amino-acid residues within the transmembrane helices to form a Ca^{2+}-binding domain. While the effects of the I-1 and I-3 mutations on G protein-independent MAP kinase activity and gene-expression events are not known, they provide important biochemical data supporting earlier genetic evidence (28,35,72) that the Ca^{2+} influx activated by cAR1 can be dissociated from G protein-dependent signaling.

Consideration of the class I mutants suggests that cAR1 undergoes a series of conformational changes during the activation process (Fig. 4). cAMP binding to the receptor triggers the formation of L-cAR1*, which is sufficient for receptor phosphorylation. Generation of an additional activated state of cAR1, L-cAR**, is indicated by the identification of mutant I-21, which only undergoes agonist-induced phosphorylation (122). In the L-cAR1** form, the receptor can interact with G proteins and the factors required to induce stimulated Ca^{2+} entry. Mutants I-1 and I-3 probably attain the cAR1** conformation but fail to interact with the appropriate downstream signaling components (121). It remains to be determined whether the L-cAR1** intermediate leads to additional conformational changes essential for one or both of these signaling pathways.

PHOSPHORYLATION MUTANTS OF cAR1

The phosphorylation sites of unstimulated and agonist-activated cAR1 have been identified through a series of serine substitution mutants. Eighteen serine residues in the C-terminal domain of cAR1 are grouped into four clusters. Serines in clusters three and four are phosphorylated in unstimulated cells; those in clusters one and two are phosphorylated in response to agonist (102). Modification of two serines within the first cluster shifts the apparent molecular mass of cAR1 from 40 kDa to

43 kDa and appears essential for loss of ligand binding (103), a desensitization process resulting from a reduction in cAR1 binding affinity for cAMP (104). Receptor phosphorylation is of physiologic importance, since a cAR1 lacking all phosphorylation sites, cm1234, shows abnormal development and arrests at the mound stage when it is expressed at high levels. (102) (J. Y. Kim, R. Soede, P. Schaap, R. Valkema, J. A. Borleis, P. J. M. Van Haastert, P. Devreotes and D. Hereld, *in preparation*). It is not clear how agonist-induced phosphorylation exerts this effect. Despite correlative evidence that phosphorylation of cAR1 may be important for the adaptation of adenylyl cyclase activity (105), synthesis of cAMP by cm1234 and wild-type cells is comparable. Other responses, including chemotaxis, gene expression (J. Y. Kim, R. Soede, P. Schaap, R. Valkema, J. A. Borleis, P. J. M. Van Haastert, P. Devreotes and D. Hereld, *in preparation*), and Ca^{2+} influx (J. L. S. Milne, *unpublished data*), are similarly unaffected in the cm1234 cells. Thus, additional adaptation signals may regulate these responses, in contrast to the β-adrenergic receptor and rhodopsin, for which agonist-stimulated phosphorylation primarily regulates adaptation of effectors (79). That others of the seven helix receptors may send phosphorylation-independent adaptation signals to downstream effectors is an exciting possibility. cAR1 phosphorylation and the subsequent loss of ligand binding may simply attenuate cAR1 signaling if the cells are exposed to high concentrations of stimulus. Removal of the receptor from the cell surface requires intracellular cAMP (106), while loss of ligand binding occurs in the absence of second-messenger production, suggesting that the two processes may be distinct. Upon exposure to agonists, others of the seven helix receptors become uncoupled from G proteins by a variety of mechanisms (107–109).

MUTAGENESIS OF ADENYLYL CYCLASE

The aggregation-specific ACA of *Dictyostelium* (110) closely resembles mammalian adenylyl cyclases that couple to G proteins (111). These enzymes have been proposed to have 12 transmembrane helices and two large (~40 kDa) intracellular domains, C1 and C2, located between helices VI and VII and at the C-terminal domain, respectively. Following cAR1 activation, ACA activity peaks at 1 to 2 min and then declines by ~5 min (16). As discussed above, the regulation of ACA is complex, involving G proteins, CRAC, and additional factors for both activation and, most likely, for adaptation. Parent and Devreotes (86) have isolated loss-of-function ACA mutants by randomly mutating transmembrane helices II to VI and the entire CI domain, and using the library for phenotypic rescue of aggregation-deficient *aca*⁻ cells. Forty clones did not rescue development, despite expressing adenylyl cyclase. When nonregulated adenylyl cyclase activity was measured *in vitro* using $MnSO_4$, 17 of these mutants were catalytically active. Four transformants expressing active and four expressing inactive isoforms of ACA were characterized further. The catalytically active enzymes were analyzed with the *in vivo* and *in vitro* assays shown in Fig. 5. These mutants show substantial $MnSO_4$-stimulated activity, but do not synthesize cAMP *in vivo* when challenged with chemoattractant. More-

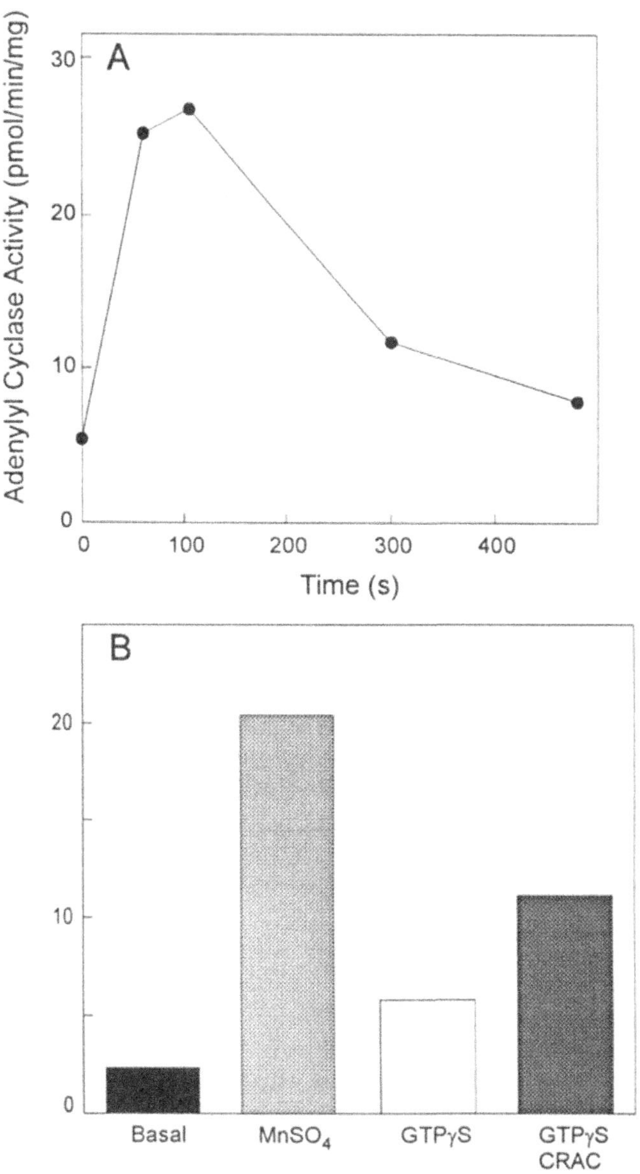

FIG. 5. *In vivo* (A) and *in vitro* (B) activation of adenylyl cyclase in *aca*⁻ cells expressing wild-type ACA. **A:** Cells starved for 5 h were resuspended in 10 mM phosphate buffer containing 2 mM $MgSO_4$ and treated with cAMP to a final concentration of 10 μM. At the indicated times, cells were removed and lysed through 5 μm filters. An aliquot of 200 μl of lysate was added to an assay mixture containing 30 nM [^{32}P]ATP. After 2 min at 22° C, the reaction was terminated and levels of [^{32}P]cAMP were determined. **B:** Basal and unregulated ACA activity of lysates of unstimulated cells was measured in the presence of 2 mM $MgSO_4$ and 5 mM $MnSO_4$, respectively. G-protein-dependent ACA activity was assessed in the presence of 40 μM GTPγS with and without the addition of exogenous CRAC from the cytosol of a CRAC-overexpressing cell line.

over, GTPγS does not stimulate ACA activity *in vitro*, even in the presence of exogenous CRAC, suggesting that these enzymes are uncoupled from the G protein-dependent signaling pathway. Sequence analysis of the corresponding plasmids indicates that each of the catalytically inactive enzymes has mutations in conserved amino-acid residues present in the C1 domain, thought to form the active site. Each of the catalytically active but uncoupled cyclases has mutations in the cytoplasmic region adjacent to helix VI and in the central region of the C1 loop. These mutations could block interactions between the cyclase and G proteins, CRAC, or other activator protein(s). Alternatively, these interactions could occur but fail to relay an activation signal to the active site. Interestingly, constitutively active ACA isoforms have been isolated with the same degenerate ACA libraries to rescue aggregation-deficient *crac*$^-$ cells (Fig. 2) (112). These gain-of-function mutants effectively bypass the requirement for G proteins and CRAC, and together with additional loss-of-function mutants will provide important information about the catalytic and regulatory domains of ACA.

CONCLUSION

The family of seven helix cAMP receptors essentially programs the sequence of events leading from single cells to a multicellular organism. Agonist association with these receptors triggers the biochemical reactions required for chemotaxis, gene expression, and morphogenesis through the modulation of classical G protein-dependent signaling pathways and a novel G protein-independent pathway. While this work has answered some of the questions posed at the outset of this chapter, it has raised others. What is the nature of G protein-independent signal transduction? How do G protein βγ subunits and the other modulators of adenylyl cyclase control its activity? Are there mammalian homologues of these regulators? Does agonist association with cAR1 trigger the formation of active intermediates by spatial reorganization of the third intracellular loop? How many active intermediates form, and which are involved in G protein-dependent and -independent signaling? Where do the proteins that mediate these responses interact with cAR1? What attenuates the activity of effectors, if not receptor phosphorylation?

Satisfactory answers to these questions will require progress at the genetic, biochemical, and structural levels, a possibility well-afforded by *Dictyostelium*. The ability to perform random mutagenesis and the potential to isolate large quantities of wild-type and mutant versions of cAR1 (see below) will effectively complement the extensive site-directed mutagenesis of rhodopsin and transducin (113,114) and the recent advances in determining the structures of the wild-type versions of these proteins (115–118). Recent work has shown that ACA (86) and cAR1 (119) associate with a 3-[(cholamidopropyl)-dimethylamino]-1-propanesulfonate (CHAPS)-resistant membrane fraction in disrupted cells that can be conveniently purified on sucrose-step gradients. Although ACA has not yet been purified, a histidine-tagged version of cAR1 can be readily isolated using these membranes as starting material (X. Xiao, Y. Yao, and P. N. Devreotes, *unpublished data*). These advances in cAR1 biochemistry, permitting the purification of wild-type and mutant cARs, will facilitate the obvious extension to this work—obtaining high-resolution structural information on the ground state of cAR1, and of different activation intermediates, in the absence and presence of G proteins.

ACKNOWLEDGMENTS

We thank Dr. C. A. Parent for providing data showing the *in vivo* and *in vitro* activation of adenylyl cyclase, Dr. M. J. Caterina for the [^{32}P]cAMP binding data in Fig. 2, and Drs. C. Gaskins, J. Segall, and R. A. Firtel for the *erk*$^-$ cells. J. L. S. M. was a recipient of a Centennial Fellowship from the Medical Research Council of Canada. This work was supported by U.S. National Institutes of Health Grant GM34933 to P. N. D.

REFERENCES

1. Gilman AG. G proteins: transducers of receptor-generated signals. *Annu Rev Biochem* 1987;56: 615–49.
2. Jelsema CL, Axelrod J. Stimulation of phospholipase A_2 activity in bovine rod outer segments by the βγ subunits of transducin and its inhibition by the α subunit. *Proc Natl Acad Sci USA* 1987;84: 3623–7.
3. Logothetis DE, Kurachi Y, Galper J, Neer EJ, Clapham DE. The βγ subunits of GTP-binding proteins activate the muscarinic K^+ channel in heart. *Nature* 1987;325:321–6.
4. Tang WJ, Gilman AG. Type-specific regulation of adenylyl cyclase by G protein βγ subunits. *Science* 1991;254:1500–3.
5. Katz A, Wu D, Simon MI. Subunits βγ of heterotrimeric G protein activate β2 isoform of phospholipase C. *Nature* 1992;360:686–9.
6. Bourne HR. Team blue sees red. *Nature* 1995;376:727–9.
7. Savarese TM, Fraser CM. *In vitro* mutagenesis and the search for structure-function relationships among G protein-coupled receptors. *Biochem J* 1992;283:1–19.
8. Baldwin JM. Structure and function of receptors coupled to G proteins. *Curr Opin Cell Biol* 1994; 6:180–90.
9. Strader CD, Fong TM, Tota MR, Underwood D. Structure and function of G protein-coupled receptors. *Annu Rev Biochem* 1994;63:101–32.
10. Devreotes PN. G protein-linked signaling pathways control the developmental program of *Dictyostelium*. *Neuron* 1994;12:235–41.
11. Pan P, Hall EM, Bonner JT. Folic acid as second chemotactic substance in the cellular slime moulds. *Nature New Biol* 1972;237:181–2.
12. Devreotes P. *Dictyostelium discoideum*: a model system for cell–cell interactions in development. *Science* 1989;245:1054–8.
13. Gross JD. Developmental decisions in *Dictyostelium discoideum*. *Microbiol Rev* 1994;58:330–51.
14. Firtel RA. Integration of signaling information in controlling cell-fate decisions in *Dictyostelium*. *Genes Dev* 1995;9:1427–44.
15. Roos W, Nanjundiah V, Malchow D, Gerisch G. Amplification of cyclic-AMP signals in aggregating cells of *Dictyostelium discoideum*. *FEBS Lett* 1975;53:139–42.
16. Dinauer MC, MacKay SA, Devreotes PN. Cyclic 3',5'-AMP relay in *Dictyostelium discoideum* III. The relationship of cAMP synthesis and secretion during the cAMP signaling response. *J Cell Biol* 1980;86:537–44.
17. Mato JM, Krens FA, Van Haastert PJM, Konijn TM. 3':5'-Cyclic AMP-dependent 3':5'-cyclic GMP accumulation in *Dictyostelium discoideum*. *Proc Natl Acad Sci USA* 1977;74:2348–51.

18. Wurster B, Schubiger K, Wick U, Gerisch G. Cyclic GMP in *Dictyostelium discoideum*: oscillations and pulses in response to folic acid and cyclic AMP signals. *FEBS Lett* 1977;76:141–4.
19. Europe-Finner GN, Newell PC. Cyclic AMP stimulates accumulation of inositol trisphosphate in *Dictyostelium. J Cell Sci* 1987;87:221–9.
20. Van Haastert PJM, De Vries MJ, Penning LC, et al. Chemoattractant and guanosine 5'-[γ-thio]triphosphate induce the accumulation of inositol 1,4,5,-trisphosphate in *Dictyostelium* cells that are labelled with [^{3}H]-inositol by electroporation. *Biochem J* 1989;258:577–86.
21. Maeda M, Aubry L, Insall R, Gaskins C, Devreotes P, Firtel R. Seven helix chemoattractant receptors transiently stimulate MAP kinase in *Dictyostelium:* role of heterotrimeric G-proteins. 1996; 271:3351–4.
22. Abe T, Maeda Y, Iijima T. Transient increase of the intracellular Ca^{2+} concentration during chemotactic signal transduction in *Dictyostelium discoideum* cells. *Differentiation* 1988;39:90–6.
23. Schlatterer C, Gollnick F, Schmidt E, Meyer R, Knoll G. Challenge with high concentrations of cyclic AMP induces transient changes in the cytosolic free calcium concentration in *Dictyostelium discoideum. J Cell Sci* 1994;107:2107–15.
24. Europe-Finner GN, Newell PC. Inositol 1,4,5-trisphosphate induces calcium release from a non-mitochondrial pool in amoebae of *Dictyostelium. Biochim Biophys Acta* 1986;887:335–40.
25. Flaadt H, Jaworski E, Schlatterer C, Malchow D. Cyclic AMP- and Ins(1,4,5)P_3-induced Ca^{2+} fluxes in permeabilised cells of *Dictyostelium discoideum*: cGMP regulates Ca^{2+} entry across the plasma membrane. *J Cell Sci* 1993;105:255–61.
26. Wick U, Malchow D, Gerisch G. Cyclic-AMP stimulated calcium influx into aggregating cells of *Dictyostelium discoideum. Cell Biol Int Rep* 1978;2:71–9.
27. Bumann J, Wurster B, Malchow D. Attractant-induced changes and oscillations of the extracellular Ca^{2+} concentration in suspensions of differentiating *Dictyostelium* cells. *J Cell Biol* 1984;98: 173–8.
28. Milne JL, Coukell MB. A Ca^{2+} transport system associated with the plasma membrane of *Dictyostelium discoideum* is activated by different chemoattractant receptors. *J Cell Biol* 1991;112: 103–10.
29. Aeckerle S, Wurster B, Malchow D. Oscillations and cyclic AMP induced changes of the K^+ concentration in *Dictyostelium discoideum. EMBO J* 1985;4:39–43.
30. Malchow D, Nanjundiah V, Wurster B, Eckstein F, Gerisch G. Cyclic AMP-induced pH changes in *Dictyostelium discoideum* and their control by calcium. *Biochim Biophys Acta* 1978;538:473–80.
31. Chen M-Y, Insall RH, Devreotes PN. Signaling through chemoattractant receptors in *Dictyostelium. Trends Genet* 1996;12:52–7.
32. Klein PS, Sun TJ, Saxe III CL, Kimmel AR, Johnson RL, Devreotes PN. A chemoattractant receptor controls development in *Dictyostelium discoideum. Science* 1988;241:1467–72.
33. Saxe III CL, Johnson R, Devreotes PN, Kimmel AR. Multiple genes for cell surface cAMP receptors in *Dictyostelium discoideum. Dev Genet* 1991;12:6–13.
34. Louis JM, Ginsburg GT, Kimmel AR. The cAMP receptor CAR4 regulates axial patterning and cellular differentiation during late development of *Dictyostelium. Genes Dev* 1994;8:2086–96.
35. Milne JL, Devreotes PN. The surface cyclic AMP receptors, cAR1, cAR2, and cAR3, promote Ca^{2+} influx in *Dictyostelium discoideum* by a $G_\alpha 2$-independent mechanism. *Mol Biol Cell* 1993;4:283–92.
36. Insall RH, Soede RDM, Schaap P, Devreotes PN. Two cAMP receptors activate common signaling pathways in *Dictyostelium. Mol Biol Cell* 1994;5:703–11.
37. Johnson RL, Vaughan RA, Caterina MJ, Van Haastert PJM, Devreotes PN. Overexpression of the cAMP receptor-1 in growing *Dictyostelium* cells. *Biochemistry* 1991;30:6982–6.
38. Sun TJ, Devreotes PN. Gene targeting of the aggregation stage cAMP receptor cAR1 in *Dictyostelium. Genes Dev* 1991;5:572–82.
39. Pupillo M, Insall R, Pitt GS, Devreotes PN. Multiple cyclic AMP receptors are linked to adenylyl cyclase in *Dictyostelium. Mol Biol Cell* 1992;3:1229–34.
40. Soede RDM, Insall RH, Devreotes PN, Schaap P. Extracellular cAMP can restore development in *Dictyostelium* cells lacking one, but not two subtypes of early cAMP receptors (cARs). Evidence for involvement of cAR1 in aggregative gene expression. *Development* 1994;120:1997–2002.
41. Johnson RL, Van Haastert PJM, Kimmel AR, Saxe III CL, Jastorff B, Devreotes PN. The cyclic nucleotide specificity of three cAMP receptors in *Dictyostelium. J Biol Chem* 1992;267:4600–7.
42. Johnson RL, Saxe III CL, Gollop R, Kimmel AR, Devreotes PN. Identification and targeted gene disruption of cAR3, a cAMP receptor subtype expressed during multicellular stages of *Dictyostelium* development. *Genes Dev* 1993;7:273–82.

43. Saxe III CL, Ginsburg GT, Louis JM, Johnson R, Devreotes PN, Kimmel AR. CAR2, a prestalk cAMP receptor required for normal tip formation and late development of *Dictyostelium discoideum*. *Genes Dev* 1993;7:262–72.
44. Pupillo M, Kumagai A, Pitt GS, Firtel RA, Devreotes PN. Multiple α subunits of guanine nucleotide-binding proteins in *Dictyostelium*. *Proc Natl Acad Sci USA* 1989;86:4892–6.
45. Hadwiger JA, Wilkie TM, Strathmann M, Firtel RA. Identification of *Dictyostelium* G_α genes expressed during multicellular development. *Proc Natl Acad Sci USA* 1991;88:8213–7.
46. Wu LJ, Devreotes PN. *Dictyostelium* transiently expresses eight distinct G-protein α-subunits during its developmental program. *Biochem Biophys Res Commun* 1991;179:1141–7.
47. Lilly P, Wu LJ, Welker DL, Devreotes PN. A G-protein β-subunit is essential for *Dictyostelium* development. *Genes Dev* 1993;7:986–95.
48. Wu L, Gaskins R, Gundersen R, et al. Signal transduction by G proteins in *Dictyostelium discoideum*. In: Dickey BF, Birnbaumer L, eds. *Handbook of experimental pharmacology, Vol 108/II, GTPases in biology II.* Heidelberg: Springer-Verlag. 1993;335–49.
49. Bominaar AA, Van Haastert PJM. Phospholipase C in *Dictyostelium discoideum*—identification of stimulatory and inhibitory surface receptors and G-proteins. *Biochem J* 1994;297:189–93.
50. Dharmawardhane S, Cubitt AB, Clark AM, Firtel RA. Regulatory role of the Gα1 subunit in controlling cellular morphogenesis in *Dictyostelium*. *Development* 1994;120:3549–61.
51. Wu LJ, Gaskins C, Zhou KM, Firtel RA, Devreotes PN. Cloning and targeted mutations of Gα7 and Gα8, two developmentally regulated G protein α-subunit genes in *Dictyostelium*. *Mol Biol Cell* 1994;5:691–702.
52. Hadwiger JA, Firtel RA. Analysis of $G_\alpha 4$, a G-protein subunit required for multicellular development in *Dictyostelium*. *Genes Dev* 1992;6:38–49.
53. Hadwiger JA, Lee S, Firtel RA. The Gα subunit Gα4 couples to pterin receptors and identifies a signaling pathway that is essential for multicellular development in *Dictyostelium*. *Proc Natl Acad Sci USA* 1994;91:10566–70.
54. Janssens PMW, Van Haastert PMJ. Molecular basis of transmembrane signal transduction in *Dictyostelium discoideum*. *Microbiol Rev* 1987;51:396–418.
55. Kesbeke F, Van Haastert PJM, De Wit RJW, Snaar-Jagalska BE. Chemotaxis to cyclic AMP and folic acid is mediated by different G-proteins in *Dictyostelium discoideum*. *J Cell Sci* 1990;96:669–73.
56. Kumagai A, Hadwiger JA, Pupillo M, Firtel RA. Molecular genetic analysis of two Gαl protein subunits in *Dictyostelium*. *J Biol Chem* 1991;266:1220–8.
57. Coukell MB, Lappano S, Cameron AM. Isolation and characterization of cAMP unresponsive (frigid) aggregation-deficient mutants of *Dictyostelium discoideum*. *Dev Genet* 1983;3:283–97.
58. Kesbeke F, Snaar-Jagalska BE, Van Haastert PJM. Signal transduction in *Dictyostelium fgdA* mutants with a defective interaction between surface cAMP receptors and a GTP-binding regulatory protein. *J Cell Biol* 1988;107:521–8.
59. Snaar-Jagalska BE, Kesbeke F, Van Haastert PJM. G-proteins in the signal-transduction pathways of *Dictyostelium discoideum*. *Dev Genet* 1988;9:215–26.
60. Wu LJ, Valkema R, Van Haastert PJM, Devreotes PN. The G protein β-subunit is essential for multiple responses to chemoattractants in *Dictyostelium*. *J Cell Biol* 1995;129:1667–75.
61. de Hostos EL, Bradtke B, Lottspeich F, Guggenheim R, Gerisch G. Coronin, an actin binding protein of *Dictyostelium discoideum* localized to cell surface projections, has sequence similarities to G-protein β-subunits. *EMBO J* 1991;10:4097–104.
62. Schiestl RH, Petes TD. Integration of DNA fragments by illegitimate recombination in *Saccharomyces cerevisiae*. *Proc Natl Acad Sci USA* 1991;88:7585–9.
63. Kuspa A, Loomis WF. Tagging developmental genes in *Dictyostelium* by restriction enzyme-mediated integration of plasmid DNA. *Proc Natl Acad Sci USA* 1992;89:8803–7.
64. Theibert A, Devreotes P. Surface receptor-mediated activation of adenylate cyclase in *Dictyostelium*. Regulation by guanine nucleotides in wild-type cells and aggregation deficient mutants. *J Biol Chem* 1986;261:15121–5.
65. Insall R, Kuspa A, Lilly PJ, et al. CRAC, a cytosolic protein containing a pleckstrin homology domain, is required for receptor and G protein-mediated activation of adenylyl cyclase in *Dictyostelium*. *J Cell Biol* 1994;126:1537–45.
66. Lilly PJ, Devreotes PN. Identification of CRAC, a cytosolic regulator required for guanine nucleotide stimulation of adenylyl cyclase in *Dictyostelium*. *J Biol Chem* 1994;269:14123–9.
67. Lilly PJ, Devreotes PN. Chemoattractant and GTPγS-mediated stimulation of adenylyl cyclase in *Dictyostelium* requires translocation of CRAC to membranes. *J Cell Biol* 1995;129:1659–65.

68. Touhara K, Inglese J, Pitcher JA, Shaws G, Lefkowitz RJ. Binding of G protein $\beta\gamma$ subunits to pleckstrin homology domains. *J Biol Chem* 1994;269:10217–20.
69. Pitcher JA, Touhara K, Payne ES, Lefkowitz RJ. Pleckstrin homology domain-mediated membrane association and activation of the β-adrenergic receptor kinase requires coordinate interaction with $G_{\beta\gamma}$ subunits and lipid. *J Biol Chem* 1995;270:11707–10.
70. Langhans-Rajasekaran SA, Wan Y, Huang X-Y. Activation of Tsk and Btk tyrosine kinases by G protein $\beta\gamma$ subunits. *Proc Natl Acad Sci USA* 1995;92:8601–5.
71. Firtel RA, van Haastert PJM, Kimmel AR, Devreotes PN. G protein linked signal transduction pathways in development: *Dictyostelium* as an experimental system. *Cell* 1989;58:235–9.
72. Milne JLS, Wu L, Caterina MJ, Devreotes PN. Seven helix cAMP receptors stimulate Ca^{2+} entry in the absence of functional G proteins in *Dictyostelium*. *J Biol Chem* 1995;270:5926–31.
73. Putney Jr JW, Bird GSJ. The signal for capacitative calcium entry. *Cell* 1993;75:199–201.
74. Segall JE, Kuspa A, Shaulsky G, et al. A MAP kinase necessary for receptor-mediated activation of adenylyl cyclase in *Dictyostelium*. *J Cell Biol* 1995;128:405–13.
75. Wu L, Hansen D, Franke J, Kessin RH, Podgorski GJ. Regulation of *Dictyostelium* early development genes in signal transduction mutants. *Dev Biol* 1995;171:149–58.
76. Schnitzler GR, Briscoe C, Brown JM, Firtel RA. Serpentine cAMP receptors may act through a G protein-independent pathway to induce postaggregative development in *Dictyostelium*. *Cell* 1995; 81:737–45.
77. Campbell PT, Hnatowich M, O'Dowd BF, Caron MG, Lefkowitz RJ. Mutations of the human β2-adrenergic receptor that impair coupling to Gs interfere with receptor down-regulation but not sequestration. *Mol Pharmacol* 1991;39:192–8.
78. Zanolari B, Raths S, Singer-Kruger B, Riezmann H. Yeast pheromone receptor endocytosis and hyperphosphorylation are independent of G protein-mediated signal transduction. *Cell* 1992;71:755–63.
79. Lefkowitz RJ. G protein-coupled receptor kinases. *Cell* 1993;74:409–12.
80. Hunyady L, Baukal AJ, Balla T, Catt KJ. Independence of type I angiotensin II receptor endocytosis from G protein coupling and signal transduction. *J Biol Chem* 1994;269:24798–804.
81. Chabry J, Botto JM, Nouel D, Beaudet A, Vincent JP, Mazella J. Thr-422 and tyr-424 residues in the carboxyl terminus are critical for the internalization of the rat neurotensin receptor. *J Biol Chem* 1995;270:2439–42.
82. Lefkowitz RJ, Cotecchia S, Samama P, Costa T. Constitutive activity of receptors coupled to guanine nucleotide regulatory proteins. *TIPS* 1993;14:303–7.
83. Felder CC, Poulter MO,Wess J. Muscarinic receptor-operated Ca^{2+} influx in transfected fibroblast cells is independent of inositol phosphates and release of intracellular Ca^{2+}. *Proc Natl Acad Sci USA* 1992;89:509–13.
84. Twitchell WA, Rane SG. Nucleotide-independent modulation of Ca^{2+}-dependent K^{+} channel current by a μ-type opioid receptor. *Mol Pharmacol* 1994;46:793–8.
85. Parent CA, Devreotes PN. Molecular genetics of signal transduction in *Dictyostelium*. *Annu Rev Biochem* 1996;65:411–440.
86. Parent CA, Devreotes PN. Isolation of inactive and G protein-resistant adenylyl cyclase mutants using random mutagenesis. *J Biol Chem* 1995;270:22693–6.
87. Lefkowitz RJ. Turned on to ill effect. *Nature* 1993;365:603–4.
88. Clapham DE. Why testicles are cool. *Nature* 1994;371:109–10.
89. Caterina MJ, Milne JLS, Devreotes PN. Mutation of the third intracellular loop of the cAMP receptor, cAR1, of *Dictyostelium* yields mutants impaired in multiple signaling pathways. *J Biol Chem* 1994;269:1523–32.
90. Yokota Y, Akazawa C, Ohkubo H, Nakanishi S. Delineation of structural domains involved in the subtype specificity of tachykinin receptors through chimeric formation of substance P/substance K receptors. *EMBO J* 1992;11:3585–91.
91. Quehenberger O, Prossnitz ER, Cavanagh SL, Cochrane CG, Ye RD. Multiple domains of the *N*-formyl peptide receptor are required for high-affinity ligand binding. Construction and analysis of chimeric *N*-formyl peptide receptors. *J Biol Chem* 1993;268:18167–75.
92. Tseng M-J, Coon S, Stuenkel E, Struk V, Logsdon CD. Influence of second and third cytoplasmic loops on binding, internalization, and coupling of chimeric bombesin/m3 muscarinic receptors. *J Biol Chem* 1995;270:17884–91.
93. Kim JY, Devreotes PN. Random chimeragenesis of G-protein-coupled receptors Mapping the affinity of the cAMP chemoattractant receptors in *Dictyostelium*. *J Biol Chem* 1994;269:28724–31.
94. Boone C, Davis NG, Sprague GF. Mutations that alter the third cytoplasmic loop of the a-factor re-

ceptor lead to a constitutive and hypersensitive phenotype. *Proc Natl Acad Sci USA* 1993;90: 9921–5.

95. Stefan CJ, Blumer KJ. The third cytoplasmic loop of a yeast G-protein-coupled receptor controls pathway activation, ligand discrimination, and receptor internalization. *Mol Cell Biol* 1994;14: 3339–49.
96. Hogger P, Shockley MS, Lameh J, Sadee W. Activating and inactivating mutations in N- and C-terminal i3 loop junctions of muscarinic acetylcholine Hm1 receptors. *J Biol Chem* 1995;270: 7405–10.
97. Franke RR, Konig B, Sakmar TP, Khorana HG, Hofmann KP. Rhodopsin mutants that bind but fail to activate transducin. *Science* 1990;250:123–5.
98. Ernst OP, Hofmann KP, Sakmar TP. Characterization of rhodopsin mutants that bind transducin but fail to induce GTP nucleotide uptake. Classification of mutant pigments by fluorescence, nucleotide release, and flash-induced light-scattering assays. *J Biol Chem* 1995;270:10580–6.
99. Prossnitz ER, Quehenberger O, Cochrane CG, Ye RD. The role of the third intracellular loop of the neutrophil *N*-formyl peptide receptor in G protein coupling. *Biochem J* 1993;294:581–7.
100. Schreiber RE, Prossnitz ER, Ye RD, Cochrane CG, Bokoch GM. Domains of the human neutrophil *N*-formyl peptide receptor involved in G protein coupling. Mapping with receptor-derived peptides. *J Biol Chem* 1994;269:326–31.
101. Ohyama K, Yamano Y, Chaki S, Kondo T, Inagami T. Domains for G-protein coupling in angiotensin II receptor type I: studies by site-directed mutagenesis. *Biochem Biophys Res Commun* 1992;189:677–83.
102. Hereld D, Vaughan R, Kim JY, Borleis J, Devreotes P. Localization of ligand-induced phosphorylation sites to serine clusters in the C-terminal domain of the *Dictyostelium* cAMP receptor, cAR1. *J Biol Chem* 1994;269:7036–44.
103. Caterina MJ, Devreotes PN, Borleis J, Hereld D. Agonist-induced loss of ligand binding is correlated with phosphorylation of cAR1, a G protein-coupled chemoattractant receptor from *Dictyostelium*. *J Biol Chem* 1995;270:8667–72.
104. Caterina MJ, Hereld D, Devreotes PN. Occupancy of the *Dictyostelium* cAMP receptor, cAR1, induces a reduction in affinity which depends upon COOH-terminal serine residues. *J Biol Chem* 1995;270:4418–23.
105. Devreotes PN, Sherring JA. Kinetics and concentration dependence of reversible cAMP-induced modification of the surface cAMP receptor in Dictyostelium. *J Biol Chem* 1985;260:6378–84.
106. Van Haastert PJM, Wang M, Bominaar AA, Devreotes PN, Schaap P. cAMP-induced desensitization of surface cAMP receptors in *Dictyostelium*: different second messengers mediate receptor phosphorylation, loss of ligand binding, degradation of receptor, and reduction of receptor mRNA levels. *Mol Biol Cell* 1992;3:603–12.
107. Hausdorff W, Caron M, Lefkowitz R. Desensitization of β-adrenergic receptor function. *FASEB J* 1990;4:2881–9.
108. Dohlman H, Thorner J, Caron, MG, et al. Model systems for the study of seven-transmembrane-segment receptors. *Annu Rev Biochem* 1991;60:653–88.
109. Roettger BF, Rentsch RU, Hadac EM, Hellen EH, Burghardt TP, Miller LJ. Insulation of a G protein-coupled receptor on the plasmalemmal surface of the pancreatic acinar cell. *J Cell Biol* 1995; 130:579–90.
110. Pitt GS, Milona N, Borleis J, Lin KC, Reed RR, Devreotes PN. Structurally distinct and stage-specific adenylyl cyclase genes play different roles in *Dictyostelium* development. *Cell* 992;69:305–15.
111. Tang WJ, Gilman AG. Adenylyl cyclases. *Cell* 1992;70:869–72.
112. Parent CA, Devreotes PN. Constitutively active adenylyl cyclase mutant requires neither G proteins nor cytosolic regulators. *J Biol Chem* 1996;271:18333–6.
113. Khorana HG. Rhodopsin, photoreceptor of the rod cell. An emerging pattern for structure and function. *J Biol Chem* 1992;267:1–4.
114. Conklin BR, Bourne HR. Structural elements of Gα subunits that interact with Gβγ, receptors, and effectors. *Cell* 1993;73:631–41.
115. Noel JP, Hamm HE, Sigler PB. The 2.2 Å crystal structure of transducin-alpha complexed with GTPγS. *Nature* 1993;366:654–63.
116. Schertler GF, Villa C, Henderson R. Projection structure of rhodopsin. *Nature* 1993;362:770–2.
117. Lambright DG, Noel JP, Hamm HE, Sigler PB. Structural determinants for activation of the α-subunit of a heterotrimeric G protein. *Nature* 1994;369:621–8.
118. Unger VM, Schertler GF. Low resolution structure of bovine rhodopsin determined by electron cyro-microscopy. *Biophys J* 1995;68:1776–86.

119. Xiao X, Devreotes PN. Identification of detergent-resistant plasma membrane microdomains in *Dictyostelium*: enrichment of signal transduction particles. *Mol Biol Cell* 1997;8:855–69.
120. Van Haastert PJM. Guanine nucleotides modulate cell surface cAMP-binding sites in membranes from *Dictyostelium discoideum*. *Biochem Biophys Res Commun* 1984;124:597–604.
121. Milne JLS, Caterina MJ, Devreotes PN. Random mutagenesis of the chemoattractant receptor, cAR1, of *Dictyostelium*: evidence for multiple states of activation. *J Biol Chem* 1997; 272:2069–76.
122. Kim J-Y, Caterina MJ, Milne JLS, Lin KC, Borleis JA, Devreotes PN. Random mutagenesis of the chemoattractant receptor, cAR1, of *Dictyostelium*: mutant classes that cause discrete shifts in agonist affinity or lock the receptor in a novel activational intermediate. *J Biol Chem* 1997;2060–8.

Signal Transduction in Health and Disease, Advances in Second Messenger and Phosphoprotein Research, Vol. 31, edited by J. Corbin and S. Francis.
Lippincott–Raven Publishers, Philadelphia © 1997.

9

Mitochondrial α-Ketoacid Dehydrogenase Kinases

A New Family of Protein Kinases

Kirill M. Popov, John W. Hawes, and Robert A. Harris

Department of Biochemistry and Molecular Biology, Indiana University School of Medicine, Indianapolis, Indiana 46202-5122

A NEW FAMILY OF PROTEIN KINASES

The mitochondrial matrix space houses three multienzyme complexes, the branched chain α-ketoacid dehydrogenase complex, the pyruvate dehydrogenase complex, and the α-ketoglutarate dehydrogenase complex. All three complexes catalyze the oxidative decarboxylation of α-ketoacid substrates and generate reduced nicotinamide adenine dinucleotide (NADH), CO_2, and coenzyme A (CoA) esters as products. Two of the complexes—the branched chain α-ketoacid dehydrogenase and the pyruvate dehydrogenase—are subject to control by phosphorylation/dephosphorylation. Specific kinases that bind tightly to these complexes, along with specific phosphatases also present in the mitochondrial matrix space, are responsible for this control. Our research effort has concentrated on gaining a better understanding of the kinases involved in the regulation of these complexes. Complementary deoxyribonucleic acids (cDNAs) for four mitochondrial kinases have been cloned from rat and human libraries (1–4; Fig. 1). One of these cDNAs encodes a kinase (branched-chain α-ketoacid dehydrogenase kinase [BCKDK]) that phosphorylates and inactivates the branched chain α-ketoacid dehydrogenase complex (1); the other three encode kinases (pyruvate dehydrogenase kinase-1 [PDK1], PDK2, and PDK3) that phosphorylate and inactivate the pyruvate dehydrogenase complex (2–4). The four proteins are closely related in terms of amino-acid sequence, and clearly belong to the same protein kinase family. Searches of DNA databases have revealed open-reading frames for genes of *Caenorhabditis elegans*, *Trypanosoma brucei*, and yeast that encode hypothetical members of this same protein kinase family.

By sequence, the mitochondrial protein kinases appear to be most closely related to the bacterial histidine protein kinases. Motifs that are normally present in eukary-

Branched-chain α-ketoacid dehydrogenase kinase

BCKDK
- **Rat cDNA cloned**
- **Encodes 43,280 dalton protein**
- **mRNA expressed in many tissues**

Pyruvate dehydrogenase kinases

PDK1
- **Rat and human cDNAs cloned**
- **Encode 46,270 dalton proteins**
- **mRNA expressed primarily in heart with lesser amounts in many other tissues**

PDK2
- **Rat and human cDNAs cloned**
- **Encode 45,031 dalton proteins**
- **mRNA expressed in many tissues**

PDK3
- **Human cDNA cloned**
- **Encodes 45,000 dalton protein**
- **mRNA expression restricted to heart and skeletal muscle**

FIG. 1. Mitochondrial proteins that have been cloned and expressed as recombinant proteins.

otic serine protein kinases (5) are not found in members of the mitochondrial protein kinases. Motifs characteristic of bacterial histidine protein kinases (6), however, are found in the mitochondrial protein kinases. These include two glycine-rich loops (GXGXG and DXGXG) located in the carboxyl terminal (C-terminal) end, as well as subdomains containing highly conserved asparagine and histidine residues. There is, however, a major difference between the mitochondrial protein kinases and the bacterial histidine protein kinases with respect to function and catalytic mechanism. Bacterial protein kinases function in a two-component system in which the kinase serves as a sensor component that phosphorylates a response regulator (6). Although the mitochondrial protein kinases might be considered sensors, mitochondrial proteins that function in a manner analogous to bacterial response regulators are not involved in regulation of the mitochondrial α-ketoacid dehydrogenase complexes. Furthermore, bacterial protein kinases autophosphorylate on highly conserved histidine residues, and transfer this high-energy phosphate to aspartate residues of response regulators (6). Although a highly conserved histidine residue exists in the mitochondrial protein kinase at a position in the primary structure similar to that in the bacterial protein kinases (1–4), efforts thus far to demonstrate the involvement of histidine autophosphorylation in the phosphorylation of the E1 components of the mitochondrial α-ketoacid dehydrogenase complexes have not been successful (K. M. Popov, *unpublished observations*). Because of this, and also because the mitochondrial protein kinases phosphorylate serine rather than aspartate residues, the mitochondrial protein kinases comprise a unique family of protein kinases, related by sequence to the bacterial histidine protein kinases but by function and mechanism to the eukaryotic serine protein kinases.

RECENT FINDINGS WITH THE BRANCHED-CHAIN α-KETOACID DEHYDROGENASE KINASE

BCKDK phosphorylates Ser^{293} (Site 1) and Ser^{303} (Site 2) of the E1α subunit of the branched chain α-ketoacid dehydrogenase (BCKDH) complex (7). Site-directed mutagenesis was used to determine the relative importance of phosphorylation of these two sites to regulation of the dehydrogenase activity of the E1 component of the complex (8). Mutation of Ser^{293} to glutamate resulted in complete loss of dehydrogenase activity, suggesting that the negative charge introduced by phosphorylation causes inactivation of E1. Mutation of this residue to alanine had no effect on dehydrogenase activity at saturating substrate concentrations, but increased K_m for the branched chain α-ketoacid substrate. Phosphorylation of Ser^{303} by BCKDK still occurred with the S293A mutant protein, but without change in dehydrogenase activity. Likewise, mutation of Ser^{303} to glutamate and also to alanine had no effect on dehydrogenase activity. Thus, Ser^{293} is the regulatory site of phosphorylation of the BCKDH E1α component. Ser^{303} appears to be a silent site of phosphorylation.

The roles of amino acids surrounding Ser^{293} in phosphorylation-site recognition by BCKDK have also been investigated (9). The five residues on each side of Ser^{293} have been mutated to alanine, and the mutant proteins have been examined with respect to dehydrogenase activity and whether they still function as substrates for BCKDK. Only Arg^{288} was found to be essential for recognition of Ser^{293} as a phosphorylation site for BCKDK. Each of the other residues could be changed to alanine without affecting phosphorylation of Ser^{293} by BCKDK.

We also found in these experiments (9) that three residues, Arg^{288}, His^{292}, and Asp^{296}, are critical to the catalytic activity of E1. Mutation of each of these residues to alanine resulted in inactive proteins. These residues are completely conserved in all mitochondrial α-ketoacid dehydrogenases, suggesting that they play important roles in the thiamine pyrophosphate-dependent decarboxylation reaction catalyzed by E1. The inactive proteins derived from mutation of these residues became bound to the E2 core of the BCKDH complex with affinity equal to that of wild-type E1, and their circular dichroism spectra were likewise identical to that of wild-type E1. Mutation of His^{292} to alanine and Ser^{293} to glutamate abolished the ability of mutant E1 proteins to reconstitute with thiamine pyrophosphate. We conclude from these findings that Arg^{288} is critical for phosphorylation of Ser^{293}, that His^{292} is critical to thiamine pyrophosphate binding, and that Arg^{288} and Asp^{296} are critical for catalytic activity but are not necessary for thiamine pyrophosphate binding (Fig. 2). We suggest from these findings that the regulatory phosphorylation site (Ser^{293}) of E1α is located within an active-site pocket where thiamine pyrophosphate interacts with ketoacid substrates to effect decarboxylation.

Starvation of animals for protein is well established as causing a decrease in the activity state (percent active) of the liver BCKDH complex (10). This is due to a greater degree of phosphorylation of the complex as the result of an increase in BCKDK tightly associated with the complex (11,12). Western blot analysis of BCKDH complexes from the livers of animals fed various diets has indicated that

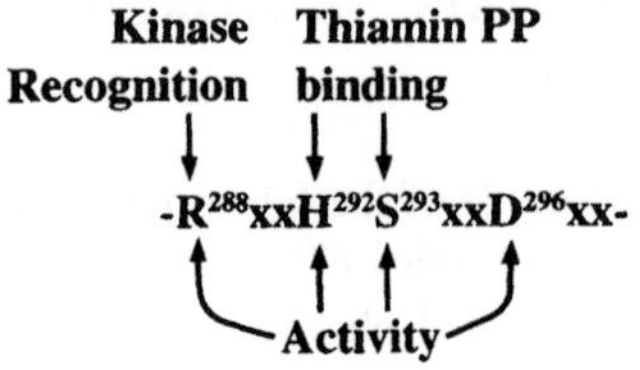

FIG. 2. Roles of important amino-acid residues in the vicinity of the regulatory phosphorylation site of the E1α component of the branched-chain α-ketoacid dehydrogenase complex. Residues necessary for kinase recognition, thiamine pyrophosphate binding, and dehydrogenase activity are indicated.

the increase in kinase activity caused by protein starvation is the result of an increase in the amount of kinase protein associated with the complex (13). Northern blot analysis revealed that an increase in message level for BCKDK also occurs in the liver of protein-starved rats (13). Thus, our current working hypothesis is that the nutritional stress induced by dietary protein deficiency promotes transcription of the BCKDK gene. This results in increased BCKDK protein and activity which, in turn, decreases the activity state of the BCKDH complex and conserves the branched-chain amino acids for protein synthesis.

RECENT FINDINGS WITH THE PYRUVATE DEHYDROGENASE KINASES

cDNAs encoding two rat PDKs (PDK1 and PDK2) and three human PDKs (PDK1, PDK2, and PDK3) have been cloned (2–4; Fig. 1). All five cDNAs encode protein kinases that are specific for phosphorylation of serine residues of the E1α subunit of the pyruvate dehydrogenase complex. Northern blot analysis has been used to characterize tissue distribution of the three isozymes of human PDK (4). The greatest amount of PDK2 message was found in heart and skeletal muscle, the lowest in placenta and lung. Brain, kidney, pancreas, and liver express intermediate amounts of PDK2. The tissue expression of PDK1 messenger ribonucleic acid (mRNA) is markedly different. Its message is expressed to the greatest extent in heart, with much less expression in other tissues (heart>skeletal muscle>liver>pancreas>brain>placenta=lung>kidney). PDK3 expression is even more tissue-specific. Its message is found almost exclusively in heart and skeletal muscle, suggesting that it may play a special role in muscle tissue. The expression of PDK2 mRNA is the greatest of all three PDKs in all tissues thus far examined, suggesting that it may be the major PDK isozyme responsible for regulation of pyruvate dehydrogenase activity in human tissues. Tissue-specific expression of PDK1 and PDK2 protein has been confirmed by Western blot analysis with extracts of heart and liver mitochondria (K. M. Popov and R. A. Harris, *unpublished observations*). Approximately equal amounts of PDK1 and PDK2 are present in heart-muscle mitochondria, whereas PDK2 is present in greater amounts than PDK1 in liver mitochondria.

The three PDK isozymes have been expressed as recombinant proteins in *Escherichia coli* (2–4). Their enzyme-specific activities vary in the order PDK3>KDK1> PDK2. Marked kinetic differences also exist among the three isozymes (K. M.

Popov, *unpublished observations*). PDK2 has the lowest K_m for adenosine triphosphate (ATP) and is the most sensitive to inhibition by adenosine diphosphate (ADP) and dichloroacetate. Both PDK1 and PDK2 are sensitive to varying degrees to activation by NADH and acetyl-CoA, and to inhibition by CoA. In contrast, PDK3 activity is virtually resistant to these compounds. Detailed studies of the kinetic parameters of PDKs have at this point been completed only with homodimers of recombinant PDK1, PDK2, and PDK3. Preliminary experiments indicate that heterodimers form when PDK1 and PDK2 are coexpressed in *E. coli*. Since PDK heterodimers are likely to be present *in vivo*, future work will include characterization of the activities and effector sensitivities of PDK heterodimers.

Long-term regulation of the pyruvate dehydrogenase complex as a result of stable changes in the activity of PDK has been demonstrated in a number of experimental systems (14). Starvation of rats causes a stable increase in PDK activity in liver, heart, and muscle that correlates with decreased activity states of the pyruvate dehydrogenase complexes present in these tissues (14,15). This is a desirable response, since conservation of three-carbon units for glucose synthesis in the starved state serves to also conserve body protein. Chemically induced diabetes also increases PDK activity and decreases the activity state of the pyruvate dehydrogenase complexes present in liver, heart, and muscle tissue. The responsible mechanism is probably the same as in starvation, but causes an undesirable response in diabetes, since hepatic glucose synthesis is promoted, peripheral tissue glucose oxidation is inhibited, and both of these effects contribute to the hyperglycemia of the diabetic state. Insulin therapy restores kinase activity and pyruvate dehydrogenase complex activity to the state observed in control animals, thereby inhibiting hepatic glucose synthesis and promoting peripheral tissue glucose utilization. Randle's laboratory has presented evidence for a stable increase in the specific activity of PDK in response to starvation and diabetes (16). The modification responsible for this change in PDK activity, which survived column-purification procedures, has not been defined. We have investigated whether an increase in PDK protein might also contribute to the increase in PDK enzyme activity in the starved and diabetic states (K. M. Popov and R. A. Harris, *unpublished observations*). Western blot analysis with antibodies specific for PDK1 and PDK2 has revealed changes in amounts of these proteins in response to starvation and diabetes. The greatest change observed thus far is a doubling of the amount of liver PDK2 in response to diabetes induced by streptozotocin (K. M. Popov and R. A. Harris, *unpublished observations*). Our current working hypothesis is that changes in both the amount of PDK protein and its specific activity may be involved in the stable increase in PDK activity induced by starvation and diabetes.

SUMMARY

Four mitochondrial protein kinases have been cloned. These proteins represent a new family of protein kinases, related by sequence to the bacterial protein kinases but by function to the eukaryotic serine protein kinases. Arg^{288} is required for recognition by BCKDK of the phosphorylation site on the E1α subunit of the BCKDH complex. BCKDK inhibits the dehydrogenase activity of the BCKDH complex by introducing a negative charge into the active-site pocket of the E1 component. Protein starvation of rats induces an increase in the amount of BCKDK bound to the BCKDH complex. This causes inactivation of the BCKDH complex and conserves branched-chain amino acids for protein synthesis in the protein-starved state. Expression of the different PDK isoenzymes is tissue specific, and the different PDK isoenzymes are unique with respect to kinetic parameters for ATP and ADP and sensitivity to allosteric effectors (NADH, NAD^+, coenzyme A, acetyl-CoA, pyruvate, and dichloroacetate). Preliminary experiments indicate that an increased amount of PDK2 protein partly explains the increase in PDK activity that occurs in rat liver in response to chemically induced diabetes.

ACKNOWLEDGMENTS

This investigation was supported by United States Public Health Service Grants DK 19259 (R. A. H.), DK 47844 (R. A. H.), GM 51262 (K. M. P.), a Grant-in-Aid from the National American Heart Association (K. M. P.), a post-doctoral fellowship from the Indiana Heart Association (J. W. H.), and the Grace M. Showalter Residuary Trust.

REFERENCES

1. Popov KM, Zhao Y, Shimomura Y, Kuntz MJ, Harris RA. Branched-chain α-ketoacid dehydrogenase kinase: molecular cloning, expression, and sequence similarity with histidine protein kinases. *J Biol Chem* 1992;267:13127–30.
2. Popov KM, Kedishvili NY, Zhao Y, Shimomura Y, Crabb DW, Harris RA. Primary structure of pyruvate dehydrogenase kinase establishes a new family of eukaryotic protein kinases. *J Biol Chem* 1993;268:26602–6.
3. Popov KM, Kedishvili NY, Zhao Y, Gudi R, Harris RA. Molecular cloning of the p45 subunit of pyruvate dehydrogenase kinase. *J Biol Chem* 1994;269:29720–4.
4. Gudi R, Bowker MM, Kedishvili NY, Zhao Y, Popov KM. Diversity of the pyruvate dehydrogenase gene family in humans. *J Biol Chem* 1995;270:28989–94.
5. Hanks SK, Quinn AM, Hunter T. The protein kinase family: conserved features and deduced phylogeny of the catalytic domains. *Science* 1988;241:42–52.
6. Stock JB, Ninfa AJ, Stock AM. Protein phosphorylation and regulation of adaptive responses in bacteria. *Microbiol Rev* 1989;53:450–90.
7. Yeaman SJ. The 2-oxo acid dehydrogenase complexes: recent advances. *Biochem J* 1989;257:625–32.
8. Zhao Y, Hawes JW, Popov KM, et al. Site-directed mutagenesis of phosphorylation sites of the branched chain α-ketoacid dehydrogenase complex. *J Biol Chem* 1994;269:18583–7.
9. Hawes JW, Schnepf RJ, Jenkins AE, Shimomura Y, Popov KM, Harris RA. Roles of amino acid residues surrounding phosphorylation site 1 of branched chain α-ketoacid dehydrogenase in catalysis and phosphorylation site recognition by BCKDH kinase. *J Biol Chem* 1995;270:31071–6.
10. Harris RA, Paxton R, Powell SM, Goodwin GW, Kuntz MJ, Han AC. Regulation of branched-chain α-ketoacid dehydrogenase complex by covalent modification. *Adv Enzyme Regul* 1986;25:219–37.
11. Espinal J, Beggs M, Patel H, Randle PJ. Effects of low protein diet and starvation on the activity of branched-chain 2-oxoacid dehydrogenase kinase in rat liver and heart. *Biochem J* 1986;237:285–88.

12. Zhao Y, Jaskiewicz, Harris RA. Effects of clofibric acid on the activity and activity state of the hepatic branched-chain 2-oxo acid dehydrogenase complex. *Biochem J* 1992:285:167–72.
13. Popov KM, Zhao Y, Shimomura Y, et al. Dietary control and tissue expression of branched-chain α-ketoacid dehydrogenase kinase. *Arch Biochem Biophys* 1995;316:148–54.
14. Denyer GS, Kerbey AL, Randle PJ. Kinase activator protein mediates longer-term effects of starvation on activity of pyruvate dehydrogenase kinase in rat liver mitochondria. *Biochem J* 1986;239: 347–54.
15. Jones BS, Yeaman SJ, Sugden MC, Holness MJ. Hepatic pyruvate dehydrogenase kinase activities during starved-to-fed transition. *Biochim Biophys Acta* 1992;1134:164–8.
16. Priestman DA, Mistry SC, Kerbey AL, Randle PJ. Purification and partial characterization of rat liver pyruvate dehydrogenase kinase activator protein (free pyruvate dehydrogenase kinase) *FEBS Lett* 1992;308:83–6.

Signal Transduction in Health and Disease, Advances in Second Messenger and Phosphoprotein Research, Vol. 31, edited by J. Corbin and S. Francis.
Lippincott–Raven Publishers, Philadelphia © 1997.

10

Phosphatases as Partners in Signaling Networks

David L. Brautigan

Departments of Microbiology and Medicine, Center for Cell Signaling, University of Virginia, Charlottesville, VA 22908

SIGNALING BY KINASE CASCADES

Considerable excitement has been generated in the past half-dozen years with the discovery of protein kinase cascades that transduce and transmit signals from plasma-membrane receptors to various sites throughout the cell (reviewed in [1]). Receptors with intrinsic protein tyrosine kinase (PTK) activity are now understood to be linked to mitogen-activated protein (MAP) kinase activation through receptor dimerization and tyrosine phosphorylation, binding of the adapter protein Grb2, and the associated guanine-nucleotide exchange factor SOS named for the *Drosophilia* son of sevenless protein. In this complex, SOS promotes the formation of guanosine triphosphate (GTP)-ras, which in turn binds to and facilitates activation of the Ser/Thr kinase Raf.

From this step the best understood pathway is the MAP kinase cascade (Fig. 1). MAP kinase, actually consisting of two isoforms called MAP kinases p42 and p44, or extracellular regulated kinase (ERK2) and ERK1, respectively, has multiple downstream targets, including cytoplasmic phospholipase A2 (cPLA2), other protein kinases, such as ribosomal S6 kinase ($p90^{rsk}$) and transcription factors within the nucleus, such as elk1. The MAP kinase is activated by phosphorylation of a threonine and a tyrosine in a T-E-Y sequence, both phosphates being transferred by a single "dual-specificity" kinase called MEK (for MAP and ERK kinase). MEK is itself activated by phosphorylation of serine residues by the protooncogene kinase Raf. Recent evidence suggests that other kinases are also capable of activating MEK, including MEK kinases and the protooncogene protein mos. These other MEK kinases may be responsible for linking the MEK–MAP kinase step to signals emanating from other cell-surface receptors, such as G-protein-linked receptors or Janus kinase (JAK)-associated receptors (Fig. 1).

Largely from genetic studies in yeast, it is now understood that several closely-related MEK–MAP kinase cascades are involved in different physiologic responses to

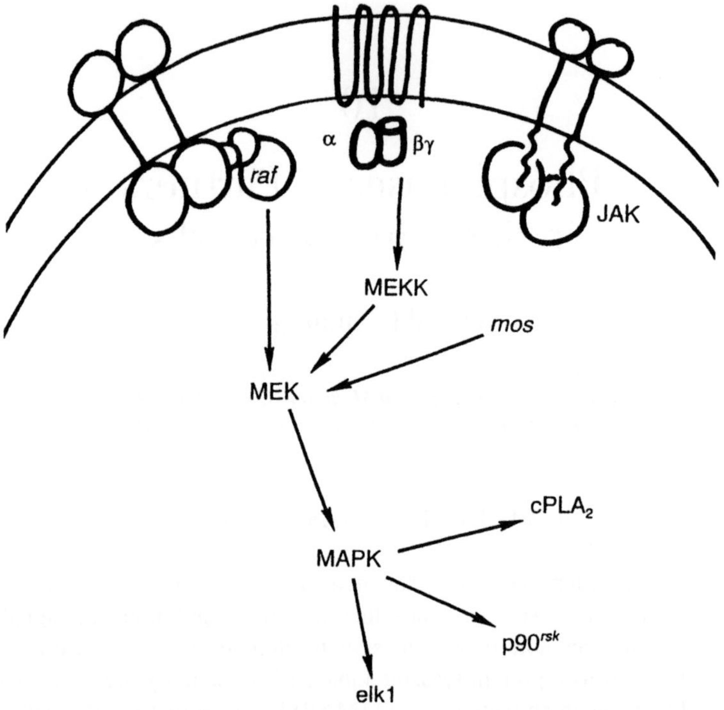

FIG. 1. Signaling pathways involve a kinase pathway. (Top) Three types of receptors spanning a membrane bilayer; a growth-factor receptor tyrosine kinase with a single transmembrane segment, a transmembrane seven-helix receptor coupled to a heterotrimeric G protein (shown as α, β, and γ), and a cytokine receptor associated with a JAK tyrosine kinase. As described in the text, and reviewed in (1), activation of the receptors leads to activation of different protein Ser/Thr kinases that phosphorylate and activate MEK, which is highly specific for MAP kinase. MAP kinase then transmits the signal to various intracellular targets.

external factors or environmental stresses (reviewed in [2]). These involve the MAP kinase homologues called JNK or SAP, for *jun* N-terminal kinase or stress-activated kinase, and p38 kinase. These pathways seem to operate in parallel as a network for information processing (3).

However, the depiction of these pathways, such as shown in Fig. 1, is fundamentally incorrect because it does not show one essential feature. Each of the steps is reversed by enzymatic dephosphorylation, catalyzed by protein phosphatases. Therefore, schemes such as that in Fig. 1, which are used essentially everywhere to discuss these signaling pathways, only show half of the story. There is an emphasis on the phosphorylation events that cause rapid activation of enzymes and factors within cells. Left out are the dephosphorylation reactions that rapidly and effectively erase these actions. Protein phosphatases inactivate both the kinases in the cascade and the downstream effects of the kinases, thereby extinguishing the signal and resetting the system to its baseline activity (see [4]).

SER/THR PHOSPHATASES AND MAP KINASE SIGNALING

Protein phosphatase type-1 (PP1) was first described 50 yr ago by Carl and Gerti Cori as the PR enzyme that inactivated phosphorylase (see [5]), and research on that enzyme continued for many years in the laboratory of Edmond Fischer in Seattle. Protein phosphatase type-2A (PP2A) was first purified and characterized as a glycogen synthase phosphatase in the laboratories of Takeda in Hiroshima (6) and Tsuiki in Sendai (7). In 1983 Tom Ingebritsen and Philip Cohen proposed a classification scheme for phosphatases based on their specificity and inhibition (8), and these two protein phosphatase enzymes have been known as PP1 and PP2A since that time.

It is remarkable that PP2A, more so than PP1, seems to play a major role in reversing all of the steps of the kinase signaling pathway shown in Fig. 1. With the purified kinases *in vitro*, PP2A was found to be highly reactive and selective in dephosphorylation and inactivation of MEK and MAP kinase, as well as of the targets of MAP kinase. Based on this information, the scheme of MAP kinase signaling needs to be revised in order to reflect the active participation of PP2A in the regulation of the system. In this revised scheme (Fig. 2), the kinases and their downstream targets are depicted as two interconvertible forms. There is the unphosphorylated form, which is less active or inactive, and there is the phosphorylated form, which is activated. In an unstimulated cell the MAP kinase and indeed the entire signaling pathway is mostly dormant. It is maintained in that state by the action of PP2A. In fact, we find that PP2A makes up some 0.25% of total protein in cells in tissue culture. By virtue of its abundance and activity, PP2A represents a formidable obstacle to signaling through this pathway. My contention is that we have to reconsider signaling in light of the fact that endogenous phosphatases actively maintain proteins within cells in a relatively unphosphorylated state. One has to imagine that in order to transmit signals from the cell surface to the nucleus and other sites in the cell, there must be mechanisms to temporarily inactivate endogenous phosphatases.

PARTNERS IN SIGNALING

The definition of a theorem is "a proposition embodying something to be proved," or "a proposition that is not self-evident but can be proved from accepted premises." For the sake of discussion, the following are three theorems to emphasize the partnership of kinases and phosphatases in protein phosphorylation in signaling.

Theorem 1. Transitions between different growth states, such as phases of the cell cycle or terminal differentiation, involve abrupt changes (those occurring in a relatively short time) in the phosphorylation levels of key regulatory proteins. This could involve an increase in phosphorylation of proteins such as MAP kinase or the tumor suppressor protein Rb, or an increase in phosphorylation of a transcription factor such as c-jun, or a decrease in phosphorylation of a cyclin-dependent

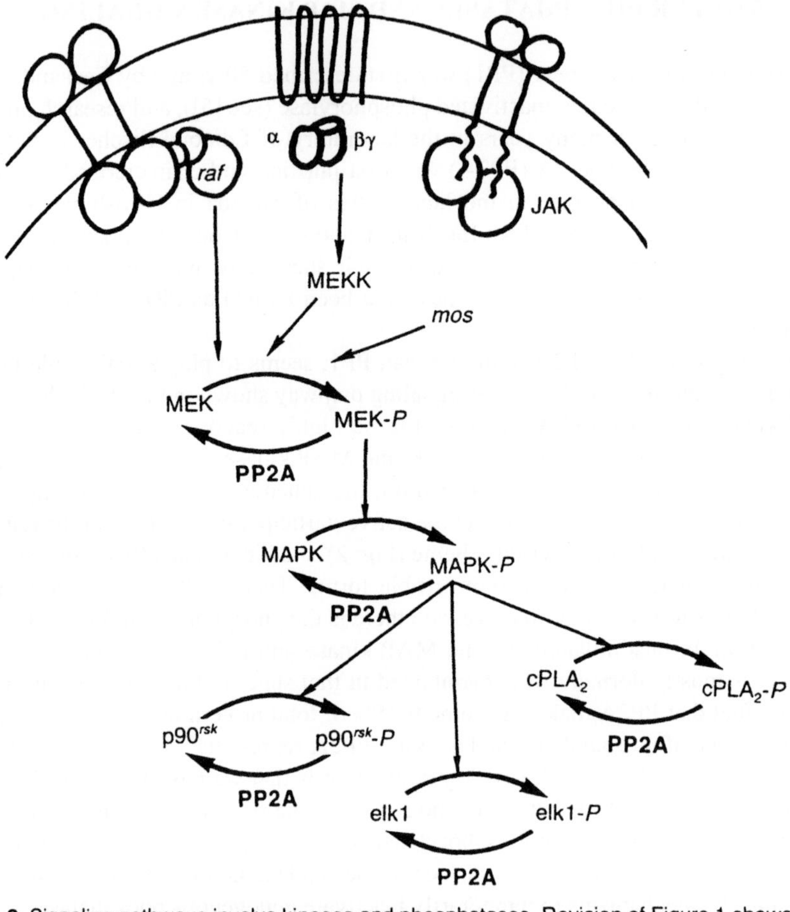

FIG. 2. Signaling pathways involve kinases and phosphatases. Revision of Figure 1 shows protein phosphatase type-2A (PP2A) is involved in control of kinases in the pathway. Also, PP2A inactivates the proteins and enzymes that produce effects of the hormone signal. The abundance and activity of PP2A raise the issue of how signals overcome this opposition.

kinase or an src family kinase. It is important to note that both increases and decreases in phosphorylation occur as signaling events.

Theorem 2. To produce an abrupt increase in the phosphorylation of an intracellular protein requires coincident activation of a protein kinase and inactivation of the corresponding protein phosphatase. These two enzymes operate on the opposing halves of a cyclic reaction scheme. Increasing one half and decreasing the other half of the cycle produces a shift in the phosphorylation state of the target. Simultaneous change in activity of the two enzymes produces a crossover, as shown in Fig. 3. This proposition implicates a common upstream signaling event or two synchronized signaling events to ensure that both the activation and inactivation

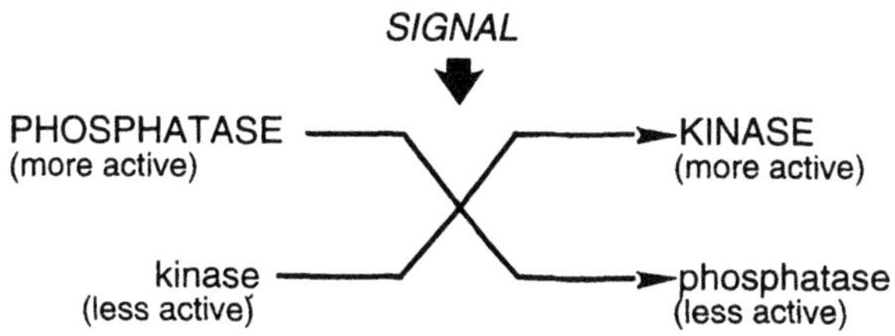

FIG. 3. Signals activate kinases and inactivate phosphatases. Scheme emphasizing the high basal activity of protein phosphatases in actively maintaining the cellular status quo. Moving in time from left to right, a cell in homeostasis has many key regulatory proteins in a mainly dephosphorylated state, because phosphatases are active (*upper case*) whereas kinases are relatively inactive (*lower case*). In response to signals, either external or cell-cycle based, there is a coincident activation of kinases and inactivation of phosphatases, producing a crossover in activities and abrupt phosphorylation of proteins to trigger the transition to a new cellular status.

occur simultaneously. A classic example of this type of crossover is the "flash activation" of glycogen phosphorylase in a glycogen protein complex upon addition of Ca^{2+} plus Mg-adenosine triphosphatase (MgATP) (9).

Theorem 3. Protein phosphatases are inactivated by two physiologic mechanisms: (i) covalent modification, specifically phosphorylation; and (ii) heterotropic interactions (i.e., the binding of inhibitors or regulatory subunits to the catalytic subunits). The regulation of the major mammalian protein phosphatases by these two mechanisms is depicted in Fig. 4. The T-shaped lines employed by geneticists to indicate a negative or inhibitory effect are used.

NEGATIVE REGULATION OF NEGATIVE REGULATORS

PP1 and PP2A collaborate to resist cell activation or transition between phases in the cell cycle by keeping many proteins relatively dephosphorylated within cells. The phosphatases are dominant regulators of the status quo. However, PP1 and PP2A are themselves under negative regulation. Turning the phosphatases off provides the opportunity to accumulate a regulatory protein in its phosphorylated state, to produce a physiologic effect. Examples of negative regulation of protein Ser/Thr phosphatases include phosphorylation (see [10]). PP1 is phosphorylated in a conserved sequence near the carboxy-terminus (C-terminus), at Thr^{320}, by the cyclin-dependent kinase (CDK), as shown by Berndt et al. (11). This occurs in fission yeast PP1 during mitosis, as shown by Yanagida et al. (12). There have also been reports of phosphorylation of Tyr residues in PP1 by Johansen and Ingebritsen using src kinase or by Villa-Moruzzi using abl kinase (13,14). These studies were done with purified proteins, a low level of phosphorylation led to partial inactivation, and the site was only identified as a protease-sensitive C-terminal segment. There is reduced specific activity of PP1 immunoprecipitated from cells transformed by temperature-sensitive forms of *v-src* (15), consistent with phosphorylation causing inactivation.

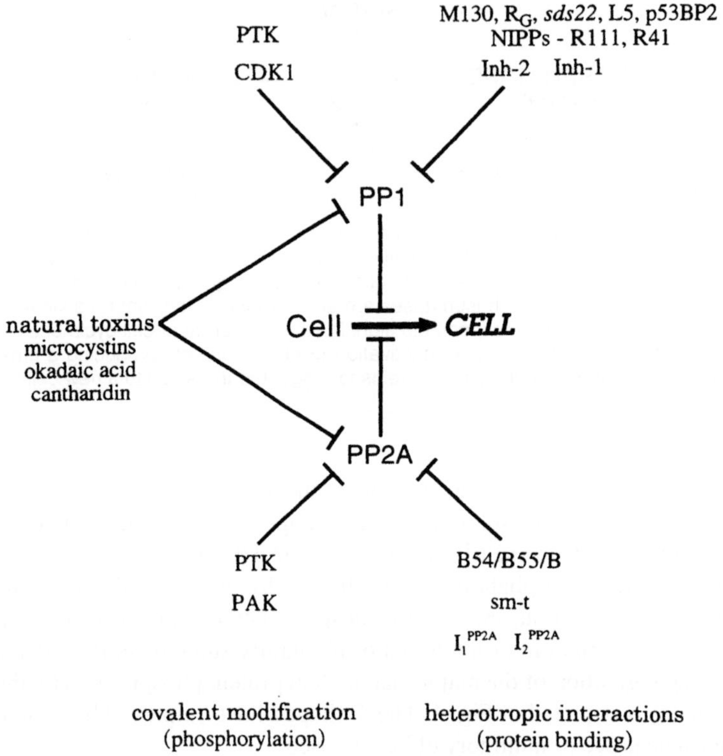

FIG. 4. Phosphatase negative regulation. As described in the text, this scheme shows how a cell is restricted from transition by PP1 and PP2A and these phosphatases themselves are under different types of negative control. Covalent modification, specifically phosphorylation, is catalyzed by protein tyrosine kinases (PTKs), cyclin-dependent kinases (CDK1), and an autophosphorylation-activated kinase said to be PAK1. In addition, heterotropic interactions between protein subunits or inhibitor proteins alter the specificity and control the activity of phosphatases and localize these enzymes to intracellular sites. Proteins are identified in the text. The phosphatases are also the targets for toxins, produced by microorganisms and insects, which bind to them with nanomolar affinity and act as inhibitors. Together, this provides a picture of negative regulation of the negative regulators of cell functions.

In parallel with PP1, phosphorylation also negatively regulates PP2A. My group has shown that PP2A is phosphorylated within a conserved sequence near the C-terminus, at Tyr^{307}, and that this causes inactivation of PP2A (16). Phosphorylation of PP2A occurs in cells transformed by *v-src* or cells stimulated by epidermal growth factor (EGF) or serum (17), and also in normal lymphocytes activated by surface cross-linking (18). The C-terminal sequence of PP2A (RRTPDYFL) has the site for Tyr phosphorylation, and is a highly conserved feature, identical in yeast and human PP2A. Therefore, it is interesting that Leu^{309} at the C-terminus of PP2A was reported separately in the same year by Clarke, Stock, and Hemmings and their coworkers to be the site of methyl esterification by a cytosolic transferase (19–21).

The consequences of this methyl esterification are not yet understood, but here is yet another covalent mechanism with a potential for regulation of PP2A function. In addition to being inactivated by Tyr phosphorylation, PP2A can also be inactivated by phosphorylation of an unidentified Thr residue, as reported by Damuni et al. (22), who used a purified autophosphorylation-activated kinase that they have now identified as PAK. Therefore, both PP1 and PP2A can be inactivated by phosphorylation of either Tyr or Thr residues. Metabolic labeling of cells shows that P-Ser is the major phosphoamino acid in PP1 and PP2A (*unpublished results*), but the position of the phosphorylation is not known. Evidence with the purified phosphatases indicates that the phosphorylation of Tyr and Thr residues (apparently not the Ser) can be readily reversed by intramolecular dephosphorylation. Phosphorylation may inactivate the phosphatase by virtue of intramolecular competitive inhibition that is relieved by the phosphatase's own catalytic action. This provides an intriguing mechanism for phosphatase inactivation and reactivation, with a specific delay time that depends on the rate of intramolecular dephosphorylation. Thus, during signaling the phosphorylation of PP1 or PP2A would protect its substrates from dephosphorylation during the period that the phosphatase was carrying out its own intramolecular dephosphorylation.

Phosphorylation as a mechanism of phosphatase regulation might account for the failure of many attempts over the past 30 yr to measure a change in phosphatase activity in cell or tissue extracts in response to hormones. To measure changes in protein phosphorylation in stimulated cells, traditional methods used homogenized cells in buffers containing inhibitors of kinases (e.g., ethylenediamine tetraacetic acid [EDTA]), phosphatases (e.g., NaF, PP_i), and proteases (e.g., phenylmethylsulfonyl fluoride [PMSF]) in order to arrest any changes in modification of the protein of interest. However, in order to measure phosphatase activity, no inhibitors could be added during extraction. Therefore, the phosphatase was free to dephosphorylate itself, with a return to full activity. As a result, there was little change in phosphatase activity in extracts in response to hormone. The conclusion from these types of experiments, which became widely accepted dogma, was that phosphatases were constitutively active in cells and were not acutely regulated by hormones, but that only the kinases were regulated. Recent experiments done to detect phosphorylation utilized potent toxin inhibitors of PP1 and PP2A (see below) to block self-dephosphorylation. To determine the effects of phosphorylation on phosphatase activity requires either an initial rate assay or the use of thioATP to make thiophosphorylated protein that resists dephosphorylation. These experimental approaches have made investigation of phosphatase phosphorylation possible. Transient phosphorylation is proposed for acute control of PP1 and PP2A.

AN ABUNDANCE OF PP1 SUBUNITS

Besides being phosphorylated, which provides a basis for acute regulation of phosphatase activity, both PP1 and PP2A bind many proteins that regulate and/or inhibit phosphatase activities. These heterotropic interactions (Fig. 4) afford a tonic form of regulation that operates in concert with phosphorylation. PP1 specifically binds two heat-stable proteins, called inhibitor-1 and inhibitor-2, that have been well studied since their discovery by Huang and Glinsman (23) 20 yr ago. Inhibitor-1 is a substrate for cyclic adenosine monophosphate (cAMP)-dependent protein kinase, and phosphorylation is required for its activity against PP1. Shenolikar et al. have data to show that the nanomolar affinity of inhibitor-1 for PP1 comes from two distinct binding interactions, one independent of phosphorylation, the other dependent on phosphorylation, which together produce the inhibition of PP1. On the other hand, inhibitor-2 is active without phosphorylation, although it is phosphorylated by multiple kinases including caseine kinase 2 (CK2), glycogen synthase kinase 3 (GSK3), cyclin-dependent kinase (CDK), and MAP kinase (see [24]). Studies in the laboratory of DePaoli-Roach, using truncated forms of recombinant inhibitor-2, indicate that there are two sites for interaction with PP1. It seems that two sites are better than one, at least for protein inhibitors!

In a collaboration with Lamb and Fernandez, we discovered the periodic synthesis and degradation of inhibitor-2 in the cell cycle of mammalian fibroblasts in tissue culture (25). Our recent studies show import of inhibitor-2 from the cytoplasm, where it accumulates during the G1 phase of the cell cycle, into the nucleus of cells during the S-phase (*unpublished results*). These results show that the synthesis and localization of inhibitor-2 is under stringent control. Following the theorems presented above, the nuclear importation of inhibitor-2 could be linked to activation of cyclin D/CDK4. Ludlow has shown that PP1 in the nucleus dephosphorylates Rb protein (26). The PP1 maintains Rb in its underphosphorylated state during the G1 phase. Simultaneous CDK activation and inhibition of PP1, through the import of inhibitor-2 into the nucleus, would produce abrupt phosphorylation of Rb protein, which occurs at the onset of S phase.

In addition to the inhibitor-1 and -2 proteins there are other nuclear inhibitor proteins (NIPPs) for PP1, first characterized by Bollen, but which more recent evidence suggests are derived from larger proteins (i.e., R41 and R111) within the nucleus (27). Genetic experiments have shown that defects in the yeast PP1 gene, called *dis2*, were suppressed by another gene, *sds22*, which encodes a phosphoprotein that functions as a PP1 regulatory subunit (28). Beyond these proteins is a diverse group of what have been called targeting subunits, which are really the regulatory subunits of PP1. Some have been copurified with PP1, such as the M130 myosin-binding subunit (29–31), or R_G, the glycogen-binding subunit (32). The yeast 2 hybrid system has been used to identify other PP1-binding proteins, such as ribosomal protein L5 (33) and p53BP2 (34). Surely there are more to come.

THE ABCS OF PP2A REGULATION

As in the case of PP1, there are proteins that exert heterotropic regulation of PP2A. Long recognized are the effects of the A and B subunits in altering substrate specificity of the C catalytic subunit in the heterotrimeric ABC structure of PP2A. The core of PP2A is composed of the A and C subunits, which together pair with a variety of B subunits (B55, B56, B54). Heterogeneity in the B subunits is responsible for producing an amazing diversity of forms of PP2A. In particular, Hemmings, DePaoli-Roach, and Mumby have cloned multiple isoforms of different B-subunit types. We can guess that the B subunits are responsible for specifying different functions and different intracellular targeting of PP2A. In cells transformed by deoxyribonucleic acid (DNA) tumor viruses, viral-encoded proteins replace the B subunits in PP2A. The polyoma virus middle T and simian virus 40 (SV40) small t proteins form heterotrimeric complexes with the AC core subunits of PP2A. Experiments by Sontag in Mumby's group showed that overexpression of small t in cells leads to its complexing with PP2A, thus reducing the activity of PP2A toward MEK and MAP kinase (35). These kinases are activated as a result of this reduction in PP2A activity. This may be one of the strategies that the viruses use to cause transformation. Recently, Damuni et al. purified two protein inhibitors of PP2A (36), which were identified from their sequences to be proteins already known as putative HLA class II-associated proteins (PHAPI and PHAPII) (37). Thus, there appear to be a variety of subunits and inhibitor proteins for both PP1 and PP2A. These serve to regulate the level of activity of these phosphatases, restrict their substrate specificity, and determine the intracellular localization of PP1 and PP2A.

PHOSPHATASES AS TARGETS OF TOXINS

Both PP1 and PP2A are the intracellular targets for various environmental toxins produced by microorganisms and insects. This includes cyclic peptides such as microcystin and nodularin, polyethers such as okadaic acid, and bridged cyclic compounds such as cantharidin. The toxicity of these compounds affirms that the phosphatases are essential for life, as had been shown independently by genetic manipulation of yeast. When administered at sublethal doses, the toxins reveal that diverse processes are under control of the phosphatases. The toxins, although structurally dissimilar from one another, all seem to bind at the active site of each phosphatase. Here they contact multiple residues near the active site, as visualized from the recently published crystallographic structures of PP1, one of which contains bound microcystin (38,39). This site involves Tyr^{272}, seen prominently in the structures of PP1 and implicated in PP1 activity by mutagenesis and comparative binding assays conducted in the laboratory of Lee (40). The toxins have been important tools in producing changes in cell functions, thus highlighting the role of PP1 and PP2A in maintaining homeostasis.

CHALLENGES AND QUESTIONS

After years of being considered constitutively active enzymes that simply reversed the action of protein kinases, phosphatases are being recognized for fulfilling an important role in signaling (e.g., see [4]). Experiments conducted in the past couple of years have exposed new possibilities for regulation of phosphatase activity. More evidence is needed to show that the covalent and heterotropic mechanisms highlighted in this chapter in fact operate in living cells in response to signals. A critical realization is that PP1 and PP2A are not just two enzymes, but rather that each catalytic subunit is distributed among a set of multisubunit phosphatase enzymes with distinctive properties. Diversity comes from the regulatory subunits available for pairing with the catalytic subunit. We have to find ways and develop reagents, such as specific antibodies, to examine these individual phosphatases. A challenge is to learn more about discrete functions of regulatory subunits in mammalian cells.

Many questions remain to be answered toward the goal of better understanding how phosphatases act as partners in signaling networks. What is the distribution of the PP1 or PP2A catalytic subunit among its possible regulatory subunits in a given cell or tissue? Do stimuli cause redistribution of the catalytic subunit among regulatory subunits? Are the sites on PP1 and PP2A for regulatory subunits independent, or overlapping and mutually exclusive? Which of these individual multisubunit phosphatases is responsive to extracellular stimuli? What are the kinases that phosphorylate the PP1 and PP2A catalytic subunits in cells? What is the timing of PP1 and PP2A phosphorylation in living cells? Which common step in a signaling pathway gives synchronous kinase activation and phosphatase inactivation? What are the kinetics (K_{cat}) of phosphatase self-dephosphorylation, and is this affected by regulatory subunits? What is the three-dimensional structure of the phosphorylated, inactive phosphatase catalytic subunit? What are the compound effects of multiple phosphorylation of PP1 or PP2A? Better understanding of the phosphatases will add complexity to the network of reactions that link signaling events together, but will bring us closer to knowledge of how living organisms respond to and adapt to their environments.

ACKNOWLEDGMENT

Research in the author's laboratory has been supported in part by Grant BE-130 from the American Cancer Society and Grant MCB9507357 from the National Science Foundation.

REFERENCES

1. Marshall CJ. Specificity of receptor tyrosine kinase signaling: transient versus sustained extracellular signal-regulated kinase activation. *Cell* 1995;80:179–85.
2. Herskowitz I. MAP kinase pathways in yeast: for mating and more. *Cell* 1995;80:187–97.
3. Cano E, Mahadevan LC. Parallel signal processing among mammalian MAPKs. *TIBS* 1995;20: 117–22.
4. Hunter T. Protein kinases and phosphatases: the yin and yang of protein phosphorylation and signaling. *Cell* 1995;80:225–36.

5. Fischer EH, Brautigan DL. A phosphatase by any other name: from prosthetic group removing enzyme to phosphorylase phosphatase. *TIBS* 1982;7:3–4.
6. Usui H, Konohara N, Yoshikawa K, Imazu M, Imaoka T, Takeda M. Phosphoprotein phosphatases in human erythrocyte cytosol. *J Biol Chem* 1983;258:10455–63.
7. Tamura S, Kikuchi H, Kikuchi K, Hirage A, Tsuiki S. Purification and subunit structure of a high molecular weight phosphoprotein phosphatase from rabbit liver. *Eur J Biochem* 1980;104:347–55.
8. Ingebritsen TS, Cohen P. The protein phosphatases involved in cellular regulation. *Eur J Biochem* 1983;132:255–61.
9. Heilmeyer LMG Jr, Meyer F, Haschke RH, Fischer EH. Control of phosphorylase activity in a muscle glycogen particle. *J Biol Chem* 1970;24:6649–56.
10. Brautigan DL. Flicking the switches: phosphorylation of serine/threonine protein phosphatases. *Semin Cancer Biol* 1995;6:211–7.
11. Dohadwala M, de Cruz e Silva EF, Hall FL, et al. Phosphorylation and inactivation of protein phosphatase 1 by cyclin-dependent kinases. *Proc Natl Acad Sci USA* 1994;91:6408–12.
12. Yamano Y, Ishii K, Yanagida M. Phosphorylation of dis2 protein phosphatase at C terminal cdc2 consensus and its potential role in cell cycle regulation. *EMBO J* 1994;13:5310–8.
13. Johansen JW, Ingebritsen TS. Phosphorylation and inactivation of protein phosphatase 1 by $pp60^{v\text{-}src}$. *Proc Natl Acad Sci USA* 1986;83:207–11.
14. Villa-Moruzzi E. Activation of type-1 protein phosphatase by cdc2 kinase. *FEBS Lett* 1992;304:211–5.
15. Belandia B, Brautigan D, Martin-Perez J. Attenuation of ribosomal protein S6 phosphatase activity in chicken embryo fibroblasts transformed by Rous sarcoma virus. *Mol Cell Biol* 1994;14:200–6.
16. Chen J, Martin BL, Brautigan DL. Regulation of protein serine-theonine phosphatase type-2A by tyrosine phosphorylation. *Science* 1994;257:1261–4.
17. Chen J, Parsons S, Brautigan DL. Tyrosine phosphorylation of protein phosphatase 2A in response to growth stimulation and v-src transformation of fibroblasts. *J Biol Chem* 1994;269:7957–62.
18. Brautigan DL, Chen J, Thompson P. Tyrosine phosphorylation of phosphatase 2A in transformed fibroblasts and normal lymphocytes. *Adv Protein Phosphatases* 1993;7:49–65.
19. Favre B, Zolnierowicz S, Turowski P, Hemmings BA. The catalytic subunit of protein phosphatase 2A is carboxyl-methylated *in vivo*. *J Biol Chem* 1994;269:16311–7.
20. Xie H, Clark S. Protein phosphatase 2A is reversibly modified by methyl esterification at its C-terminal leucine residue in bovine brain. *J Biol Chem* 1994;269:1981–4.
21. Lee VM-Y, Balin BJ, Otvos L Jr, Trojanowski JQ. A68: a major subunit of paired helical filaments and derivatized forms of normal tau. *Science* 1991;251:675–8.
22. Guo H, Damuni Z. Autophosphorylation-activated protein kinase phosphorylates and inactivates protein phosphatase 2A. *Proc Natl Acad Sci USA* 1993;90:2500–4.
23. Huang FL, Glinsmann WH. Separation and characterization of two phosphorylase phosphatase inhibitors from rabbit skeletal muscle. *Eur J Biochem* 1976;70:419–26.
24. Wang M, Guan K-L, Roach PJ, DePaoli-Roach AA. Phosphorylation and activation of the ATP-Mg-dependent protein phosphatase by the mitogen-activated protein kinase. *J Biol Chem* 1995;270:18352–8.
25. Brautigan DL, Sunwoo J, Labbe JC, Fernandez A, Lamb NJC. Cell cycle oscillation of phosphatase inhibitor-2 in rat fibroblasts coincident with $p34^{CDC2}$ restriction. *Nature* 1990;344:74–8.
26. Ludlow JW, Glendening DM, Linvingston DM, DeCaprio JA. Specific enzymatic dephosphorylation of the retinoblastoma protein. *Mol Cell Biol* 1993;13:367–72.
27. Jagiello I, Buellens M, Stalmans W, Bollen M. Subunit structure and regulation of protein phosphatase-1 in rat liver nuclei. *J Biol Chem* 1995;270:17257–63.
28. Stone EM, Yamano H, Kinoshita N, Yanagida M. Mitotic regulation of protein phosphatases by the fission yeast sds22 protein. *Curr Biol* 1993;3:13–26.
29. Alessi D, MacDougall LK, Sola MM, Ikebe M, Cohen P. The control of protein phosphatase-1 by targeting subunits. The major myosin phosphatase in avian smooth muscle is a novel form of protein phosphatase-1. *Eur J Biochem* 1992;210:1023–35.
30. Shimizu H, Ito M, Miyahara K, et al. Characterization of the myosin-binding subunit of smooth muscle myosin phosphatase. *J Biol Chem* 1994;269:30407–11.
31. Shirazi A, Iizuka K, Fadden P, et al. Purification and characterization of the mammalian myosin light chain phosphatase holoenzyme. The differential effects of the holoenzyme and its subunits on smooth muscle. *J Biol Chem* 1994;269:31598–606.
32. Stralfors P, Hiraga A, Cohen P. The protein phosphatases involved in cellular regulation. *Eur J Biochem* 1985;149:295–303.

33. Hirano K, Ito M, Hartshorne DJ. Interaction of the ribosomal protein, L5, with protein phosphatase type 1. *J Biol Chem* 1995;270:19786–90.
34. Helps N, Barker HM, Elledge SJ, Cohen PTW. Protein phosphatase 1 interacts with p53BP2, a protein which binds to the tumour suppressor p53. *FEBS Lett* 1995;377:295–300.
35. Sontag E, Fedorov S, Kamibayashi C, Robbins D, Cobb M, Mumby M. The interaction of SV40 small tumor antigen with protein phosphatase 2A stimulates the MAP kinase pathway and induces cell proliferation. *Cell* 1993;75:887–97.
36. Li M, Guo H, Damuni Z. Purification and characterization of two potent heat-stable protein inhibitors of protein phosphatase 2A from bovine kidney. *Biochemistry* 1995;34:1988–96.
37. Vaesen M, Barnikol-Watanabe S, Gotz H, et al. Purification and characterization of two putative HLA class II associated proteins: PHAPI and PHAPII. *Biol Chem Hoppe-Seyler* 1994;357:113–26.
38. Goldberg J, Hsien-bin H, Young-guen K, Greengard P, Nairn AC, Kuriyan J. Three-dimensional structure of the catalytic subunit of protein serine/threonine phosphatase-1. *Nature* 1995;376:745–53.
39. Egloff M-P, Cohen PTW, Reinemer P, Barford D. Crystal structure of the catalytic subunit of human protein phosphatase 1 and its complex with tungstate. *J Mol Biol* 1995;254:942–59.
40. Zhang L, Zhang Z, Long F, Lee EYC. Tyrosine-272 is involved in the inhibition of protein phosphatase-1 by multiple toxins. *Biochemistry* 1996;35:1606–11.

Signal Transduction in Health and Disease, Advances in Second Messenger and Phosphoprotein Research, Vol. 31, edited by J. Corbin and S. Francis.
Lippincott–Raven Publishers, Philadelphia © 1997.

11

The cAMP in Thyroid

From the TSH Receptor to Mitogenesis and Tumorigenesis

N. Uyttersprot, A. Allgeier, M. Baptist, D. Christophe, F. Coppee, K. Coulonval, S. Deleu, F. Depoortere, S. Dremier, F. Lamy, C. Ledent, C. Maenhaut, F. Miot, V. Panneels, J. Parma, M. Parmentier, I. Pirson, V. Pohl, P. Roger, V. Savonet, M. Taton, M. Tonacchera, J. van Sande, F. Wilkin, G. Vassart, and J. E. Dumont

Institute of Interdisciplinary Research (I.R.I.B.H.N.), Free University of Brussels, B-1070 Brussels, Belgium

INTRODUCTION

The thyroid cell, as an object of study, is interesting in the following several respects:

1. It synthesizes and secretes the thyroid hormones, which control and are necessary for maintaining the metabolism and level of activity of most organs, and ensure brain and body development. The sorry state of the cretin (i.e., the child who has been hypothyroid before or since birth) proves the importance of the thyroid gland.
2. It is a model for specialized epithelial cells, differentiated, yet still able to proliferate (a non-stem-cell tissue).
3. Its very specialization and unique role in iodine metabolism allow the easy investigation of its function.
4. It responds to the major proliferation cascades.

In this short review, we intend to describe the regulation of the thyroid cell and to show how its study has led to results that may be of general and medical relevance.

Thyroid function *in vivo* is controlled mainly by thyrotropin (TSH) and iodide. Pituitary TSH activates the thyroid, and the secretion of TSH is inhibited by thyroid hormone in a classical negative feedback mechanism. Iodide, on the other hand, tonically inhibits the function of the gland, so that at low iodide (i.e., thyroid substrate) concentrations, the thyroid is stimulated. As is the case for most differentiated tissues, chronic or intense stimulation of thyroid function is also followed by hypertrophy (i.e., an increase in capacity). Thus, both chronic stimulation by TSH and chronic iodide deficiency induce thyroid hyperplasia and cell hypertrophy. This effect of TSH, as well as those of the thyroid-stimulating antibodies (TSAbs) of Graves disease, show that there is little *in vivo* desensitization and downregulation of the TSH receptor, as it has also been demonstrated *in vitro* (1).

Thyroid growth is also influenced by other factors. For instance, thyroid weight and cell population are increased in acromegaly and decreased in growth-hormone deficiency. Growth hormone presumably acts on the thyroid through locally induced insulin-like growth factor-1 (IGF-1) (2).

CASCADES REGULATED BY THYROID-STIMULATING HORMONE AND IODIDE

Very early after its discovery, cyclic adenosine monophosphate (cAMP) was shown by Sutherland and Butcher to be generated by thyroid adenylate cyclase in response to TSH (3). Gilman, at the time a doctoral candidate of Rall, demonstrated that TSH enhances cAMP accumulation in beef thyroid slices (4,5). Using dog thyrocytes as a model, our group performed the exhaustive task of testing the validity of Sutherland's famous criteria for the various effects of TSH. TSH was found to activate adenylate cyclase in acellular preparations and the accumulation of cAMP in whole cells. All of the functional effects of TSH that were studied turned out to be mimicked by cAMP, cAMP analogues, cholera toxin, and later, forskolin. These effects included the oxidation and binding to proteins of iodide, the secretion of thyroid hormones, the oxidation of glucose through the hexose monophosphate pathway, and others. Moreover, the effects of TSH were reproduced or enhanced by inhibition of phosphodiesterases (6–8). Contrary to our expectations, most effects of TSH were caused by cAMP in this system. Later, in other systems, Knopp et al. (9) showed that cAMP also caused the induction of iodide transport. We confirmed this in dog thyrocytes. Thus, all of the main steps of iodide metabolism in the dog thyroid were shown to be activated by TSH through cAMP (for review, see reference [6]). Later, similar results were obtained in mouse thyroid and in Fisher rat thyroid cell line (FRTL) cells. The effects of TSH were reproduced by the TSAbs of Graves disease patients (10). Effects of TSAbs on iodide uptake or cAMP accumulation in FRTL-5 cells or, more recently, in TSH receptor-expressing Chinese hamster ovary (CHO) cells, became the basis for the functional assay of these TSAbs. However, even in dog thyrocytes, there remains a few effects of TSH that are not accounted for by cAMP: calcium release from slices, phosphatidylinositol synthesis . . . (11).

We had predicted that calcium must also play a role in thyroid cell regulation (6).

Indeed, it was shown in human thyroid gland that TSH also stimulates the PIP_2–phospholipase C (PLC) cascade, releasing inositol trisphosphate (IP_3) and presumably diacylglycerol. This effect was obtained for higher concentrations of TSH than was the effect on cAMP accumulation (by a factor of 10). It was not reproduced by TSAb at concentrations even higher than those measured in the serum of very hyperthyroid patients. In human thyroid, stimulation of the PIP_2–PLC cascade accounts for the activation of protein iodination and thyroid hormone synthesis. These metabolic steps are controlled by H_2O_2 generation, which is stimulated by both Ca^{2+} and diacylglycerol. It is interesting that in the dog thyroid, in which TSH does not activate the PIP_2–PLC cascade, a positive control by cAMP exists, whereas it does not exist in the human cell, in which TSH activates both cascades (12).

Iodide inhibits both adenylate cyclase- and PIP_2–PLC-mediated pathways and their effects. In all cases these inhibitions are relieved by drugs that inhibit iodide uptake or iodide oxidation. This gave rise to the XI concept, according to which the effects of iodide are mediated by an unknown intracellular derivative of iodine, XI. The postulated XI could in fact be an oxidized form of iodide or iodinated protein targets (13). A search for a well-defined XI has led to two main candidates: an iodolactone and an iodohexadecanal (iHDA). The former reproduces several effects of iodine but is not synthesized in significant amounts in thyroid cells in the absence of exogenous arachidonate. The latter is formed in the absence of exogenous precursor (presumably from endogenous plasmalogens) and reproduces some effects of iodide (14).

The thyroid cell is also regulated by various neurotransmitters that modulate adenylate cyclase and/or PLC through their respective receptors. The nature of the existing receptors and therefore of the regulatory neurotransmitters varies greatly from one species to another. A physiologic role of any of these neurotransmitters remains to be demonstrated.

THE THYROID-STIMULATING HORMONE RECEPTOR

From the biologic data, it was quite conceivable that two different TSH receptors existed. In fact, when we finally first cloned the TSH receptor it turned out to be unique. It was able to activate both guanine nucleotide-binding proteins G_s and G_q and thus adenylate cyclase and PLC, respectively (15). This was demonstrated in CHO and African green monkey SV40-transformed kidney cells (COS cells) transfected with a plasmid encoding this receptor downstream from a simian virus 40 (SV40) promoter (16).

The TSH receptor is a classical seven-transmembrane-domain-type receptor. In fact, its cloning was based on homologies of its transmembrane domains II, III, IV, and VI with the corresponding domains of already known receptors. These homologies allowed cloning of the TSH receptor gene through the polymerase chain reaction (PCR) on multiple degenerate primers, which was later used by our group and others to clone the majority of the known seven-transmembrane-domain-type receptors (17). The TSH receptor itself is characterized by a very long extracellular N terminal domain and a short third intracellular loop. It is very homologous to the luteinizing hormone (LH) and follicle-stimulating hormone (FSH) receptors. The long N-terminal domain is sufficient to bind TSH (18). Expression of the receptor, although modulated positively and negatively to some extent by TSH and downregulated by dedifferentiating treatments, is very robust. This makes sense, since this protein is the main intermediate in the physiologic control of the thyroid cell (19).

It is rather striking that, when expressed in transfected cells, both the dog and the human TSH receptors activate adenylate cyclase and activate PLC only at higher TSH concentrations and for higher TSH receptor densities (20). TSH thus behaves as a full agonist for the TSH receptor effect on adenylate cyclase and as a partial agonist for PLC. The lack of effect of dog TSH receptor on PLC in dog thyroid cells is therefore not due to a defect of this receptor itself. Moreover, in isolated dog thyroid membranes, the TSH receptor activates Gs, Gq, Gi, and even other G proteins. There is no present explanation for this discrepancy, but we hypothesize that membrane microdomains may limit the accessibility of G proteins to the TSH receptor.

The TSAbs can activate the PIP_2–PLC cascade in CHO cells expressing TSH receptor, but this is true only for the most active TSAbs in cells that overexpress the receptor by a factor of 50 to 100. The coupling of TSAbs to the receptor is therefore much less efficient than the coupling of TSH for activating Gq and PLC. This *in vitro* effect therefore has little relevance to the situation *in vivo* (21).

CONTROL OF GENE EXPRESSION

The expression of genes encoding protein required for specialized thyroid functions, including those for the presumed but not cloned iodide transporter, the still unknown H_2O_2 generating system, thyroperoxidase, and thyroglobulin, is stimulated *in vivo* and *in vitro* by TSH. In some species, such as the rat, thyroglobulin gene expression is at its maximum at normal TSH levels and can be reduced only by thyroid hormone treatment and the consequent decrease in TSH serum levels. While stimulation of thyroperoxidase gene expression is rapid and does not require prior protein neosynthesis, the stimulation of thyroglobulin gene expression requires prior new protein synthesis and is slower in cells in culture. All of these positive modulations are reproduced *in vitro* by agents mimicking cAMP (cAMP analogues), or enhancing cAMP accumulation (cholera toxin, forskolin). These conclusions apply to all species studied so far. Thus, with regard to specific gene expression, the thyroid is highly controlled by the cAMP cascade (22).

CONTROL OF THYROID CELL GROWTH AND DIFFERENTIATION

It has long been known that chronic stimulation of the thyroid, either by repeated administration of TSH, by TSH-secreting pituitary adenomas, or by TSH secreted in response to a blockade of thyroid hormone synthesis, led to growth of the thyroid and the generation of a goiter. When our work on thyroid cell growth began in the 1980s, the prevalent dogma in the cell-proliferation field was that cAMP was a mediator of differentiation and growth inhibition. This concept certainly applies to fibroblasts and other types of cells of mesenchymal origin, but its generalization was certainly excessive. Moreover, the first attempts to stimulate thyroid cell proliferation with TSH failed in human cells. Thus, when we found that production of ornithine decarboxylase (ODC) was induced in dog thyroid slices by TSH- and cAMP-enhancing agents and was decreased by acetylcholine (Ach), which activates the PIP_2–PLC cascade, our findings were neglected and the use of ODC as a growth marker was disparaged (23). Later, using the methodology created by Kerkof et al. (24) and Fayet and Hovsépian (25), and the culture medium used by Ambesi-Impiombato et al. (26) to cultivate their FRTL-5 cells, we succeeded in establishing growing primary cultures of dog thyroid cells. The proliferation of these cells was stimulated by TSH, cAMP analogues, and cAMP enhancers. However, we had considerable difficulty in publishing our results. A referee at one journal thought that our work was "not original because everybody knows that TSH stimulates the growth of the thyroid," while another referee dismissed it because "it is well known that cAMP inhibits growth" (27). The difficulty was increased when many publications claimed that the TSH proliferative effect on FRTL-5 cells was not mediated by cAMP. It therefore took quite a few years to establish the concept that in some thyroid cells (dog, human, FRTL-5 cells), TSH acting through cAMP enhances both thyroid cell proliferation and the expression of differentiation (2).

We now know that the thyroid cell responds to three families of mitogenic cascades: the TSH–cAMP cascade, the growth-factor–protein tyrosine kinase (PTK) family of cascades, and the phorbol ester–protein kinase C (PKC) cascade (Fig. 1). Each of these cascades involves successive steps of intracellular signal generation, protein phosphorylation, gene induction, and protein synthesis, with a progressive overlap as one progresses downstream. The main distinction between these cascades is the end effect: proliferation and differentiation for the cAMP cascade, proliferation and loss of differentiation for the other two cascades. Differentiation is meant here as gene expression of the thyroid-specific proteins thyroglobulin (TG), thyroperoxidase (TPO), and iodide transporter, and also of the signal-transduction proteins including the TSH receptor, thyroid transcription factor-1 (TTF1), and cell–cell interaction proteins such as E-cadherin. The effects of each cascade on mitogenesis and differentiation can be dissociated in confluent cultures in which proliferative effects are repressed. The effects on differentiation are also fully reversible (2).

The first family of mitogenic cascades to consider in thyroid cell growth and differentiation is the growth-factor–PTK family, which is common to most cells. A first, simplistic schema of these families is now roughly defined, with a first cascade of protein tyrosine phosphorylations triggered by receptor dimerization, followed by the activation of Ras guanosine trisphosphate (GTP)-binding protein, a serine kinase cascade leading to the activation of the mitogen-activated protein (MAP) kinases, activation by phosphorylation of transcription factors, immediate early gene induction of the classical protooncogenes c-*jun*, c-*myc*, c-*fos*, and others, a cascade of induction of cyclins and cyclin-dependent kinases (CDKs), and the phosphorylation and inactivation of Rb and Rb-like proteins, releasing the E2F transcription factors, which then induce the repression of proteins necessary for deoxyribonucleic acid (DNA) synthesis.

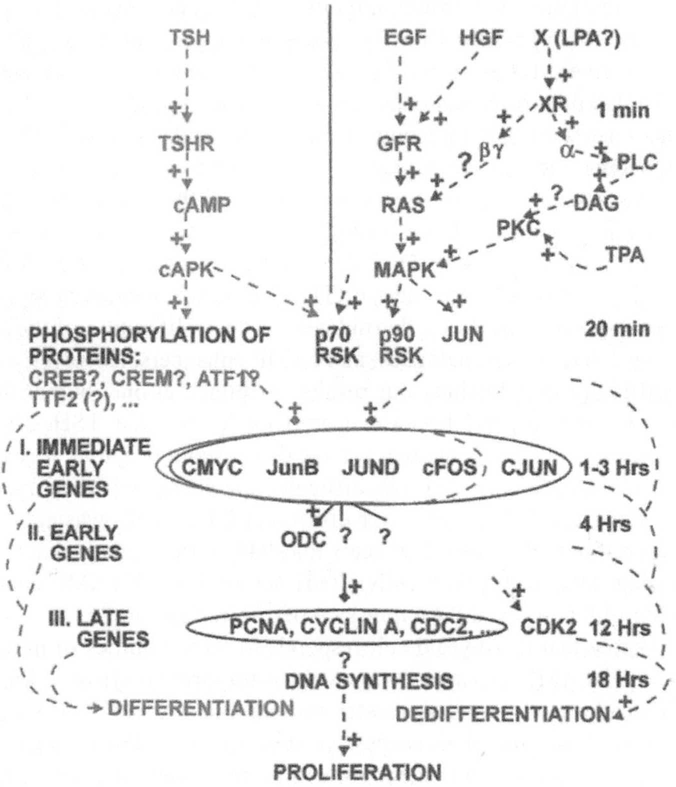

FIG. 1. Mitogenic pathways in the thyroid cell.

Constitutive activation of any of the positive factors in the cascade will lead to uncontrolled cell proliferation. Examples of such factors in thyroid tumors include Ret, which is a protein tyrosine kinase not normally present in thyroid cells, and activated Ras. Inactivation of a negative factor in the cascade (i.e., of an antioncogene) will give the same result. Known examples in thyroid tumors are Rb and p53, which induces CIP21, an inhibitor of CDKs.

Members of the G protein-coupled receptor family connect themselves on the growth factor cascade. Receptors activating Gi release βγ subunits, which through an unknown mechanism stimulate Ras. Additionally, receptors activating PLC release diacylglycerol (DAG); this activates PKC, which then stimulates Raf immediately downstream of Ras.

The TSH–cAMP cascade, which is the cascade that has been most extensively studied by our group, involves largely distinct steps until cyclin and CDK induction and activation. It involves the TSH receptor, Gs, adenylate cyclase, cAMP, cAMP-dependent protein kinases (PKA), and still unknown phosphorylated targets (perhaps transcription factors like cAMP-reponse element binding proteins [CREM], but not TTF- or the paired-domain transcription factor Pax8). This leads to some proto-oncogene expression and, through still unknown steps, to the common cyclin–CDK complexes (28).

Clearly, this outline is oversimplistic. For instance, several growth factors activate similar but not identical growth-factor cascades leading to very different results: growth and complete loss of differentiation for epidermal growth factor (EGF), growth and partial loss of differentiation for hepatocyte growth factor (HGF), and some differentiation but no mitogenesis for IGF-1 (2).

In our work, the mitogenic action of the cascades was demonstrated by cell-growth curves (DNA content) or by the autoradiographic counting of cells that incorporated tritiated thymidine. The mitogenic effect was confirmed in each biochemical experiment. In general, in dog thyroid, HGF is the strongest mitogen, followed by TSH and forskolin, and then by EGF and phorbol esters. Combinations of EGF, serum, and TSH are strongly synergistic, bringing almost all of the cells into DNA synthesis within 30 hr.

Expression of differentiation is evaluated by Northern blotting, *in situ* hybridization of thyroglobulin or thyroperoxidase messenger ribonucleic acid (mRNA), or iodide transport. In cultured cells, specific gene expression is greatly enhanced by TSH and greatly depressed by EGF and tetradecanoyl phorbol acetate (TPA), thus demonstrating the effects of these respective cascades on differentiation expression. It should be emphasized that the mitogenic and differentiating effects of TSH occur in the same cells. After pretreatment with EGF, the same cells that enter into DNA synthesis also begin to express thyroglobulin mRNA in the presence of TSH.

All of the effects of TSH on dog thyroid cells are reproduced by cAMP analogues, forskolin, and cholera toxin, thus demonstrating that they result from activation of the cAMP cascade (2).

Since the TSH receptor couples to Gi in cell membranes and, to some extent, in intact cells, stimulation of growth by TSH could be due to activation of Gi, release of $\beta\gamma$ subunits, and subsequent activation of Ras and its cascade. However, pretreatment of dog thyroid cells with pertussis toxin, which completely inhibits the pure Gi effect of norepinephrine and further increases TSH-stimulated cAMP accumulation, fails to inhibit the TSH-induced proliferative effect. Moreover TSH does not activate MAP kinase in these cells (29).

Thus, *in vitro*, the TSH–cAMP cascade stimulates both thyroid cell function and proliferation, which led us to propose in 1989 that "it is quite possible that overactivity of the cAMP system will be found responsible for hyperfunctioning benign tumors or adenomas as opposed to dedifferentiated malignant tumors where the classical oncogenes are involved" (30). This was a daring extrapolation: from results with cells in culture to effects in cells *in vivo*; and from results with experiments done on a time scale of a few days to the behavior of cells over a span of at least 34 divisions and 20 yr . . .

To test our hypothesis *in vivo*, we relied on the adenosine A2a receptor, which we had cloned (17). Because cells release adenosine, this receptor is chronically stimulated in most systems. Preliminary experiments showed that the microinjection of adenosine A2a receptor mRNA into dog thyrocytes induced DNA synthesis (31). Expression of the receptor in the thyroid *in vivo* could therefore tell us about the results of a constitutive activation of the cAMP cascade *in vivo*. We therefore expressed the A2a receptor specifically in the thyroid glands of transgenic mice. As expected, mice expressing the adenosine A2a receptor specifically in the thyroid are hyperthyroid: thyroxine (T4) and triiodothyronine (T3) serum concentrations and iodide concentration in the thyroid are much higher than in control mice (31). Transgenic mice develop a goiter: the thyrocytes have a high labeling index, whereas it is negligible in control mice. Qualitatively similar results have been obtained in mice expressing constitutively activated Gαs specifically in the thyroid (32). Such mutations downstream from the receptor in the cascade further narrow the specificity in the cAMP pathway. Thus, constitutive activation of the cAMP cascade leads *in vivo* to a phenotype of hyperfunctioning adenoma involving the whole thyroid. There is considerable evidence that the cAMP pathway may have similar roles in other cell types (33).

THE cAMP PATHWAY IN THYROID TUMORS AND DISEASE

What about human hyperfunctioning thyroid adenomas? When a seven-transmembrane-domain-type receptor binds its agonist, it opens up, allowing activation of the G protein downstream. Lefkowitz et al. (34) had shown by directed mutagenesis that specific mutations in the adrenergic α1b or β2 receptors have the same result as agonist binding. Could a similar phenomenon occur in human hyperfunctioning adenomas? Investigating the same region (the third intracellular loop that Lefkowitz et al. had pinpointed) in the TSH receptor in thyroid adenomas, we found a mutation at the corresponding position (35).

Sequencing gel studies demonstrate such a mutation in toxic adenoma. The two nucleotide bands at the same position are similar in intensity. This suggests that: (i) only one allele is mutated, and that the mutation must therefore be dominant; and (ii) the two alleles are equally present, which means that all of the cells of the adenoma express the mutation (i.e., that all of these cells derive from a single cell in which this mutation took place, or that the tumor has a monoclonal origin).

Can the mutation explain the tumor? First it confers on the receptor a constitutive activation: cAMP accumulation is much greater in COS-7 cells expressing the mutated TSH receptor than in those that do not express it. Most of the mutations do not affect the PIP_2–PLC cascade (36). Moreover, when plasmids expressing the mutated receptor are microinjected into dog thyroid cells, many of these cells begin incorporating tritiated thymidine in their nuclei.

Thirteen such gain-of-function mutants have now been described, and such mutations explain nine cases in our series of 11 hyperfunctioning adenomas. This illustrates the interest in identifying mutations conferring a selective advantage on affected cells in pathologic tissues rather than in systematic mutagenesis. Thus, TSH-receptor-activating mutations account for the majority of hyperfunctioning thyroid adenomas. Some other cases of such adenomas are probably caused by mutations conferring constitutive activation to Gαs, as demonstrated by other workers

(37). Such mutations also explain congenital hyperthyroidism (Leclere's disease) (38,39). In the hereditary form of this disease, the defect is inherited as an autosomal dominant character. All patients presenting the mutation sooner or later exhibit the phenotype of goiter and hyperthyroidism. Thus, as predicted from the *in vitro* data, constitutive activation of one element of the cAMP cascade leads to thyroid cell growth and hyperfunction. It will now be interesting to seek mutations of adenylate cyclase in otherwise unexplained hyperfunctioning adenomas. The pathologic role of the TSH–cAMP cascade is also demonstrated in nongenetic diseases. TSH-secreting pituitary adenomas cause hyperthyroidism and thyroid enlargement. Moreover, Graves disease, also characterized by hyperthyroidism and goiter, is caused by activation of the TSH receptor by TSAbs (10).

Do TSH receptor mutations have a role in cancer? They may. We have crossed mice expressing the adenosine A2 receptor in their thyroid cells with mice expressing the E7 oncogene in this tissue. As mentioned earlier, the first mice exhibit a goiter and thyroid hyperfunction, but no carcinoma. The second mice, in which E7 relieves the inhibitory effect of Rb downstream in the different mitogenic cascades, develop a euthyroid colloid goiter and later a localized tumor (40,41). In crosses between the two types of mice, tumorigenesis and metastases are observed (Fig. 2).

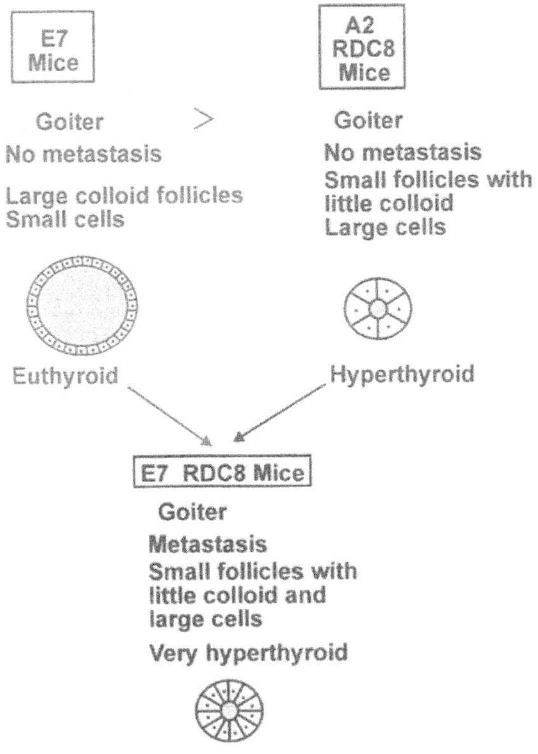

FIG. 2. Complementation of thyroid partial oncogenes (for the adenosine A2a receptor and HPV16–E7) on thyroid tumorigenesis.

Thus, two incomplete oncogenes complement each other and lead to invasive papillary carcinoma with lung metastases.

Do such mutations exist in human thyroid cancers? Studying exon 10 of the TSH receptor through the sequencing of tumor DNA of 15 thyroid cancers from Chernobyl, we have not so far found such a mutation. Complete sequencing is in progress. Exon 10 covers the entire seven-transmembrane domain of the receptor. Positive results have been obtained in thyroid cancers selected for increased adenylate cyclase activity in homogenates (Suarez, *unpublished results*).

Studies of mutated TSH receptors led to the investigation of the possible effects of normal receptors. Wild-type dog or human TSH receptors, when expressed transiently in COS-7 or permanently in CHO cells, increase cAMP accumulation. Thus, the normal receptor has a basal constitutive activity on Gs and adenylate cyclase (20). This is not true for all such receptors: the α-melanocyte-stimulating hormone (α-MSH) receptor has no such activity. Nevertheless, for receptors having such activity this raises interesting physiopathologic and therapeutic possibilities. Simple overexpression of the receptor could lead to the same disease as its constitutive activation. The thyroid could be inhibited by negative agonists of the TSH receptor.

THE THYROID-STIMULATING HORMONE–cAMP CASCADE IN HUMAN DISEASE

As seen from the findings described earlier, the cAMP cascade is implicated in normal mitogenesis and in tumorigenesis in human thyroid cells. Clearly, any activation at any step of the TSH–cAMP cascade, whether by a mutational event or through a physiologic or pathologic stimulation, will have the same consequences of thyroid functional stimulation and goiter (Fig. 3). Hypophyseal thyrotroph adenomas secreting TSH, as well as stimulating antibodies directed against the TSH receptor produced in Graves disease, are good examples. Activating mutations of the TSH receptor and Gαs have been described, and together should account for most hyperfunctioning adenomas (42). Similar mutations of adenylate cyclase or of cAMP-dependent kinases should be sought. Such mutations are dominant gain-of-function mutations. When hereditary they are autosomal dominant.

Inactivation of a negative controlling element of the cascade (e.g., the regulatory subunit of cAMP-dependent kinase, proteins of the iodide inhibitory pathway) would be a loss-of-function recessive mutation. To cause a hyperthyroid phenotype, double mutations would be necessary.

On the other hand, any inactivation of a positive element of the cascade, whether genetic or otherwise, should lead to hypofunction and hypotrophy (Fig. 3), or at least, if compensated by pituitary thyroid hormone feedback, to high TSH levels. Two complementary hereditary inactivation defects of the TSH receptor have been described, both indeed leading to euthyroid high TSH levels (43). Blockade of the TSH receptor by autoimmune antibodies leads to hypothyroidism and atrophy. Thus, the majority of thyroid diseases are in fact signal-transduction diseases.

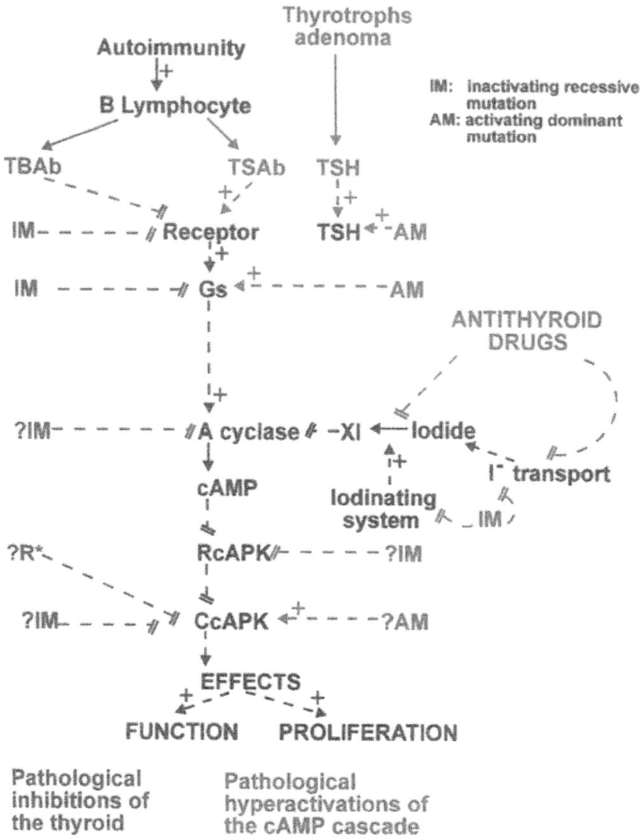

FIG. 3. The TSH–cAMP cascade and human thyroid disease. (———>) Positive control; (——| |) negative control.

THE MITOGENIC cAMP PATHWAY: PROTEINS AND GENES INVOLVED

The TSH–cAMP mitogenic pathway involves the TSH receptor, Gs, adenylate cyclase, and cAMP. Stimulation of the cascade at the level of Gs by cholera toxin or mutations, at the level of adenylate cyclase by forskolin, or by cAMP analogues has the same stimulatory effect. Evidence for the involvement of cAMP-dependent kinases in the pathway is the correspondence in the specificity of pairs of cAMP analogues in specifically stimulating PKA I and (to a lesser extent) PKA II, and in the mitogenic actions of these analogues (44). Also, microinjection of Walsh protein-kinase-inhibitor peptide decreases the TSH-induced mitogenic stimulation. However, we do not know about the elements involved downstream from protein kinase A (PKA). Certainly, dominant negative CREB-expressing plasmids inhibit cAMP action, but they also inhibit HGF-induced mitogenesis, and overexpression of CREB has the same effect (Dremier, *unpublished results*). Excess of transcription factors probably squelches the activity of other factors. Moreover, CREB-minus mice exhibit no obvious thyroid phenotype. If we do not know which protein phosphorylation may be involved in the TSH–cAMP mitogenic effect, we do know that this pathway does not induce phosphorylations identical to those resulting from activation of the growth-factor or tumor-promoter PKC cascades (45). In particular, the TSH–cAMP cascade does not lead either to p42 or p44 MAP kinases, or to p90 S6 kinase phosphorylation and activation (29). Similarly, immediate early gene inductions are different in the cAMP cascade from those in the EGF- and phorbol ester-activated cascades. If *jun* D and *jun* B are induced in both categories of cascades, c-*fos* is much less induced, c-*myc* induced only very transiently, and c-*jun* not at all in the cAMP cascade (2). Finally, the cAMP cascade is much more sensitive to the inhibitory effect of transforming growth factor-β (TGF-β) than are the EGF and phorbol ester pathways. The three pathways converge, at least qualitatively, at the cyclin/CDK induction and activation levels. They are therefore completely distinct at the level of receptor, signal generation, and protein phosphorylation, partially overlapping at the level of immediate early gene expression, and convergent at the level of the cyclin/CDK fundamental late G_1 mechanism. We have therefore defined, by exclusion, the steps in other cascades that are not involved in the cAMP mitogenic pathway. To identify the steps of this cascade, we are now looking at phosphorylated proteins and newly expressed or repressed genes (46).

FUNCTIONAL ROLES OF THE THYROID-STIMULATING HORMONE–cAMP PATHWAY AND INSULIN–INSULIN-LIKE GROWTH FACTOR-1 CASCADE

To cultivate their FRTL-5 cells, Ambesi-Impiombato et al. (26) had shown that insulin was required for cell multiplication. We later showed (47), and Williams (48) and collaborators confirmed, that insulin at high concentration or IGF-I was necessary for the mitogenic effect of TSH or forskolin, EGF, or phorbol esters on dog and human thyroid cells. In some cells (e.g., the FRTL-5 rat thyroid cell line), but not in dog or human cells, IGF-I was sufficient by itself to induce mitogenesis. All of these effects are believed to be mediated by the IGF-I receptor, since they are induced by low concentrations of IGF-I or high concentrations of insulin. Studies by our group and others of the effects of TSH or other growth factors on the steps of the mitogenic cascade have always been conducted in cells incubated with insulin or IGF-I (2).

In order to ascertain the nature of the permissive IGF-I action in thyroid cell proliferation, we first investigated the effects of TSH or forskolin on the one hand and of high insulin concentrations on the other hand on the growth and proliferation of dog thyrocytes in primary culture. As expected, TSH or forskolin separately had no effect on protein or DNA accumulation. Also as expected, insulin had no effect on DNA accumulation. However, much to our surprise, insulin induced a marked pro-

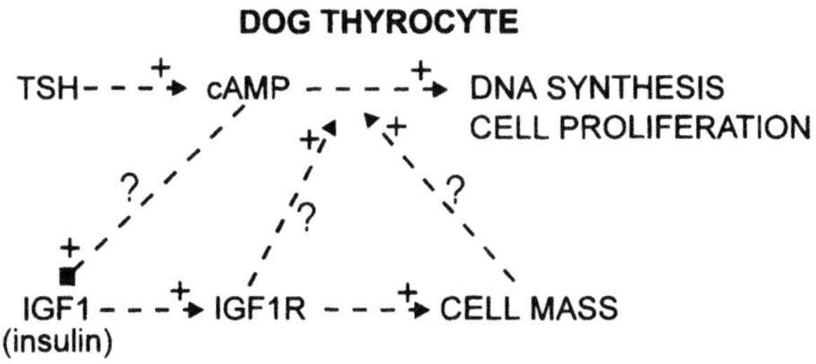

FIG. 4. Respective roles of TSH and the cAMP cascade on the one hand and insulin and/or IGF-1 on the other hand on the growth and division of the dog thyroid cell. (— — →±) Positive control; (— — ♦) induction; (?) Inferred but not demonstrated relations.

tein accumulation, with the protein content of our cells increasing within a few days by a factor of 3- to 4-fold. The protein/DNA ratio in cells of similar size should remain constant whether these cells proliferate or not. Strikingly, this ratio increased markedly in insulin-treated cells. If such cells were also treated with TSH or forskolin, both DNA and protein accumulated, but the protein/DNA ratio did not increase. These data suggest that insulin and IGF-I are responsible for the growth in size of the thyrocytes. TSH and cAMP can only stimulate DNA synthesis and cell division in such cells. Thus, we have dissociated in our model the system responsible for cell growth and the cascade responsible for the induction of proliferation in such cells (Fig. 4). The latter cascade cannot operate without the action of the first (Taton, *unpublished results*). In yeast, mutations that allow cell division regardless of cell size lead to death. Most work on cell-proliferation cascades uses cell proliferation or DNA synthesis as an endpoint, and therefore does not distinguish between events leading to cell growth and events which in the same cells induce DNA synthesis and cell division. Our finding should therefore allow us to delineate both cascades.

CONCLUSION

We have demonstrated the mitogenic role of the TSH receptor–adenylate cyclase cascade *in vitro* in dog and human thyroid cells, *in vivo* in transgenic mice, and in human disease in autonomous adenomas and congenital hyperthyroidism. We have now outlined a new model of proliferative control involving the permissive action of IGF-1, which stimulates growth and the action of the cAMP cascade, with the latter then stimulating the division of the hypertrophic cells. However many questions remain. What is the relation between structure and function in the normal and mutated TSH receptors? Which are the steps downstream of cAMP-dependent protein kinases in the mitogenic and differentiating cascade? What limits the growth of autonomous adenomas? By which mechanism does iodide affect proliferation? These and other questions are currently being addressed.

ACKNOWLEDGMENTS

This research was supported by the Belgian Programme on Interuniversity Poles of Attraction (PAI); Fonds National de la Recherche Scientifique (FNRS); Télévie (FNRS); European Union (EU) Biomed Cancer Programme and Radioprotection Programme; Association contre le Cancer; Association Sportive contre le Cancer; Fonds pour la Formation à la Recherche dans l'Industrie et l'Agriculture (FRIA); Loterie Nationale; and Academic Programs and Medicine Overseas (APMO).

REFERENCES

1. Dumont JE, Vassart G. Thyroid regulation. In: Degroot JJ, ed. *Endocrinology*. Philadelphia: W.B. Saunders, 1995:543–59.
2. Dumont JE, Lamy F, Roger PP, Maenhaut C. Physiological and pathological regulation of thyroid cell proliferation and differentiation by thyrotropin and other factors. *Physiol Rev* 1992;72:667–97.
3. Klainer LM, Chi YM, Freidberg SC, Rall TW, Sutherland EW. The effect of neurohormones on the formation of adenosine 3',5' phosphate by preparations from brain and other tissues. *J Biol Chem* 1962;237:1239–43.
4. Gilman AG, Rall TW. Factors influencing adenosine 3',5' phosphate accumulation in bovine thyroid slices. *J Biol Chem* 1968;243:5867–71.
5. Gilman AG, Rall TW. The role of adenosine 3',5'-phosphate in mediating effects of thyroid stimulating hormone on carboxydrate metabolism of bovine thyroid slices. *J Biol Chem* 1968;243: 5872–81.
6. Dumont JE. The action of thyrotropin on thyroid metabolism. *Vitam Horm* 1971;29:287–412.
7. Dumont JE, Willems C, Van Sande J, Neve P. Regulation of the release of thyroid hormones. *Ann NY Acad Sci* 1971;185:291–316.
8. Rodesch F, Neve P, Willems C, Dumont JE. Stimulation of thyroid metabolism by thyrotropin, cyclic 3',5' AMP dibutyryl cyclic 3',5'-AMP and prostaglandin E1. *Eur J Biochem* 1969;8:26–32.
9. Knopp J, Stolc V, Tong W. Evidence for the induction of iodide transport in bovine thyroid cells treated with thyroid stimulating hormone or dibutyryl cyclic adenosine 3',5'-monophosphate. *J Biol Chem* 1970;245:4403–8.
10. Zakarija M, Jin S, McKenzie JM. Evidence supporting the identity in Graves' disease of thyroid-stimulating antibody and thyroid growth-promoting immunoglobulin G as assayed in FRTL5 cells. *J Clin Invest* 1988;81:879–84.
11. Dumont JE, Boeynaems JM, Decoster C, et al. Biochemical mechanisms in the control of thyroid function and growth. *Adv Cyclic Nucl Res* 1978;9:723–34.
12. Corvilain B, Laurent E, Lecomte M, Van Sande J, Dumont JE. Role of the cyclic adenosine 3',5'-monophosphate and the phosphatidylinositol-Ca^{2+} cascades in mediating the effects of thyrotropin and iodide on hormone synthesis and secretion in human thyroid slices. *J Clin Endocrinol Metab* 1994;79:152–9.
13. Van Sande J, Grenier G, Willems C, Dumont JE. Inhibition by iodide of the activation of the thyroid cyclic 3′,5′-AMP system. *Endocrinology* 1975;96:781–6.
14. Panneels V, Van Sande J, Van den Bergen H, et al. Inhibition of human thyroid adenylyl cyclase by 2-iodoaldehydes. *Mol Cell Endocrinol* 1994;106:41–50.
15. Parmentier M, Libert F, Maenhaut C, et al. Molecular cloning of the thyrotropin (TSH) receptor. *Science* 1989;296:1620–2.
16. Van Sande J, Raspe E, Perret J, et al. Thyrotropin activates both the cyclic AMP and the PIP2 cascades in CHO cells expressing the human cDNA of TSH receptor. *Mol Cell Endocrinol* 1990; 74:R1–R6.
17. Libert F, Parmentier M, Lefort A, et al. Selective amplification and cloning of four new members of the G protein-coupled receptor family. *Science* 1989;244:569–72.
18. Vassart G, Dumont JE. The thyrotropin receptor and the regulation of thyrocyte function and growth. *Endocrine Rev* 1992;13:596–611.

19. Maenhaut C, Brabant G, Vassart G, Dumont JE. *In vitro* and *in vivo* regulation of thyrotropin receptor mRNA levels in dog and human thyroid cells. *J Biol Chem* 1992;267:3000–7.
20. Van Sande J, Swillens S, Gerard C, et al. In Chinese hamster ovary K1 cells dog and human thyrotropin receptors activate both the cyclic AMP and the phosphatidylinositol 4,5-bisphosphate cascades in the presence of thyrotropin and the cyclic AMP cascade in its absence. *Eur J Biochem* 1995;229:338–43.
21. Van Sande J, Lejeune C, Ludgate M, et al. Thyroid stimulating immunoglobulins, like thyrotropin activate both the cyclic AMP and the PIP2 cascades in CHO cells expressing the TSH receptor. *Mol Cell Endocrinol* 1992;88:R1–R5.
22. Christophe D, Gerard C, Hansen C, et al. Control of thyroglobulin gene expression. *Hormones and Cell Regulation—Colloque INSERM* 1987;153:205–13.
23. Mockel J, Decaux G, Unger J, Dumont JE. *In vitro* regulation of ornithine decarboxylase in dog thyroid slices. *Eur J Biochem* 1980;104:297–304.
24. Kerkof PR, Long PJ, Chaikoff IL. *In vitro* effects of thyrotropin hormone. I. On the pattern of organization of monolayer cultures of isolated sheep thyroid gland. *Endocrinology* 1964;74:170–9.
25. Fayet G, Hovsépian S. Demonstration of growth in porcine thyroid cell culture. *Biochimie* 1979;61: 923–30.
26. Ambesi-Impiombato FS, Parks LAM, Coon HG. Culture of hormone-dependent functional epithelial cells from rat thyroids. *Proc Natl Acad Sci USA* 1980;77:3455–9.
27. Roger PP, Hotimsky A, Moreau C, Dumont JE. Stimulation by thyrotropin, cholera toxin and dibutyryl cyclic AMP of the multiplication of differentiated thyroid cells *in vitro*. *Mol Cell Endocrinol* 1982;26:165–72.
28. Baptist M, Dumont JE, Roger PP. Demonstration of cell cycle kinetics in thyroid primary culture by immunostaining of proliferating cell nuclear antigen: differences in cyclic AMP-dependent and -independent mitogenic stimulations. *J Cell Sci* 1993;105:69–80.
29. Lamy F, Wilkin F, Baptist M, Posada J, Roger PP, Dumont JE. Phosphorylation of mitogen-activated protein kinases is involved in the epidermal growth factor and phorbol ester, but not in the thyrotropin/cAMP, thyroid mitogenic pathways. *J Biol Chem* 1993;268:8398–401.
30. Dumont JE, Jauniaux JC, Roger PP. The cyclic AMP-mediated stimulation of cell proliferation. *Trends Biochem Sci* 1989;14:67–71.
31. Ledent C, Dumont JE, Vassart G, Parmentier M. Thyroid expression of an A2 adenosine receptor transgene induces thyroid hyperplasia and hyperthyroidism. *EMBO J* 1992;11:537–42.
32. Michiels FM, Caillou B, Talbot M, et al. Oncogenic potential of guanine nucleotide stimulatory factor alpha subunit in thyroid glands of transgenic mice. *Proc Natl Acad Sci USA* 1994;91:10488–92.
33. Roger PP, Reuse S, Maenhaut C, Dumont JE. Multiple facets of the modulation of growth by cAMP. *Vitam Horm* 1995;51:59–191.
34. Kjelsberg MA, Cotecchia S, Ostrowski J, Caron MG, Lefkowitz RJ. Constitutive activation of the alpha 1B-adrenergic receptor by all amino acid substitutions at a single site. Evidence for a region which constrains receptor activation. *J Biol Chem* 1992;267:1430–3.
35. Parma J, Duprez L, Van Sande J, et al. Somatic mutations in the thyrotropin receptor gene cause hyperfunctioning thyroid adenomas. *Nature* 1993;365:649–51.
36. Van Sande J, Parma J, Tonacchera M, Swillens S, Dumont JE, Vassart G. Somatic and germline mutations of the TSH receptor gene in thyroid diseases. *J Clin Endocrinol Metab* 1995;80:2577–85.
37. Landis CA, Masters SB, Spada A, Pace AM, Bourne HR, Vallar L. GTPase inhibiting mutations activate the delta chain of Gs and stimulates adenylate cyclase in human pituitary tumors. *Nature* 1989; 340:692–6.
38. Duprez L, Parma J, Van Sande J, et al. Germline mutations in the thyrotropin receptor gene cause non-autoimmune autosomal dominant hyperthyroidism. *Nature Genetics* 1994;7:396–401.
39. Kopp P, Van Sande J, Parma J, et al. Congenital hyperthyroidism caused by a mutation in the thyrotropin-receptor gene. *N Engl J Med* 1995;19:150–4.
40. Ledent C, Marcotte A, Dumont JE, Vassart G, Parmentier M. Differentiated carcinomas develop as a consequence of the thyroid specific expression of a thyroglobulin-human papillomavirus type 16 E7 transgene. *Oncogene* 1995;10:1789–97.
41. Ledent C, Parmentier M, Vassart G, Dumont JE. Models of thyroid goiter and tumors in transgenic mice. *Mol Cell Endocrinol* 1994;100:167–9.
42. Vassart G, Parma J, Van Sande J, Dumont JE. The thyrotropin receptor and the regulation of thyrocyte function and growth: update 1994. *Endocrine Rev* 1994;3:77–80.

43. Sunthornthepvarakul T, Gottschalk ME, Hatashi Y, Refetoff S. Resistance to thyrotropin caused by mutations in the thyrotropin-receptor gene. *N Engl J Med* 1995;332:155–60.
44. Van Sande J, Lefort A, Beebe S, et al. Pairs of cyclic AMP analogs, that are specifically synergistic for type I and type II cAMP-dependent protein kinases, mimic thyrotropin effects on the function, differentiation, expression, and mitogenesis of dog thyroid cells. *Eur J Biochem* 1989;183:699–708.
45. Contor L, Lamy F, Lecocq R, Roger PP, Dumont JE. Differential protein phosphorylation in induction of thyroid cell proliferation by thyrotropin, epidermal growth factor or phorbol ester. *Mol Cell Biol* 1988;8:2494–503.
46. Miot F, Wilkin F, Dremier S, et al. Cloning of cDNA specifically involved in the thyroid cAMP mitogenic pathway. *Horm Res* 1994;42:27–30.
47. Roger PP, Servais P, Dumont JE. Stimulation by thyrotropin and cyclic AMP of the proliferation of quiescent canine thyroid cells cultured in a defined medium containing insulin. *FEBS Lett* 1983; 157:323–9.
48. Wynford-Thomas D, Stringer BMJ, Williams ED. Proliferative response to cyclic AMP elevation of thyroid epithelium in suspension culture. *Mol Cell Endocrinol* 1987;51:163–6.

Signal Transduction in Health and Disease, Advances in Second Messenger and Phosphoprotein Research, Vol. 31, edited by J. Corbin and S. Francis. Lippincott–Raven Publishers, Philadelphia © 1997.

12

Ca^{2+}/Calmodulin-Dependent Myosin Light-Chain Kinases

James T. Stull, Kristine E. Kamm, Joanna K. Krueger, Pei-ju Lin, Katherine Luby-Phelps, and Gang Zhi

Department of Physiology, University of Texas Southwestern Medical Center at Dallas, Dallas, TX 75235-9040

INTRODUCTION

Ca^{2+}/calmodulin-dependent myosin light-chain kinases play a central role in the regulation of smooth-muscle contractility and in a variety of nonmuscle functions (Fig. 1). Increases in cytosolic Ca^{2+} concentration result in the formation of a Ca^{2+}/calmodulin complex that binds to and activates myosin light-chain kinase (1). Subsequent to this activation, a serine residue in the N-terminus of the regulatory light chain of myosin is phosphorylated. In skeletal muscle, this phosphorylation moves the myosin head away from the backbone of the thick filament, and increases the rate at which myosin cross-bridges move to the force-bearing state (2). This movement potentiates submaximal skeletal-muscle contractions initiated by Ca^{2+} binding to the thin-filament regulatory protein troponin. Phosphorylation of skeletal-muscle myosin does not lead to a marked increase in actin-activated myosin adenosine triphosphatase (ATPase) activity, presumably because of lack of MgATPase inhibition by the nonphosphorylated regulatory light chain (RLC) (i.e., skeletal-muscle myosin is always in an activatable state) (3). Conversely, for smooth-muscle and nonmuscle myosins, the nonphosphorylated regulatory light chain inhibits myosin ATPase activity even in the presence of actin. This inhibitory state is reversed by RLC phosphorylation, leading to a marked increase in actin-activated myosin ATPase activity, which is thought to play central roles in the initiation of smooth-muscle contraction (3,4), platelet contraction (5), endothelial-cell retraction (6), exocytosis (7), receptor capping (8), and other nonmuscle functions (Fig. 1).

MYOSIN LIGHT-CHAIN KINASE PHOSPHORYLATION MODULATES ACTIVATION OF THE ENZYME BY CA^{2+}/CALMODULIN *IN VIVO*

Investigations in a number of laboratories have provided experimental evidence to satisfy the criteria of Krebs and Beavo (1979) for establishing the physiologic sig-

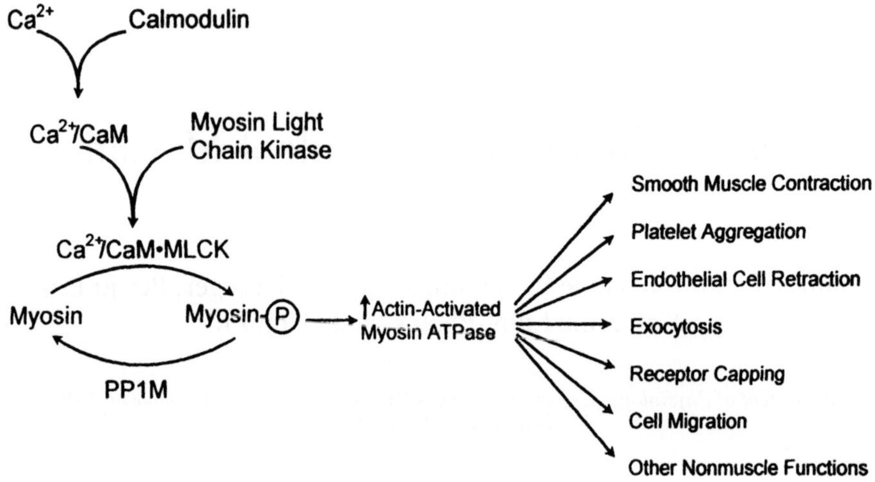

FIG. 1. A general scheme for the activation of myosin light-chain kinase and its role in various physiologic processes.

nificance of myosin light-chain kinase phosphorylation in smooth muscle *in vivo* (Table 1)(9). The primary conclusion from these studies is that Ca^{2+}/calmodulin-dependent protein kinase II (CaMK II) phosphorylates myosin light-chain kinase at a specific serine residue in the C-terminus of the calmodulin-binding domain (Site A) in smooth-muscle tissues and cells in culture. Since this phosphorylation results in desensitization of myosin light-chain kinase to activation by Ca^{2+}/calmodulin (increases K_{CaM}), it is physiologically predicted that regulatory light-chain phosphorylation will be desensitized with respect to cytosolic Ca^{2+} concentrations. This prediction has been confirmed experimentally (10). Although other sites in myosin light-chain kinase are phosphorylated in response to the activation of different signaling pathways, recent data show that the introduction, via site-directed mutagenesis, of a negative charge in place of the serine at Site A is sufficient to desensitize myosin light-chain kinase to activation by Ca^{2+}/calmodulin (11). Introduction of a negative charge at another site phosphorylated *in vivo* does not result in desensitiza-

TABLE 1. *Criteria for establishing the significance of myosin light-chain kinase phosphorylation* in vivo

1. Purified myosin light-chain kinase is phosphorylated by Ca^{2+}/calmodulin-dependent protein kinase II and dephosphorylated by protein phosphatases PP1 and PP2A (12,13,30,31).
2. The affinity of myosin light-chain kinase for Ca^{2+}/calmodulin decreases with phosphorylation of Site A in the enzyme (32–35).
3. Myosin light-chain kinase is phosphorylated and dephosphorylated in intact cells, with accompanying changes in sensitivity to Ca^{2+} (10,33,35–38).
4. The cytoplasmic Ca^{2+} concentration correlates with the extent of myosin light-chain kinase phosphorylation at Site A (35–39).

(From Krebs and Beavo, ref. [9], with permission.)

tion. Thus, Site A, through phosphorylation, appears to be the primary site for modulating activation of myosin light-chain kinase by Ca^{2+}/calmodulin *in vivo*.

How is the activity of myosin light-chain kinase regulated physiologically? Based on evidence obtained *in vivo*, a crucial point for understanding the complexity of this regulatory system involves defining the differences in Ca^{2+} sensitivities of the phosphorylation reactions of: (i) myosin light-chain kinase by Ca^{2+}/calmodulin-dependent protein kinase II; and (ii) myosin regulatory light-chain by Ca^{2+}/calmodulin-dependent myosin light-chain kinase. The Ca^{2+} concentration required for half-maximal phosphorylation of myosin light-chain kinase is approximately 500 nM (10). When CaMK II activity is inhibited *in vivo* and myosin light-chain kinase is not phosphorylated, the Ca^{2+} concentration required for half-maximal phosphorylation of the myosin regulatory light chain is 170 nM. However, when phosphorylation of the light chain is allowed to proceed with increases in cytosolic Ca^{2+} concentrations, this value increases to 300 nM. Thus, there is a marked difference in the Ca^{2+} sensitivities of myosin light chain and myosin light-chain kinase phosphorylation *in vivo*. This desensitization mechanism comes into play under conditions in which cytosolic Ca^{2+} concentrations rise to high levels.

THERE IS LITTLE FREE CALMODULIN IN SMOOTH-MUSCLE CELLS

Initially there were two aspects of the biochemical properties of the myosin light-chain kinase/myosin regulatory light-chain system that made it difficult to understand mechanistically how myosin light-chain kinase was phosphorylated in smooth-muscle cells. When Ca^{2+}/calmodulin is bound to purified myosin light-chain kinase, phosphorylation by Ca^{2+}/calmodulin-dependent protein kinase II at the serine in Site A is inhibited (12,13). Biochemically, the Ca^{2+}/calmodulin-independent form of this protein kinase phosphorylates myosin light-chain kinase only in the absence of Ca^{2+}/calmodulin. Consequently, it is predicted that when cytosolic Ca^{2+} concentrations increase in a cell and Ca^{2+}/calmodulin binds to myosin light-chain kinase, its phosphorylation would be inhibited. Another theoretical problem is related to the high concentrations of myosin light-chain kinase and calmodulin in smooth-muscle cells relative to the value of K_{CaM} (concentration of Ca^{2+}/calmodulin required for half-maximal activation *in vitro* = 1 nM). The total calmodulin and myosin light-chain kinase concentrations in smooth muscle are 40 and 4 μM, which are four and three orders of magnitude, respectively, greater than the value of K_{CaM} (10,14,15). Based on activation properties in solution in relation to the high concentrations of these two proteins, it is predicted that an increase in the value of K_{CaM} from 1 nM to 10 nM by phosphorylation at Site A would have minimal effects on the Ca^{2+} dependence of regulatory light-chain phosphorylation (16). However, in both situations the assumption is made that the free calmodulin concentration available for activation is in excess of the myosin light-chain kinase concentration and similar to the total calmodulin concentration present in cells. If Ca^{2+}/calmodulin were limiting, not all of the myosin light-chain kinase would be in the bound state when cytosolic Ca^{2+}

concentrations increase inside a cell, and a fraction would therefore be available for phosphorylation by CaMK II.

Recent experimental observations in our laboratory demonstrate that calmodulin may in fact be limiting for the activation of myosin light-chain kinase *in vivo*. Smooth-muscle tissues were permeabilized by soaking in 4 mM ethylene glycol-bis-(β-aminoethyl ether)-N,N,N′,N′-tetraacetic acid (EGTA) and 1% Triton-X 100, and were stored in 50% glycerol at −20°C for 2 wk. After removal of glycerol, this skinned tissue preparation contracts in the presence of Ca^{2+} and MgATP, and the addition of 2 μM calmodulin is sufficient to potentiate the force response. Measurements of the tissue content of myosin light-chain kinase showed no loss in the permeable fiber (3.2 ± 0.4 μM) as compared with the intact tissue (3.4 ± 0.2 μM). Recent measurements of the mobility of myosin light-chain kinase in smooth-muscle cells, through fluorescence recovery after photobleaching (FRAP), are consistent with the high-affinity binding of the enzyme, presumably to the contractile protein system (Table 2). Surprisingly, the high calmodulin content in the intact fiber (39 ± 2 μM) decreased by only 50% with the extensive permeabilization procedure (16 ± 2 μM). Most of this calmodulin fraction is unavailable for the activation of myosin light-chain kinase, as evidenced by the potentiation produced with only 2 μM added calmodulin. Thus, it appears that a major portion of the calmodulin endogenous to smooth-muscle tissue is bound, even in the absence of Ca^{2+}. This observation is consistent with a previous report (17) and has been recently confirmed (18).

Two additional experimental approaches were used to investigate calmodulin availability in smooth-muscle cells in culture (10). Coinjection of rhodamine-labeled calmodulin with fluorescein-labeled dextrans of various sizes was used to estimate the size at which a freely diffusible molecule was lost after a short incubation in the presence of the detergent β-escin and 4 mM EGTA. Fifteen minutes of incubation showed that both calmodulin (16.5 kDa) and a 40-kDa dextran were retained; however, 3-kDA and 10-kDa dextrans were not. With a longer incubation period (45 min), the 40-kDa dextran fluorescence disappeared without a significant change in calmodulin concentration. Thus, most of the calmodulin in smooth-muscle cells appears bound to cellular structures in the absence of Ca^{2+}.

TABLE 2. *Diffusion coefficients for proteins in bovine tracheal smooth-muscle cells in primary culture.*

	Size (kDa)	Mobility (%)	Diffusion Coefficient (cm^2/sec)
Dextran	20	100	1.6×10^{-8}
	40	100	7.0×10^{-9}
Calmodulin	16.5	5	2.2×10^{-9}
MLCK	128	?	Less than 10^{-10}

Data are from Tansey *et al.*, ref. 10, with permission. Except for smooth-muscle myosin light-chain kinase. Proteins were labeled with fluorescent probes and microinjected into cells (10). The fastest recovering component of the recovery curve after photobleaching was used to estimate the diffusion coefficients.

FRAP was used to measure the mobility of the labeled calmodulin following its microinjection into smooth-muscle cells in culture. The recovery curves were complex and did not fit a single component, indicating that there were multiple compartments of calmodulin in cells. By measuring the rate of calmodulin recovery during the first few seconds, an apparent mean diffusion coefficient was obtained for the most rapidly recovering component (Table 2). A striking feature of these results is that even for the fastest mobile fraction of calmodulin, its apparent diffusion coefficient is 7-fold lower than that of a similar sized dextran (Table 2). These and other recent experiments indicate that no more than 5% of the total endogenous calmodulin in resting smooth-muscle cells is unbound (10,18,19). When the Ca^{2+} concentration is increased to 450 nM and 3 μM, some of the bound calmodulin is released; however, at higher Ca^{2+} concentrations calmodulin becomes increasingly immobilized (19). These results emphasize the complexity of understanding calmodulin distribution in cellular subcompartments. However, they provide experimental support for the hypothesis that calmodulin may be limiting for myosin light-chain kinase activation in smooth-muscle cells.

MYOSIN LIGHT-CHAIN KINASE ACTIVITY IS REGULATED BY AN INTRASTERIC MECHANISM

The characterization of myosin light-chain kinases in vertebrate tissues has led to the identification of two distinct enzymes, found in skeletal and smooth muscles, respectively (1,20). There are important molecular and biochemical differences between these two enzymes; however, the organization of their functional domains is similar. They contain a catalytic core in the central portion that is highly homologous to that of other protein kinases. The crystal structures of several protein kinases show that the catalytic cores of these enzymes contain two lobes, with the smaller lobe binding MgATP, leaving the γ-phosphate positioned for transfer (21). Phosphorylation occurs at a cleft between the lobes, and the larger lobe provides binding sites for protein substrates and catalysis. The C-terminus of the catalytic core of myosin light-chain kinase is connected to a shorter, regulatory domain that contains a linker region followed by a calmodulin-binding domain (22). Evidence has been presented that specific residues in the linker region (four basic and two hydrophobic residues) may bind to the catalytic core, thereby effecting autoinhibition (23–25). The calmodulin-binding domain, by homology to twitchin kinase, may also bind to the catalytic core (26). However, there is no direct experimental evidence that within the intact enzyme the calmodulin-binding domain makes contact with residues in the catalytic core.

An analysis of mutated myosin light-chain kinases has focused on the possible role of acidic residues in the catalytic core binding to basic residues in the autoinhibitory domain. Mutations in eight acidic residues predicted to be on the surface of the catalytic core had no effect on the Ca^{2+}/calmodulin activation or catalytic properties of the enzyme (Table 3). Mutations in seven residues showed effects only on the Ca^{2+}/calmodulin activation properties, without any changes in the enzyme's K_m

TABLE 3. *Putative binding properties of acidic residues in the catalytic cores of smooth-muscle myosin light-chain kinase*

Regulatory Domain	Light Chain	Both	Neither
E780		E777	E700
E785		E821	D784
D786			D896
E788			D898
E793			E900
E858			D911
D923			D914
			E915

Site-directed mutations were made by standard procedures, and enzyme activities were measured in COS cell lysates (24,40). K_m and K_{CaM} values were determined to identify residues that may bind to the light chain, regulatory domain, or both.

or V_{max} values. These specific residues are distributed in a linear fashion across the surface of the large lobe, to the active site between the clefts of the two lobes of the catalytic core, and are predicted to bind to basic residues in the regulatory Domain 6 of the enzyme (24). Mutagenesis has also indicated that two acidic residues (E777 and E821) may be important for binding of both the autoinhibitory domain and the smooth-muscle regulatory light chain of the enzyme, respectively. Based on the positions of these residues, it is predicted that they will bind to the arginine in the regulatory light chain at the P-3 position. Thus, exposure of the E777 and E821 residues upon calmodulin binding may be sufficient for light-chain binding and phosphorylation.

We were surprised to find that mutations of other acidic residues did not produce significant increases in the K_m value for light-chain substrate. Synthetic peptides have been used to define the consensus phosphorylation sequences of both smooth- and skeletal-muscle myosin light-chain kinases, and have revealed substrate determinants that are distinct for the two enzymes (27). With the chicken skeletal-muscle regulatory light chain, two groups of basic residues, at P-6 to P-8 and P-10 to P-11, are important for low K_m values. However, for recognition by the smooth-muscle myosin light-chain kinase, arginine at P-3 was an important residue, in addition to the three basic residues at the P-6 to P-8 positions. When this arginine was changed to an alanine in the peptide substrate, the K_m value increased from 12 μM to 200 μM and the V_{max} value decreased 5-fold. However, when this change was made in the intact regulatory light chain, the K_m value increased to only 71 μM and the V_{max} value decreased 3-fold. A charge-reversal mutation (R16E) resulted in greater changes (Fig. 2). The K_m value increased to 99 μM with about a 28-fold decrease in the V_{max} value. This mutation resulted in a 340-fold change in the V_{max}/K_m ratio. In the skeletal-muscle light chain there is a glutamate at the P-3 position, and predictably, the skeletal-muscle light chain is a poor substrate for the smooth-muscle kinase (Fig. 2). When this glutamate residue at P-3 is mutated to an arginine in the skeletal-muscle light chain, the mutant protein is still a poor substrate for the smooth-muscle kinase. These results, and the fact that the synthetic peptide substrate has a V_{max} value that is

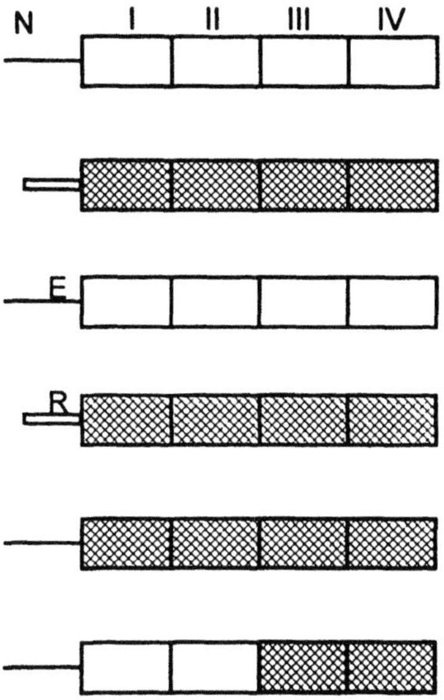

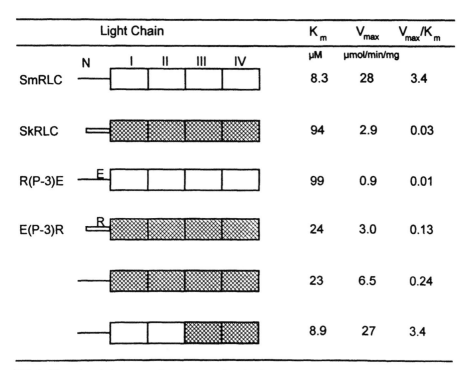

Light Chain	K_m (μM)	V_{max} (μmol/min/mg)	V_{max}/K_m
SmRLC	8.3	28	3.4
SkRLC	94	2.9	0.03
R(P-3)E	99	0.9	0.01
E(P-3)R	24	3.0	0.13
	23	6.5	0.24
	8.9	27	3.4

FIG. 2. Phosphorylation properties of mutated and chimeric myosin regulatory light chains. Mutated and chimeric proteins were expressed in *Escherichia coli*, purified, and used as substrates for smooth-muscle myosin light-chain kinase (29). SmRLC and SkRLC refer to smooth- and skeletal-muscle regulatory light chains, respectively, with specific designations for the N-terminus and the four subdomains. R(P-3)E and E(P-3)R indicate site-directed mutations at the P-3 position, where R and E were respectively mutated to E and R.

only 10% of the value for the regulatory light chain, suggest that higher-ordered structures may play a role in determining the substrate specificity of myosin light-chain kinase. We therefore made chimeras between the smooth- and skeletal-muscle regulatory light chains to identify potential positions for these important structures (Fig. 2).

Myosin light chains are structurally related to Ca^{2+}-binding proteins. For the vertebrate smooth- and skeletal-muscle regulatory light chain, Ca^{2+}-specific binding properties have been lost. However, there are subdomains that are similar to the helix-loop-helix motif of the EF Ca^{2+}-binding hand. Light chains contain four such subdomains that follow the more extended N-terminus where the phosphorylatable serine is located (28). Our chimeras between smooth- and skeletal-muscle regulatory light chain were based on this subdomain organization (29). The addition of the smooth-muscle N-terminus to skeletal muscle Subdomains 1, 2 , 3, and 4 resulted in a chimeric light chain that had a V_{max}/K_m ratio that was 14-fold smaller than that of the smooth-muscle regulatory light chain (Fig. 2). Thus, not all of the determinants for substrate specificity of the enzyme are located in the N-terminus of the light chain, which contains the consensus phosphorylation sequence. When the N-terminus and Subdomains 1 and 2 of the smooth-muscle light chain were added to Subdomains 3 and 4 of the skeletal-muscle light chain, the catalytic properties for phosphorylation were identical to those of the smooth-muscle regulatory light chain. Based on these results, we conclude that the arginine at P-3 in the smooth-muscle light chain is an important substrate determinant, in addition to some undefined structures in Subdomains 1 and 2. These results, combined with the previous identification of important hydrophobic residues at positions P+1 through P+3, show that the specificity of myosin light-chain kinase is not restricted to determinants located in a primary sequence immediately surrounding the phosphorylatable serine (29). Additionally, residues determined to be important in the synthetic peptide substrate (basic residues at P-6 to P-8) may not be important in the context of the protein substrate. Future experiments will focus on the identification of specific structures in Subdomains 1 and 2 that contribute to substrate specificity, and on determining the mechanism by which they function.

ACKNOWLEDGMENTS

The authors express appreciation to the National Institutes of Health for support through Grants HL06290 and HL26043, the National Science Foundation for Grant MCB-9304603, and to the American Heart Association for a Grant-in-Aid.

REFERENCES

1. Stull JT, Nunnally MH, Michnoff CH. Calmodulin-dependent protein kinases. In: Krebs EG, Boyer PD, eds. *The enzymes*. Orlando, FL: Academic Press; 1986:113–66.
2. Sweeney HL, Bowman BF, Stull JT. Myosin light chain phosphorylation in vertebrate striated muscle: regulation and function. *Am J Physiol* 1993;264:C1085–95.
3. Trybus KM. Role of myosin light chains. *J Muscle Res Cell Motil* 1994;15:587–94.
4. Kamm KE, Stull JT. Regulation of smooth muscle contractile elements by second messengers. *Annu Rev Physiol* 1989;51:299–313.
5. Kawamoto S, Bengur AR, Sellers JR, Adelstein RS. *In situ* phosphorylation of human platelet myosin heavy and light chains by protein kinase C. *J Biol Chem* 1989;264:2258–65.
6. Goeckeler ZM, Wysolmerski RB. Myosin light chain kinase-regulated endothelial cell contraction: the relationship between isometric tension, actin polymerization, and myosin phosphorylation. *J Cell Biol* 1995;130:613–27.
7. Ludowyke RI, Peleg I, Beaven MA, Adelstein RS. Antigen-induced secretion of histamine and the phosphorylation of myosin by protein kinase C in rat basophilic leukemia cells. *J Biol Chem* 1989; 264:12492–501.
8. Kerrick WGL, Bourguignon LY. Regulation of receptor capping in mouse lymphoma T cells by Ca^{2+}-activated myosin light chain kinase. *Proc Natl Acad Sci USA* 1984;81:165–9.
9. Krebs EG, Beavo JA. Phosphorylation-dephosphorylation of enzymes. *Annu Rev Biochem* 1979; 48:923–59.
10. Tansey MG, Luby-Phelps K, Kamm KE, Stull JT. Ca^{2+}-dependent phosphorylation of myosin light chain kinase decreases the Ca^{2+} sensitivity of light chain phosphorylation within smooth muscle cells. *J Biol Chem* 1994;269:9912–20.

11. Kamm KE, Grange RW. Ca^{2+} sensitivity of contraction. In: Barany M, ed. *Biochemistry of smooth muscle contraction*. Orlando, FL: Academic Press, 1996: 355–65.
12. Hashimoto Y, Soderling TR. Phosphorylation of smooth muscle myosin light chain kinase by Ca^{2+}/calmodulin-dependent protein kinase-II. Comparative study of the phosphorylation sites. *Arch Biochem Biophys* 1990;278:41–5.
13. Ikebe M, Reardon S. Phosphorylation of smooth myosin light chain kinase by smooth muscle Ca^{2+}/calmodulin-dependent multifunctional protein kinase. *J Biol Chem* 1990;265:8975–8.
14. Hartshorne DJ. Biochemistry of the contractile process in smooth muscle. In: Johnson LR, ed. *Physiology of the gastrointestinal tract*. New York: Raven Press; 1987:423–82.
15. Rüegg JC, Pfitzer G, Zimmer M, Hofmann F. The calmodulin fraction responsible for contraction in an intestinal smooth muscle. *FEBS Lett* 1984;170:383–6.
16. Kamm KE, Stull JT. The function of myosin and myosin light chain kinase phosphorylation in smooth muscle. *Annu Rev Pharmacol Toxicol* 1985;25:593–620.
17. Rüegg JC, Meisheri K, Pfitzer G, Zeugner C. Skinned coronary smooth muscle: calmodulin, calcium antagonists, and cAMP influence contractility. *Basic Res Cardiol* 1983;78:462–71.
18. Zimmerman UJ, Schlaepfer WW. Characterization of calcium-activated neutral protease (CANP)-associated protein kinase from bovine brain and its phosphorylation of neurofilaments. *Biochem Biophys Res Commun* 1995;129:804–11.
19. Luby-Phelps K, Hori M, Phelps JM, Won D. Ca^{2+}-regulated dynamic compartmentalization of calmodulin in living smooth muscle cells. *J Biol Chem* 1995;270:21532–8.
20. Stull JT, Krueger JK, Zhi G, and Gao Z-H. Molecular properties of myosin light chain kinases. In: *Calcium as Cell Signal*, K. Maruyama, Y. Nonomura, and K. Kohama (eds.), Igaku-Shoin Ltd., Tokyo, 1995:175–184.
21. Taylor SS, Radzio-Andzelm E. Three protein kinase structures define a common motif. *Structure* 1994;2:345–55.
22. Knighton DR, Pearson RB, Sowadski JM, et al. Structural basis of the intrasteric regulation of myosin light chain kinases. *Science* 1992;258:130–5.
23. Fitzsimons DP, Herring BP, Stull JT, Gallagher PJ. Identification of basic residues involved in activation and calmodulin binding of rabbit smooth muscle myosin light chain kinase. *J Biol Chem* 1992; 267:23903–9.
24. Krueger JK, Padre RC, Stull JT. Intrasteric regulation of myosin light chain kinase. *J Biol Chem* 1995;270:16848–53.
25. Tanaka M, Ikebe R, Matsuura M, Ikebe M. Pseudosubstrate sequence may not be critical for autoinhibition of smooth muscle myosin light chain kinase. *EMBO J* 1995;14:2839–46.
26. Hu S-H, Parker MW, Lel JY, Wilce MCJ, Benian GM, Kemp BE. Insights into autoregulation from the crystal structure of twitchin kinase. *Nature* 1994;369:581–4.
27. Kemp BE, Stull JT, Kemp BE, eds. *Peptides and protein phosphorylation*. Boca Raton, FL: CRC Press; 1990:115–33.
28. Collins JH. Myosin light chains and troponin C: structural and evolutionary relationships revealed by amino acid sequence comparisons. *J Muscle Res Cell Motil* 1991;12:3–25.
29. Zhi G, Herring BP, Stull JT. Structural requirements for phosphorylation of myosin regulatory light chain from smooth muscle. *J Biol Chem* 1994;269:24723–7.
30. Pato MD, Adelstein RS, Crouch D, Safer B, Ingebritsen TS, Cohen P. The protein phosphatases involved in cellular regulation. 4: Classification of two homogeneous myosin light chain phosphatases from smooth muscle as protein phosphatase-2A1 and and a homogeneous protein phosphatase from reticulocytes active on protein synthesis initiation factor eIF-2 as protein phosphatase-2A2. *Eur J Biochem* 1983;132:283–7.
31. Nomura M, Stull JT, Kamm KE, Mumby MC. Site-specific dephosphorylation of smooth muscle myosin light chain kinase by protein phosphatases 1 and 2A. *Biochemistry* 1992;31:11915–20.
32. Conti MA, Adelstein RS. The relationship between calmodulin binding and phosphorylation of smooth muscle myosin kinase by the catalytic subunit of 3':5'-cAMP-dependent protein kinase. *J Biol Chem* 1981;256:3178–81.
33. Miller JR, Silver PJ, Stull JT. The role of myosin light chain kinase phosphorylation in β-adrenergic relaxation of tracheal smooth muscle. *Mol Pharmacol* 1983;24:235–42.
34. Nishikawa M, de Lanerolle P, Lincoln TM, Adelstein RS. Phosphorylation of mammalian myosin light chain kinases by the catalytic subunit of cyclic AMP-dependent protein kinase and by cyclic GMP-dependent protein kinase. *J Biol Chem* 1984;259:8429–36.

35. Stull JT, Hsu L-C, Tansey MG, Kamm KE. Myosin light chain kinase phosphorylation in tracheal smooth muscle. *J Biol Chem* 1990;265:16683–90.
36. Tang D-C, Stull JT, Kubota Y, Kamm KE. Regulation of the Ca^{2+} dependence of smooth muscle contraction. *J Biol Chem* 1992;267:11839–45.
37. Word RA, Tang D-C, Kamm KE. Activation properties of myosin light chain kinase during contraction/relaxation cycles of tonic and phasic smooth muscles. *J Biol Chem* 1994;269:21596–602.
38. Van Riper DA, Weaver BA, Stull JT, Rembold CM. Myosin light chain kinase phosphorylation in swine carotid artery contraction and relaxation. *Am J Physiol* 1995;268:H1–10.
39. Stull JT, Silver PJ, Miller JR, Blumenthal DK, Botterman BR, Klug GA. Phosphorylation of myosin light chain in skeletal and smooth muscles. *Fed Proc* 1983;42:21–6.
40. Gallagher PJ, Herring BP, Trafny A, Sowadski J, Stull JT. A molecular mechanism for autoinhibition of myosin light chain kinases. *J Biol Chem* 1993;268:26578–82.

Signal Transduction in Health and Disease, Advances in Second Messenger and Phosphoprotein Research, Vol. 31, edited by J. Corbin and S. Francis.
Lippincott–Raven Publishers, Philadelphia © 1997.

13

Cascade Activation of the Calmodulin Kinase Family

Hisayuki Yokokura, Osamu Terada, Yasuhito Naito, Ryotaro Sugita, and Hiroyoshi Hidaka

Department of Pharmacology, Nagoya University School of Medicine, Nagoya 466, Japan

Ca^{2+} is widely recognized as an essential intracellular second messenger in eukaryotic systems regulating processes such as muscle contraction, neurotransmitter release, gene expression, and cell proliferation (reviewed in [1,2]). In a number of cases the effects of Ca^{2+} are believed to be mediated by the ubiquitously distributed Ca^{2+} receptor, calmodulin (CaM). Strong evidence in turn indicates that the effects of Ca^{2+}/CaM are often achieved through the regulation of protein phosphorylation (reviewed in [3–5]). A family of Ca^{2+}/CaM-dependent protein kinases (CaM kinases; CaMK) has been identified, consisting of phosphorylase kinase, myosin light-chain kinase (MLCK), and elongation factor-2 (EF-2) kinase (CaMK-III), which are highly specific enzymes, whereas CaMK-II and -IV are multifunctional enzymes. CaMK-I was first purified from bovine brain on the basis of its ability to phosphorylate site 1 of the neuronal protein synapsin I (6). Since then, the enzyme has been found to be expressed in both neuronal and nonneuronal tissues (7,8), thus necessitating a reevaluation of the kinase as a potential multifunctional protein kinase. In this respect, two other good substrates for CaMK-I have been identified *in vitro*: cAMP-response element-binding protein (CREB) (9), and CF-2, a portion of the R-domain of the cystic fibrosis transmembrane conductance regulator (CFTR) (10). The recent findings of multiple isoforms of CaMK-I and the regulation of the enzyme will be briefly reviewed in this chapter.

MULTIPLE ISOFORMS OF CA^{2+}/CaM KINASE I

Purification of Ca^{2+}/CaM Kinase I from Animal Brain

CaMK-I is active as a monomer (6), although various molecular weights of 37 to 43 kDa have been assigned to the enzyme. A preparation of bovine brain, purified with synapsin I as a substrate, was found to consist of two major polypeptides of

37,000 *M*r and 39,000 *M*r and a minor polypeptide of 42,000 *M*r (6). Two other Ca^{2+}/CaM-dependent protein kinases, termed CaMK-Ia (43,000 *M*r) and CaMK-Ib (39,000 *M*r), have been purified from rat brain using a synthetic peptide based on Site 1 of synapsin I as a substrate (11). Another preparation, termed CaMK-V (41,000 *M*r), was purified from rat brain using the synthetic peptide syntide-2 as a substrate (12). The partial amino-acid sequences of the enzyme were determined and compared with all the primary structures of the proteins discovered in the past. CaMK-V was believed to be a distinct enzyme with a specific sequence, and was designated as CaMK-V because it was discovered after CaMK-I and IV. A complementary deoxyribonucleic acid (cDNA) encoding CaMK-I has been cloned from rat brain (13), and CaMK-V seems likely to be identical or closely related to CaMK-I based on amino-acid sequence identity. Taken together, these results suggest the possibility that CaMK-I comprises a family of isoforms. However, the exact relationship between CaMK-I, -Ia, -Ib, and -V remains to be elucidated.

Immunodetection of Ca^{2+}/CaM Kinase I in Rat Brain

Immunoblot analysis of CaMK-I has also supported the possibility that this enzyme consists of multiple isoforms. A polyclonal antibody to CaMK-V stained two polypeptides, of 41 kDa and 40 kDa, in preparations of partially purified CaMK-V from rat cerebrum (7). This antibody was raised against a synthetic peptide based on the partial amino acid sequences of CaMK-V and was found to immunoreact with the recombinant CaMK-I expressed in *Escherichia coli* (14). These findings are consistent with the finding that the polyclonal antibodies to CaMK-I, termed CC76 and 77, stained polypeptides of 42,000 *M*r and ~38,000 *M*r in rat-cortex lysates (8). Interestingly, in rat cortex, but not in any other brain regions or in nonneuronal tissues, only one of the antibodies, CC77, recognized a polypeptide of ~65,000 *M*r (8). Although not clearly certain, this antibody seems likely to also recognize the latter polypeptide in some cell lines, including NS204 and COS (8). Antisera have now been obtained by immunizing rabbits with the recombinant CaMK-I expressed in *E. coli* as a glutathione S-transferase fusion protein. Immunoblot analysis of rat-brain lysates using one such antiserum, termed N3, revealed three major immunoreactive bands, of 42,800 *M*r, 39,500 *M*r, and 38,000 *M*r, which were not recognized by preimmune serum (Fig. 1). This antiserum also recognized a polypeptide of 74,000 *M*r in rat-brain lysates (Fig. 1), while other antisera commonly recognized the three major polypeptides and varied in their abilities to recognize other polypeptides (data not shown). Based on the recognition of the ~65,000 *M*r and 74,000 *M*r polypeptides only by the specific antibodies, it is likely that these polypeptides are not functionally related to CaMK-I. In rat brain, immunocytochemistry with the CaMK-I or -V antibody revealed strong staining in cortex, hippocampus, amygdala, hypothalamus, caudatoputamen, brain stem, and choroid plexus (7,8). The labeling was mainly observed in neuropil, but clusters of intensely labeled neuronal cell bodies were also detected all along the neuraxis. Neuronal nuclei and glial cells did not appear to be

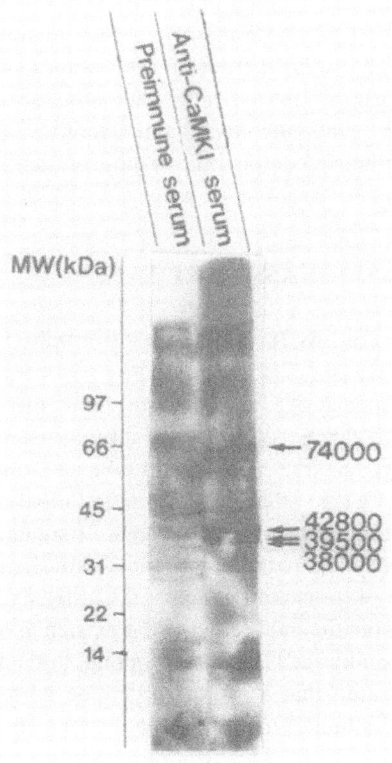

FIG. 1. Immunoblot analysis of CaM kinase-I isoforms in rat brain. Rat brain was homogenized in a buffer containing 20 mM Tris-HCl (pH 7.4), 0.1% sodium dodecyl sulfate (SDS), 1% Triton X-100, and 1% sodium deoxycholate, and the supernatant was collected by centrifugation. Aliquots (150 µg protein/lane) were applied on SDS-polyacrylamide gel electrophoresis (PAGE) using a 7.5% to 15% gradient gel. The proteins were transferred to nitrocellulose membrane and immunoblotted with preimmune serum or antiserum to recombinant CaMK-I. Protein bands were visualized with the ECL detection kit.

stained. Subcellular fractionation studies confirmed the cytosolic localization of CaMK-I in the brain (8). The antibodies used in immunohistochemical studies are likely to recognize all of the isoforms of CaMK-I. The specific localization of each isoform in neuronal tissues has to be further examined.

Complementary DNA Cloning of Ca^{2+}/CaM Kinase Isoforms from Rat Brain

To clarify the isoforms of CaMK-I, we thought it best to work toward cloning them with cDNA. A cDNA encoding CaMK-I was isolated from rat brain and a 37,545 *M*r was deduced for the resulting protein from the amino acid sequence (13). Another cDNA of a predicted rat CaMK-I from fetal lung was cloned, and a protein of 41,643 *M*r was deduced (15). Analysis of the two cDNAs suggested that the original DNA sequence was incorrect. To date, it has been found that there was indeed a sequencing artifact in the original DNA sequence (16,17). Notably, a frameshift occurred at residue 322, resulting in an extended carboxyl-terminus (C-terminus) (17).

Taken together, the two cDNAs from rat brain and rat fetal lung encode the same isoform of CaMK-I. In our recent work, cDNAs for isoforms of CaMK-I distinct from the previously reported one may have been isolated from rat fetal brain (18). The cDNAs seems likely to encode the isoforms of CaMK-I, since they are closely related to that of CaMK-I, with >70% DNA sequence identity. The transcription of the cDNAs in rat brain, and the properties of their recombinant proteins, are under investigation.

THE REGULATION OF CA^{2+}/CaM KINASE I ACTIVITY

Intrasteric Regulation of Ca^{2+}/CaM Kinase I by Its Regulatory Domain

Comparison of the deduced amino acid sequence of CaMK-I with that of the α subunit of rat-brain CaMK-II revealed a high degree of identity (approximately 42%) throughout the catalytic domain of the two enzymes (13). In addition, a second domain was identified near the C-terminus of CaMK-I that showed limited identity (approximately 32%) with the Ca^{2+}/CaM-binding domain of the CaMK-II subunit (13). The structural basis for the Ca^{2+}/CaM-dependent regulation of the CaMK-I has been established (17,19). Studies with a series of truncation mutants of CaMK-I revealed the location of the Ca^{2+}/CaM-binding domain and predicted the existence of an autoinhibitory domain either near or overlapping the Ca^{2+}/CaM-binding domain. These conclusions were supported by the finding that a synthetic peptide corresponding to residues 294 to 321 of CaMK-I inhibited the wild-type enzyme in a manner that was competitive with respect to Ca^{2+}/CaM. This peptide also inhibited the constitutively active mutant (residues 1 to 293) in a manner that was competitive with the peptide substrate syntide-2 (Ki = 1.2 μM). CaMK-I is likely to be regulated through an intrasteric mechanism common to other protein kinases, protein kinase A (PKA), PKC, MLCK, and CaMK-II.

Purification of Activators of Ca^{2+}/CaM Kinase I from Animal Brain

It has recently been appreciated that in addition to its Ca^{2+}/CaM-dependence, CaMK-I has to be in the phosphorylation state to be maximally active. The activating phosphorylation was initially interpreted to be autocatalytic on the basis of its Ca^{2+}/CaM-dependence (20). However, the pure protein activator of CaMK-Ia has been obtained from porcine brain (21). The activator is a monomer with a ~52,000 Mr, and has itself been recently identified as a CaMK (19,22). Concurrently, similar results have been obtained from the study of CaMK-V (23–25). Incubation of a preparation of CaMK-V under phosphorylating conditions resulted in enzyme phosphorylation in a Ca^{2+}/CaM-dependent manner, and a protein of 64,000 Mr (p64) was also phosphorylated (23). Notably, the enzyme, when separated from p64, was phosphorylated at a significantly decreased level, and its catalytic activity was similarly decreased, suggesting that p64 is a CaMK-V activator (23). The maximally active state of the enzyme could be restored by the addition of p64. These findings led us to hypothesize that the activator itself may be a Ca^{2+}/CaM-dependent protein kinase with an ability to phosphorylate and activate CaMK-V. Recently, the activator has been isolated from rat brain and highly purified, and has been found by the gel phosphorylation method to be autophosphorylated and to phosphorylate CaMK-I, exerting both effects in a Ca^{2+}/CaM-dependent manner (25). CaMK-Ia and -V are likely to be identical and to constitute one of the CaMK-I isoforms, and their corresponding activators with different *M*r values have been identified. However, it is also possible that CaMK-Ia and -V are different isoforms that are regulated by their specific activators.

Cascade Activation of Ca^{2+}/CaM Kinases by Ca^{2+}/CaM

CaMK-Ia and -V kinases appear to be enzymes of restricted substrate specificity. In addition to phosphorylating CaMK-Ia and -V, they phosphorylate the related enzyme CaMK-IV. Recently, two additional reports have described the purification from rat brain of CaMK-IV kinase, which had an *M*r of ~68,000 (26,27). Tokumitsu et al. have isolated a cDNA of the enzyme encoding a 505-amino-acid protein with a calculated *M*r of ~56,000 (28). Like CaMK-Ia and -V kinase, CaMK-IV kinase is apparently regulated by Ca^{2+}/CaM and also phosphorylates and activates CaMK-I as well as CaMK-IV (28). With regard to their regulation by phosphorylation, CaMK-I and -IV were phosphorylated by CaMK-Ia kinase at Thr^{177} and Thr^{310}, respectively (29), which correspond to the activating phosphorylation sites in PKA, mitogen-activated protein (MAP) kinases, MAP kinase kinases, and cyclin-dependent protein kinases (CDKs) (30). Recent resolution of the crystal structure of PKA, ERK2, and CDK2 suggests that all of the phosphorylatable residues are part of a "lip" structure at the entrance to each enzyme active site (31–33). Phosphorylation in this activation loop is central to the regulation of these three kinases (30). It has been proposed that negative charges on this loop align the residues of the active site that participate in catalysis (30). Notably, CaMK-I isoforms that we have cloned have Thr at the corresponding position and could be regulated by CaMK-Ia, -V, or -IV kinase, an issue that remains to be resolved. (An inactive mutant of residues 1 to 306 of CaMK-I, presumably devoid of a partial CaM-binding domain, is not phosphorylated or activated by CaMK-Ia kinase [17,19]). Thus, Thr^{177} must be unavailable for phosphorylation in the autoinhibited conformation of CaMK-I, and must become accessible through removal of the autoinhibitory domain by CaM-binding to the full-length enzyme. It has also been found that even after phosphorylation, Ca^{2+}/CaM is completely required for CaMK-I or -V to be active. Taken together, the foregoing findings indicate that Ca^{2+}/CaM plays three roles in the regulation of CaMK-I (Fig. 2): (i) Ca^{2+}/CaM activates CaMK-I kinase (22,23); (ii) Ca^{2+}/CaM induces conformational changes in CaMK-I, resulting in accessibility of Thr^{177} to the activated CaMK-I kinase (19); and (iii) Ca^{2+}/CaM also exposes the active site of CaMK-I, so that substrates can bind to it (17,19). The theoretical "calcium/calmodulin-depen-

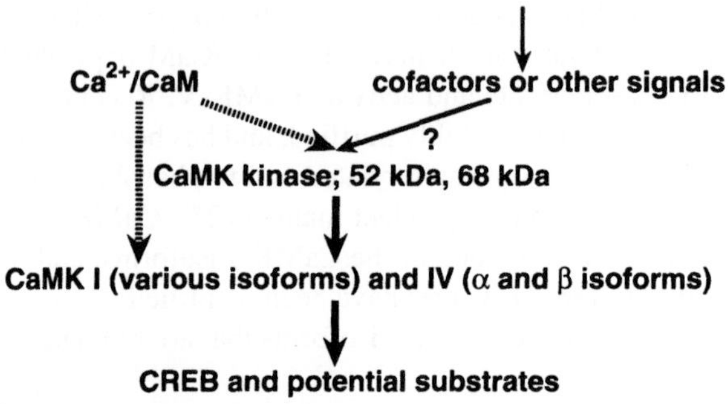

Fig. 2. Cascade activation of CaMK-I and -IV. Note that it is possible to postulate that CaMK kinases are also regulated by factors distinct from Ca^{2+}/CaM, or by posttranslational modifications such as phosphorylation and dephosphorylation.

dent protein kinase kinase cascade" that we proposed with extraordinary courage in 1994 (24) is now becoming a reality.

REFERENCES

1. Campbell AK. *Intracellular Calcium: Its Universal Role as Regulator.* New York: John Wiley & Sons, 1983.
2. Davis TN. What's new with calcium? *Cell* 1992;71:557–62.
3. Nairn AC, Hemmings HC, Jr, Greengard P. Protein kinases in the brain. In: Richardson CC, ed. *Annual Review of Biochemistry.* Palo Alto, CA; 1985:931–76.
4. Hanson PI, Schulman H. Neuronal Ca^{2+}/calmodulin-dependent protein kinases. In: Richardson CC, eds. *Annual Review of Biochemistry.* Palo Alto, CA; 1992:559–601.
5. Nairn AC, Picciotto MR. Calcium/calmodulin-dependent protein kinases. *Semin Cancer Biol* 1994;5: 295–303.
6. Nairn AC, Greengard P. Purification and characterization of Ca^{2+}/calmodulin-dependent protein kinase I from bovine brain. *J Biol Chem* 1987;262:7273–81.
7. Ito T, Mochizuki H, Kato M, et al. Ca^{2+}/calmodulin-dependent protein kinase V: tissue distribution and immunohistochemical localization in rat brain. *Arch Biochem Biophys* 1994;312:278–84.
8. Picciotto MR, Zoli M, Bertuzzi G, Nairn AC. Immunochemical localization of calcium/calmodulin-dependent protein kinase I. *Synapse* 1995;20:75–84.
9. Sheng M, Thompson MA, Greenberg ME. CREB: A Ca^{2+}-regulated transcription factor phosphorylated by calmodulin-dependent kinases. *Science* 1991;252:1427–30.
10. Picciotto MR, Cohn JA, Bertuzzi G, Greengard P, Nairn AC. Phosphorylation of the cystic fibrosis transmembrane conductance regulator. *J Biol Chem* 1992;267:12742–52.
11. DeRemer MF, Saeli RJ, Edelman AM. Ca^{2+}-calmodulin-dependent protein kinase Ia and Ib from rat brain. I. Identification, purification, and structural comparisons. *J Biol Chem* 1992;267:13460–5.
12. Mochizuki H, Ito T, Hidaka H. Purification and characterization of Ca^{2+}/calmodulin-dependent protein kinase V from rat brain. *J Biol Chem* 1993;268:9143–7.
13. Picciotto MR, Czernik AJ, Nairn AC. Calcium/calmodulin-dependent protein kinase I-cDNA cloning and identification of autophosphorylation site. *J Biol Chem* 1993;268:26512–21.
14. Ito T, Yokokura H, Nairn AC, Nimura Y, Hidaka H. Ca^{2+}/calmodulin-dependent protein kinase V and I may form a family of isoforms. *Biochem Biophys Res Commun* 1994;201:1561–6.
15. Cho FS, Phillips KS, Bogucki B, Weaver TE. Characterization of a rat cDNA clone encoding calcium/calmodulin-dependent protein kinase I. *Biochem Biophys Acta* 1994;1224:156–60.
16. Correction. *J Biol Chem* 1995;270:10358.

17. Yokokura H, Picciotto MR, Nairn AC, Hidaka H. The regulatory region of calcium/calmodulin-dependent protein kinase I contains closely associated autoinhibitory and calmodulin-binding domains. *J Biol Chem* 1995;270:23851–9.
18. Yokokura H, Terada O, Naito Y, Hidaka H. Isolation and comparison of rat cDNAs encoding Ca^{2+}/calmodulin-dependent protein kinase I isoforms. *Biochim Biophys Acta* 1997;1338:8–12.
19. Haribabu B, Hook S, Selbert MA, et al. Human calcium-calmodulin dependent protein kinase I: cDNA cloning, domain structure and activation by phosphorylation at threonine-177 by calcium-calmodulin dependent protein kinase I kinase. *EMBO J* 1995;14:3679–86.
20. DeRemer MF, Saeli RJ, Brautigan DL, Edelman AM. Ca^{2+}-calmodulin-dependent protein kinase Ia and Ib from rat brain. II. Enzymatic characteristics and regulation of activities by phosphorylation and dephosphorylation. *J Biol Chem* 1992;267:13466–71.
21. Lee JC, Edelman AM. A protein activator of Ca^{2+}-calmodulin-dependent protein kinase Ia. *J Biol Chem* 1994;269:2158–64.
22. Lee JC, Edelman AM. Activation of Ca^{2+}-calmodulin-dependent protein kinase Ia is due to direct phosphorylation by its activator. *Biochem Biophys Res Commun* 1995;210:631–7.
23. Mochizuki H, Sugita R, Ito T, Hidaka H. Phosphorylation of Ca2+/calmodulin-dependent protein kinase V and regulation of its activity. *Biochem Biophys Res Commun* 1993;197:1595–1600.
24. Sugita R, Mochizuki H, Ito T, Yokokura H, Kobayashi R, Hidaka H. Ca^{2+}/calmodulin-dependent protein kinase kinase cascade. *Biochem Biophys Res Commun* 1994;203:694–701.
25. Inoue S. Mizutani A, Sugita R, Sugita K, Hidaka H. Purification of characterization of a novel protein activator of Ca^{2+}/calmodulin dependent protein kinase I. *Biochem Biophys Res Commun* 1995;215: 861–867.
26. Okuno S, Kitani T, Fujisawa H. Purification and characterization of Ca^{2+}/CaM-dependent protein kinase IV kinase from rat brain. *J Biochem* 1994;116:923–30.
27. Tokumitsu H, Brickey DA, Glod J, Hidaka H, Sikela J, Soderling TR. Activation mechanisms for Ca^{2+}/calmodulin-dependent protein kinase IV. *J Biol Chem* 1994;269:28640–7.
28. Tokumitsu H, Enslen H, Soderling TR. Characterization of a Ca^{2+}/calmodulin-dependent protein kinase cascade. *J Biol Chem* 1995;270:19320–4.
29. Selbert MA, Anderson KA, Huang QH, Goldstein EG, Means AR, Edelman AM. Phosphorylation and activation of Ca^{2+}-calmodulin-dependent protein kinase IV by Ca^{2+}-calmodulin-dependent protein kinase Ia kinase. *J Biol Chem* 1995;270:17616–21.
30. Taylor SS, Radzio-Andzelm E. Three protein kinase structures define a common motif. *Structure* 1994;2:345–55.
31. Knighton DR, Wheng J, Ten Eyck LF, et al. Crystal structure of the catalytic subunit of cyclic adenosine monophosphate-dependent protein kinase. *Science* 1991;253:407–14.
32. DeBondt HL, Rosenblatt J, Jancarik J, Jones HD, Morgan DO, Kim SH. Crystal structure of cyclin-dependent protein kinase 2. *Nature* 1993;363:595–602.
33. Zhang F, Strand A, Robbins D, Cobb MH, Goldsmith EJ. Atomic structure of the MAP kinase ERK2 at 2.3 Å resolution. *Nature* 1994;367:704–11.

Signal Transduction in Health and Disease, Advances in Second Messenger and Phosphoprotein Research, Vol. 31, edited by J. Corbin and S. Francis.
Lippincott–Raven Publishers, Philadelphia © 1997.

14

Modulation of Sodium and Calcium Channels by Protein Phosphorylation and G Proteins

William A. Catterall

Department of Pharmacology, University of Washington School of Medicine, Seattle, Washington 98195

INTRODUCTION

Most animal cells maintain large ionic gradients across their surface membranes, such that the intracellular fluid contains a high concentration of potassium ions and low concentrations of sodium ions and calcium ions relative to the extracellular fluid. These large ionic gradients are maintained by the action of energy-dependent ion pumps specific for Na^+ and K^+ or for Ca^{2+}. In addition, nearly all cells maintain an internally negative membrane potential of the order of -60 mV because their surface membranes are specifically permeable to K^+, allowing K^+ to leak out of cells faster than Na^+ and Ca^{2+} can leak in. The ion gradients and the resting membrane potential provide the electrochemical driving force for ion movements mediated by ionic channels involved in electrical signaling.

Ion channels mediating electrical signaling are intrinsic membrane proteins that form ion-selective pores through which ions can move down their electrochemical gradients into or out of cells. The ionic channels can be divided into two classes: voltage-regulated and ligand-regulated. The primary regulatory or "gating" processes are intrinsic to the ion-channel molecules themselves, occur on a millisecond time scale, and allow voltage-sensitive ion channels to respond rapidly to changes in membrane voltage and ligand-gated ion channels to respond rapidly to synaptically released neurotransmitters. Voltage-sensitive ion channels are regulated by two experimentally separable processes: activation, which controls the voltage- and time-dependence of opening of the ion channel after depolarization; and inactivation, which controls the voltage- and time-dependence of channel closure during maintained depolarization.

The responsiveness of voltage-gated ion channels to membrane potential is also regulated on the time scale of seconds to minutes by G-protein-coupled receptors, and these regulatory processes are crucial in the control of hormone secretion, neurotransmitter release, muscle contraction, and gene transcription. Both direct binding of G proteins and phosphorylation of the ion-channel proteins themselves are now implicated as important effectors of this second-order regulation of ion-channel function. G-protein regulation of inward rectifying potassium channels is now well-documented to involve the direct binding of Gβγ subunits (1,2). Many hormones and neurotransmitters also act through G-protein mediators by altering intracellular levels of the second messengers cyclic adenosine monophosphate (cAMP), cyclic guanosine monophosphate (cGMP), diacylglycerols, and calcium. cAMP and cGMP often produce cellular responses through activation of specific cyclic-nucleotide-dependent protein kinases (3–5). Many of the effects of Ca^{2+} are mediated by stimulation of Ca^{2+}-dependent protein kinases including Ca^{2+}/calmodulin protein kinase (CaMK) and protein kinase C (PKC) (6,7). Numerous examples of ion-channel modulation by phosphorylation events initiated by neurotransmitter and hormone action are now documented (8–10). This chapter focuses on the regulation of two classes of voltage-gated ion channels: the Na^+ channels and Ca^{2+} channels, by protein phosphorylation and G proteins, with emphasis on the molecular mechanisms that may be responsible for these regulatory processes in three representative cases—skeletal-muscle L-type Ca^{2+} channels, brain N-type Ca^{2+} channels, and brain Na^+ channels.

MODULATION OF L-TYPE Ca^{2+} CHANNELS

Voltage-gated Ca^{2+} channels mediate the influx of extracellular Ca^{2+} into excitable cells during membrane depolarization, and provide the link between electrical signals at the cell surface and intracellular physiologic events (11–13). Elevation of the intracellular Ca^{2+} concentration initiates such specialized responses as secretion in neuronal and endocrine tissue and contraction of muscle. The intracellular free Ca^{2+} concentration influences many cellular processes, and the level of cytoplasmic Ca^{2+} is regulated in a complex manner. Not surprisingly, voltage-sensitive Ca^{2+} channels, which help control the concentration of this important intracellular messenger, are themselves subject to modulation.

Multiple Ca^{2+}-channel subtypes have been identified in different tissues (13–15). L-type Ca^{2+} channels responsible for long-lasting Ca^{2+} currents have been described in smooth, cardiac, and skeletal muscle and in many types of neurons and endocrine cells. This type of Ca^{2+} channel is distinguished by its high sensitivity to blockade by organic Ca^{2+}-channel antagonists such as the dihydropyridines, verapamil, and diltiazem.

Modulation of L-type channels by a wide variety of hormones and neurotransmitters has been demonstrated in both cardiac muscle and neuronal tissue (15,16). The sensitivity of channels to modulation by different hormones and neurotransmitters varies with tissue source. The mechanism underlying modulation of L-type Ca^{2+} channels has not been elucidated in most cases that have been described.

THE SKELETAL-MUSCLE L-TYPE Ca^{2+} CHANNEL AS A MOLECULAR MODEL FOR Ca^{2+}-CHANNEL STRUCTURE AND REGULATION BY PROTEIN PHOSPHORYLATION

Skeletal-muscle L-type Ca^{2+} channels in the transverse tubule membrane play critical roles in initiating contraction and regulating Ca^{2+} entry (17). A voltage-driven conformational change in the α1 subunit of this Ca^{2+} channel is thought to trigger the release of Ca^{2+} from the sarcoplasmic reticulum (18,19) via protein–protein interactions at the transverse tubule–sarcoplasmic reticulum junction (20–22). The resulting Ca^{2+} release through the sarcoplasmic reticulum Ca^{2+} channel initiates muscle contraction (23).

L-type Ca^{2+} channels in skeletal muscle mediate long-lasting Ca^{2+} currents and are localized to the transverse tubule membrane. Like cardiac Ca^{2+} channels, the activation of skeletal-muscle L-type Ca^{2+} channels is enhanced by phosphorylation by cAMP-dependent protein kinase (24). In cultured muscle cells, repetitive depolarization causes a dramatic enhancement of Ca^{2+} currents (25). This enhancement can increase calcium currents 10-fold in the critical membrane potential range near -20 mV. This enhancement of calcium currents is strongly voltage dependent and is also entirely dependent on the activity of cAMP-dependent protein kinase. This probably results from a voltage-dependent phosphorylation of the Ca^{2+} channel itself. This novel regulatory mechanism greatly increases Ca^{2+}-channel activity during tetanic stimulation of skeletal-muscle cells, and may play a critical role in regulation of the contractile force of skeletal muscle in response to the frequency of stimulation of the motor nerve in a way analogous to regulation of the contractile force of cardiac muscle through regulation of the cardiac Ca^{2+} channel.

Molecular Properties of Skeletal-Muscle Ca^{2+} Channels

T-tubule membranes contain 50- to 100-fold more dihydropyridine (DHP) receptor sites than other tissues, and hence have been the principal tissue for molecular studies of the L-type Ca^{2+} channel. The most abundant form of the rabbit skeletal-muscle L-type Ca^{2+} channel is a complex of five subunits (26,27): α1 (175 kDa), α2 (143 kDa), β (55 kDa), γ (30 kDa), and δ (24-27 kDa), associated as illustrated in Fig. 1. The α1, β, and γ subunits are products of distinct genes (28–30). Molecular biologic and protein chemistry studies indicate that the α2 and δ subunits are encoded by the same gene (31), whose protein product is proteolytically processed and disulfide-bonded to form the α2δ complex (32,33). Although α2 subunit has hydrophobic segments (31), it appears that only δ is a transmembrane protein (34).

The α1 subunit alone can function as a voltage-gated ion channel when expressed in mammalian cells (35). The cDNA sequence of the α1 subunit predicts a protein of 1,873 amino acids, whose structure is similar to that of the Na^+ channel α subunit, with four internally homologous domains, each of which contains six transmem-

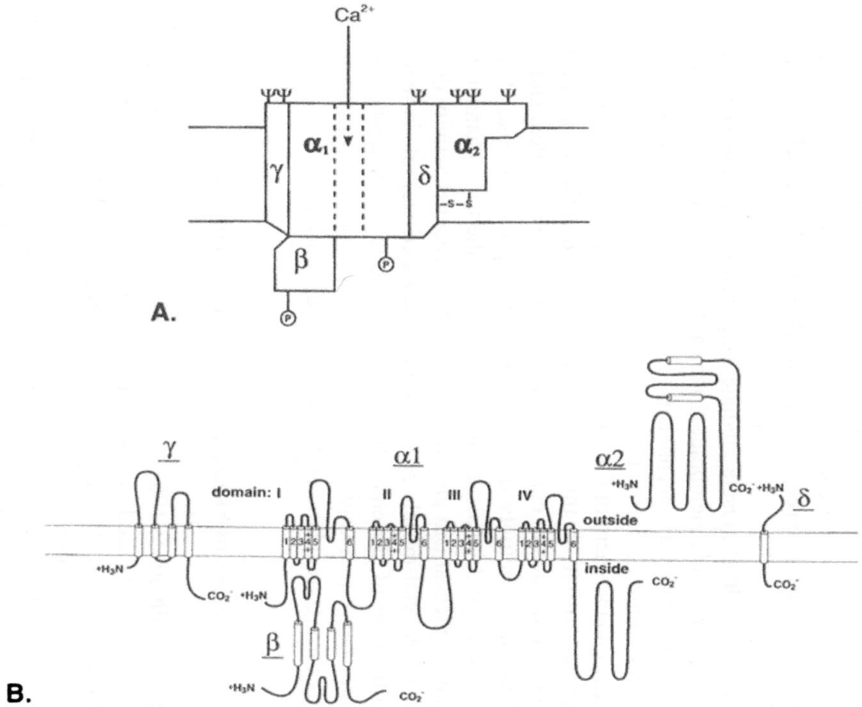

FIG. 1. Subunit structure of skeletal-muscle Ca^{2+} channels. **A:** A model of the subunit structure of the skeletal-muscle Ca^{2+} channel derived from biochemical experiments. P: sites of cAMP-dependent protein phosphorylation; Ψ: N-linked glycosylation. **B:** Transmembrane folding models of the Ca^{2+}-channel subunits derived from primary-structure determination and hydropathy analysis. *Cylinders* represent predicted α-helical segments in the transmembrane regions of the α1, δ, and γ subunits, and in the peripherally associated α2 and β subunits. The transmembrane folding patterns are derived from hydropathy analysis and analogy with the current models for the structures of Na^+ and K^+ channels for α1. The transmembrane arrangement of α2 is not well defined by hydropathy analysis, and the indicated structure therefore reflects the results of *in vitro* translation experiments showing that the α2 subunit is an extracellular protein (34).

brane helices, which have been named S1 to S6 (Fig. 1; see [28]). The Ca^{2+} influx through the skeletal-muscle L-type Ca^{2+} channel is modulated by several classes of compounds including the dihydropyridines, phenylalkylamines, and benzothiazepines, which bind to the channel at distinct receptor sites, just as they do in L-type Ca^{2+} channels in the heart (reviewed in [36]). Inhibition of Ca^{2+} channels by these compounds is important in the treatment of a number of cardiovascular disorders, such as hypertension, angina, myocardial infarction, and arrhythmias. The $\alpha 1$ subunit contains the receptor sites for these various channel modulators, as determined by photoaffinity labeling and antibody mapping of labeled peptide fragments.

The $\alpha 1$ and β subunits of the L-type Ca^{2+} channel are substrates for phosphorylation by a number of protein kinases (37–40). It is well established that the Ca^{2+} flux through the purified skeletal-muscle Ca^{2+} channel is regulated by phosphorylation-dependent events (41–44). Ion-flux studies in reconstituted phospholipid vesicles show that phosphorylation of the $\alpha 1$ and β subunits can greatly increase the number of functional Ca^{2+} channels in purified preparations (43,44). Single-channel recording experiments in planar bilayer membranes detect both increases in the number of functional Ca^{2+} channels and increases in the activity of single Ca^{2+} channels after phosphorylation by cAMP-dependent protein kinase (41,42). Thus, the $\alpha 1$ and β subunits of the purified Ca^{2+} channel contain sites at which cAMP-dependent protein phosphorylation modulates channel function. Knowledge of the location of the sites responsible for channel regulation will give insight into the molecular basis of the regulatory process.

Molecular Properties of Two Size Forms of the $\alpha 1$ Subunit

Two size forms of the $\alpha 1$ subunit are present in purified preparations of skeletal-muscle Ca^{2+} channels, T-tubule membranes, and intact skeletal-muscle cells in culture (45–47). Antibodies directed against a peptide corresponding to residues 1,505 to 1,522 of the $\alpha 1$ subunit recognize a form of the $\alpha 1$ subunit with an apparent molecular mass of 175 kDa (Fig. 2A, B). Antibodies directed against a peptide that represents the carboxy-terminal (C-terminal) 18 amino acids of the $\alpha 1$ subunit, as predicted by cloning and sequence analysis, recognize a 212-kDa form of the $\alpha 1$ subunit (Fig. 2B), which is present at a lower level than the 175-kDa form. The minor 212-kDa form ($\alpha 1_{212}$) contains the complete amino-acid sequence encoded by the $\alpha 1$ messenger ribonucleic acid (mRNA), while the major form of the $\alpha 1$ subunit, corresponding to the 175-kDa protein observed in purified preparations, is truncated at its C terminus. A more accurate estimate of the molecular mass of this truncated subunit by Ferguson analysis, which compensates for anomalous effects of acrylamide concentration in sodium dodecyl sulfate–polyacrylamide gel electrophoresis (SDS–PAGE), indicated that the mass was approximately 190 kDa, and we refer to this form of the $\alpha 1$ subunit as $\alpha 1_{190}$. Antibody mapping of the C-terminal region of $\alpha 1_{190}$ placed the C-terminus between residues 1,685 and 1,699, in agreement with the size predicted by Ferguson analysis (Fig. 2A, arrowhead; see [46]).

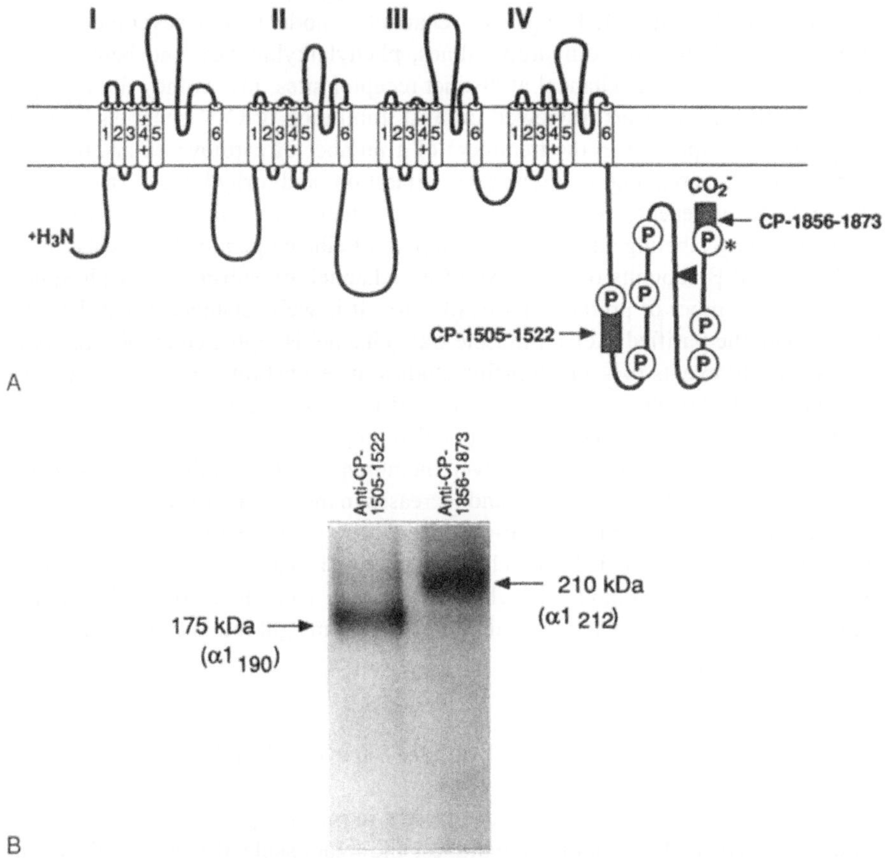

FIG. 2. A: Phosphorylation sites of the α1 subunit of the rabbit skeletal-muscle L-type Ca^{2+} channel predicted from the amino-acid sequence. The recognition sites of antibodies directed against residues 1,505 to 1,522 (anti-CP-1505–1522) and 1,856 to 1,873 (anti-CP-1856–1873) are shown. The *arrowhead* indicates the region in which the C-terminus of $\alpha1_{190}$ occurs. Ⓟ indicates potential cAMP-dependent phosphorylation sites in the C-terminal region, with * indicating the mostly rapidly phosphorylated site ($Ser^{1,854}$). **B:** Immunoprecipitation of purified ^{32}P-labeled Ca^{2+} channels and analysis by SDS–PAGE were done with the antibodies indicated above. Note that both $\alpha1_{190}$ and $\alpha1_{212}$ are phosphorylated by cAMP-dependent protein kinase.

The skeletal-muscle L-type Ca^{2+} channel has been proposed to serve dual roles as a voltage-gated ion channel mediating slow Ca^{2+} currents and as a voltage sensor for initiation of excitation–contraction coupling. Microinjection of cDNA encoding the α1 subunit into muscle cells cultured from mice with the muscular dysgenesis mutation indicate that a single α1 subunit gene is responsible for both of these functions in skeletal muscle (48). Calcium-channel isoforms in skeletal muscle may serve these two functional roles *in vivo* (46), but cDNA encoding $\alpha1_{190}$ is sufficient to produce Ca^{2+} channels that can function in both ion conductance and excitation–contraction coupling in skeletal muscle cells (49).

Differential Phosphorylation of the Two Forms of the α1 Subunit

Both forms of the α1 subunit are phosphorylated in response to physiologic stimuli that activate cAMP-dependent protein phosphorylation in intact muscle cells (47). Three phosphopeptides are phosphorylated in the full-length form of the α1 subunit which are not observed in the truncated 190-kDa form. These results confirm that the two size forms of the α1 subunit are present in intact muscle cells, and suggest the possibility that the two forms may be subject to differential regulation by protein phosphorylation.

The α1 subunit contains a number of consensus sequences for phosphorylation by cAMP-dependent protein kinase, most of which occur in the C-terminal region (Fig. 2A). The full-length α1 subunit contains three potential cAMP-dependent phosphorylation sites that are missing in $\alpha 1_{190}$ (Fig. 2A). Since phosphorylation of these sites in the C terminus of α1 may be important for regulation of ion-conductance activity, it is of interest to examine the phosphorylation of the two forms of the α1 subunit by cAMP-dependent protein kinase. The most rapidly phosphorylated site in the truncated form of the α1 subunit in purified Ca^{2+}-channel preparations is Ser^{687}, located in the intracellular loop between domains II and III (Fig. 1; see [50,51]). In contrast, time-course experiments indicated that two sites in the C-terminal portion of $\alpha 1_{212}$ are the most intensely and rapidly phosphorylated, and that phosphorylation of these peptides was maximal after 10 s (51,52). These sites were phosphorylated about 100-fold more rapidly than Ser^{687}. Further studies comparing phosphorylation under native and denaturing conditions indicated that phosphorylation occurs almost exclusively on these sites unless the channel is first denatured to expose other phosphorylation sites within the remainder of α1 (51). Immunoprecipitation of the labeled phosphopeptides with antipeptide antibodies identified these phosphorylation sites as $Ser^{1,854}$, the C-terminal-most cAMP-dependent phosphorylation consensus sequence in the α1 subunit (Fig. 2A), and $Ser^{1,757}$, one of two nearby consensus sequences (serine residues 1,757 and 1,772). Phosphorylation of $Ser^{1,757}$ and/or $Ser^{1,854}$ in the C-terminus of the full-length α1 may be important in regulation of Ca^{2+}-channel function. If phosphorylation of this residue is required for some aspect of Ca^{2+}-channel regulation, the function of Ca^{2+} channels containing the two size forms of the α1 subunit would be differentially affected.

Phosphorylation of the β Subunit of Skeletal-Muscle Ca^{2+} Channels

Like the α1 subunit of the skeletal-muscle Ca^{2+} channel, the β subunit is stoichiometrically phosphorylated by cAMP-dependent protein kinase *in vitro* in detergent-solubilized preparations (37) and in reconstituted Ca^{2+} channels that are modulated by cAMP-dependent protein kinase (41,43). Ser^{182}, Thr^{205}, and Ser^{501} have been shown to be phosphorylated *in vitro* (29,45,53), but only the phosphorylation site at Thr^{205} is conserved in β subunits. β subunits are also phosphorylated in intact cells, but the sites have not been identified (53). Since phosphorylation of both α1 and β subunits has been shown to be correlated with regulation of the ion-conductance activity of skeletal-muscle Ca^{2+} channels, both are candidates for sites of channel modulation by phosphorylation.

Role of Kinase Anchoring in Voltage-Dependent Modulation of Skeletal-Muscle Ca^{2+} Channels

The Ca^{2+} conductance of the skeletal-muscle Ca^{2+} channel activates slowly, and only a small fraction of the channels respond to a single depolarizing stimulus (54). Trains of prepulses to positive membrane potentials that mimic action potentials greatly increase Ca^{2+} current in response to a subsequent depolarization (see [25]; Fig. 3A, top traces). This Ca^{2+}-channel potentiation requires the activity of protein kinase A (PKA), and may result from rapid phosphorylation by PKA at positive membrane potentials, causing a negative shift in the voltage dependence of channel activation and a slowing of channel deactivation. The potentiation is reversed within seconds at repolarized membrane potentials by the action of phosphoprotein phosphatases. The force of vertebrate skeletal-muscle contraction is substantially increased in response to high-frequency stimulation (55), and this effect is dependent upon extracellular $Ca^{2+}$and on the ion-conductance activity of L-type Ca^{2+} channels (56–58). We have hypothesized that increased contractile force may result from increased Ca^{2+} entry through L-type Ca^{2+} channels whose activity is potentiated by voltage-dependent phosphorylation during and following depolarization (25). This increased Ca^{2+} entry would increase the store of Ca^{2+} for release from the sarcoplasmic reticulum during subsequent action potentials. Similar L-type Ca^{2+}-channel potentiation by depolarizing prepulses is observed in cardiac myocytes (59–63) and chromaffin cells (64), and also involves voltage-dependent protein phosphorylation (64,65).

A surprising feature of skeletal-muscle Ca^{2+}-channel potentiation by voltage-dependent PKA phosphorylation is its rapid response time. Trains of positive prepulses as short as 3 msec delivered at 200 Hz are effective in potentiating Ca^{2+}-channel activity (25). Many PKA-mediated events occur on a time scale of seconds, owing to concentration and diffusion limits. PKA is a soluble protein consisting of a central dimer of regulatory subunits with two associated catalytic subunits. Binding of cAMP to the regulatory subunits reversibly releases active catalytic subunits (3). The RII regulatory subunit of PKA also binds to PKA anchoring proteins (AKAPs) through a conserved binding domain (66), which can be disrupted by a 24-amino-acid region from a human thyroid-anchoring protein, Ht 31. Ht 31 peptide prevents RII–AKAP binding in most tissues, and disrupts phosphorylation required for the basal activity of glutamate receptors in hippocampal neurons (67). This interaction occurs through a putative amphipathic helix, and is disrupted by substitution of proline residues at critical positions in the helix (68). We have examined the require-

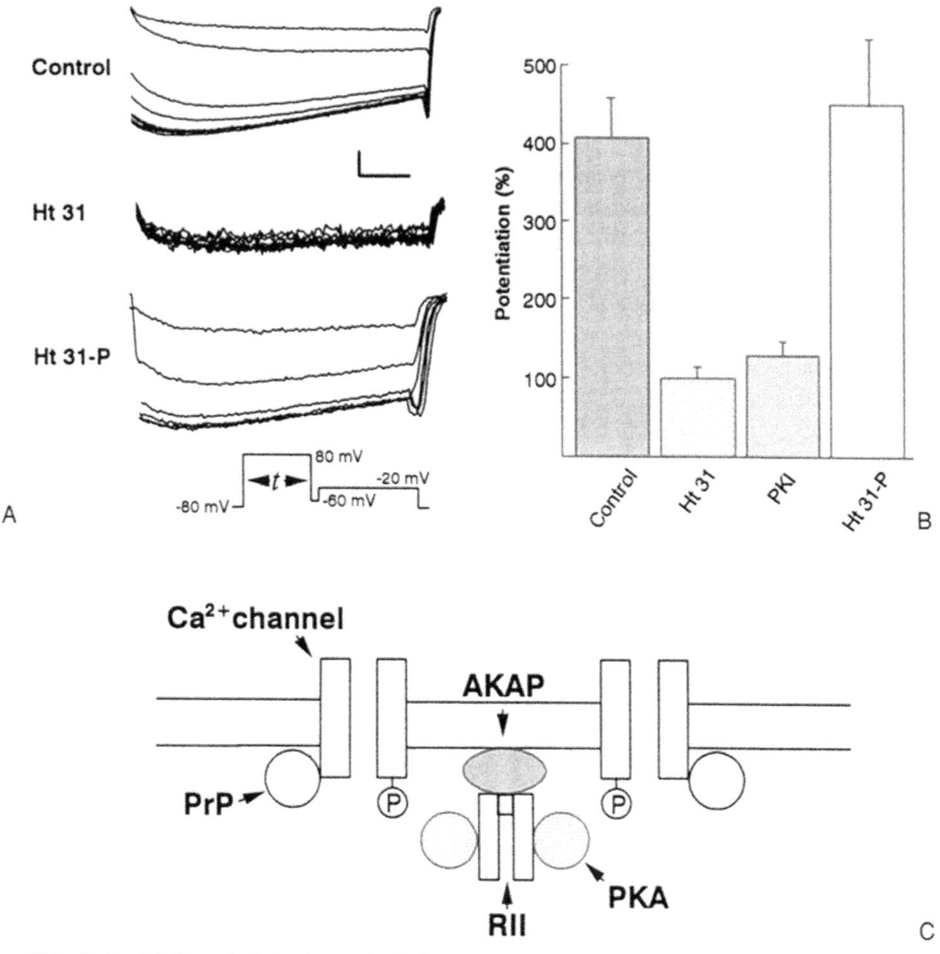

FIG. 3. Modulation of skeletal-muscle Ca^{2+} channels requires anchored PKA. Small cultured mouse skeletal myotubes of the CB_3 cell line were patch-clamped in the whole-cell configuration with electrodes containing one of three synthesized peptides: (i) Ht 31, containing 24 critical amino acids from the human thyroid kinase anchoring protein Ht 31; (ii) Ht 31-P, in which the putative amphipathic helix formed by Ht 31 was disrupted by replacing two isoleucine residues with prolines; and (iii) PKI, a 20-residue inhibitor of cAMP-dependent protein kinase. **A:** Potentiation of Ca^{2+}-channel current (carried by Ba^{2+}) was tested with the protocol shown in the *inset*, with repetitive stimulations applied at 3-s intervals. The duration of the conditioning prepulse to 80 mV was varied from 50 to 400 ms in 50-ms increments, and was followed by a 300-ms test pulse to -20 mV. Ca^{2+}-channel current during the test pulse is presented. Ca^{2+}-channel current increased with conditioning pulse duration in the control (peptide-free) condition, but showed only a slight increase when 100 μM Ht 31 was present. The modifed control peptide Ht 31-P (500 μM) had no apparent effect on potentiation. Cell capacitance was used to normalize current in each cell, and averaged 134 pF. Scale bar is 0.5 A/F (*top*), 0.6 A/F (*middle*), 0.8 A/F (*bottom*), and 50 ms. **B:** Mean potentiation (± SEM; 200-ms conditioning pulse) measured at the end of the test pulse (n = 11–15) under each of these three conditions and in 10 μm PKI is shown. **C:** Model for Ca^{2+}-channel voltage-dependent potentiation. The catalytic subunit of PKA is physically associated with the Ca^{2+} channel through its interaction with the cAMP-binding regulatory subunit (RII) and an A-kinase anchoring protein (AKAP) that binds to RII at a putative amphipathic helical domain. A protein phosphatase (PrP) is also closely associated with the channel.

ment for anchoring of PKA during voltage-dependent potentiation, using Ht 31 peptides as probes to competitively inhibit the anchoring process. Ht 31 peptide prevents potentiation by endogenous PKA as effectively as inhibition of PKA by a specific peptide inhibitor or by omission of ATP from the intracellular solution (Fig. 3A, B; see [69]). In contrast, a proline-substituted mutant AKAP peptide has no effect (Fig. 3A, B). Potentiation in the presence of 2 μM exogenous catalytic subunit of PKA is unaffected, indicating that kinase anchoring is specifically blocked by the AKAP peptide, whereas the effect of saturating kinase concentrations is unaffected. No effects of these agents were observed on the level or voltage dependence of basal Ca^{2+}-channel activity before potentiation, suggesting that close physical proximity between the skeletal-muscle Ca^{2+} channel and PKA is critical for voltage-dependent potentiation of Ca^{2+}-channel activity but not for basal activity. Our results support the hypothesis (Fig. 3C) that PKA must be anchored within approximately 150 nm of the Ca^{2+} channel to produce rapid phosphorylation in response to brief depolarizing stimuli, but is not required to maintain the basal activity of the channel.

MODULATION OF N-TYPE AND P/Q-TYPE CA^{2+} CHANNELS IN NEURONS BY G PROTEINS

In neurons, L-type Ca^{2+} channels are expressed along with N-type and P/Q-type Ca^{2+} channels. These distinct channel types are distinguished by their physiologic properties, and especially by their blockage by specific drugs and toxins (13–15). N-type Ca^{2+} channels contain α_{1B} subunits, while P/Q-type Ca^{2+} channels are thought to contain α_{1A} subunits. Both types of channels are localized at low density on dendrites and at high density in nerve terminals of many classes of central neurons (70), while L-type Ca^{2+} channels containing α_{1C} and α_{1D} are localized primarily in cell bodies and proximal dendrites of the same neurons (71). The differential distribution of these channels at the subcellular level is a fundamental determinant of their physiologic roles. N-type and P/Q-type Ca^{2+} channels are the primary channels involved in providing Ca^{2+} to trigger rapid release of neurotransmitters at fast synapses (72–75), while L-type Ca^{2+} channels are the primary channels that provide Ca^{2+} in cell bodies for modulation of gene expression (76,77).

Although the N-type and P/Q-type Ca^{2+} channels of neurons are modulated by protein phosphorylation, the most prominent mechanism of modulation of these channels is through membrane-delimited G-protein pathways that do not require soluble second messengers or protein phosphorylation (78,79). For example, in the well-studied rat sympathetic ganglion neuron, five different pathways impinge on N-type Ca^{2+} channels and modulate their function in subtly different ways. The most widespread of these pathways is observed in many peripheral and central neurons. Ca^{2+} currents can be inhibited in many cell types by neurotransmitters acting through G proteins via a membrane-delimited pathway—that is, a pathway without soluble intracellular messengers (1,2,78–80). Inhibition is typically caused by a positive shift in the voltage dependence and a slowing of channel activation, and is relieved by strong depolarization resulting in facilitation of Ca^{2+} currents (81,82). This membrane-delimited pathway regulates the activity of N-type and P/Q-type Ca^{2+} channels (1,78,79,83), which are located in presynaptic terminals and participate in neurotransmitter release (70,72–75). Synaptic transmission is inhibited by neurotransmitters through this mechanism (79,80). G-protein α subunits are thought to confer specificity in receptor coupling (1,2,79,80,84–86), but until recently it was not known whether the Gα or Gβγ subunits are responsible for modulation of Ca^{2+} channels. Surprisingly, we found that transfection of Gβγ into cells expressing P/Q-type Ca^{2+} channels induces modulation like that caused by activation of G-protein-coupled receptors, but that Gα subunits do not induce such modulation (see [87]; Fig. 4). Cotransfection of tsA-201 cells with the Ca^{2+} channel α1, α2δ, and β subunits and Gβγ causes a shift in the voltage dependence of Ca^{2+}-channel activation to more positive membrane potentials and reduces the steepness of voltage-dependent activation (Fig. 4A), effects that closely mimic the actions of neurotransmitters and guanyl nucleotides on N-type and P/Q-type Ca^{2+} channels. In contrast, transfection with a range of Gα subunits does not have this effect (Fig. 4A). This voltage shift can be reversed by strong positive prepulses resulting in voltage-dependent facilitation of the Ca^{2+} current in the presence of Gβγ (Fig. 4B), again closely mimicking the effects of neurotransmitters and guanyl nucleotides on Ca^{2+} channels. Similarly, injection or expression of Gβγ subunits in sympathetic ganglion neurons induces facilitation and occludes modulation of N-type channels by norepinephrine, but Gα subunits do not (87). In both cases, the Gγ subunit is ineffective by itself, but overexpression of exogenous Gβ subunits is sufficient to cause channel modulation. These results surprisingly point to the Gβγ subunits as the primary regulators of presynaptic Ca^{2+} channels. Binding of Gβγ induces voltage-dependent facilitation of presynaptic N-type and P/Q-type Ca^{2+} channels, resulting in a mode of regulation similar to that induced by voltage-dependent protein phosphorylation of skeletal-muscle L-type Ca^{2+} channels. Thus, quite different molecular mechanisms yield formally similar potentiation or facilitation of the function of skeletal-muscle L-type and neuronal N- and P/Q-type Ca^{2+} channels.

MOLECULAR MECHANISMS OF MODULATION OF Na^+ CHANNELS

Molecular Properties of Sodium Channel Subunits

Na^+ channels consist of a large glycosylated α subunit of about 260 kDa, expressed in nerve and muscle tissues as a complex with a single glycosylated β1 subunit of about 36 kDa or a β1 subunit and a glycosylated β2 subunit of 33 kDa (88–90). The α subunits of nerve and muscle Na^+ channels are proteins of about 2,000 amino acids, organized in four homologous domains, with approximately 50% amino acid identity among them (Fig. 5) (91–96). As in Ca^{2+} channel α1 subunits, each homologous domain contains six probable transmembrane α helices (S1 through S6) and an additional membrane-associated segment between transmem-

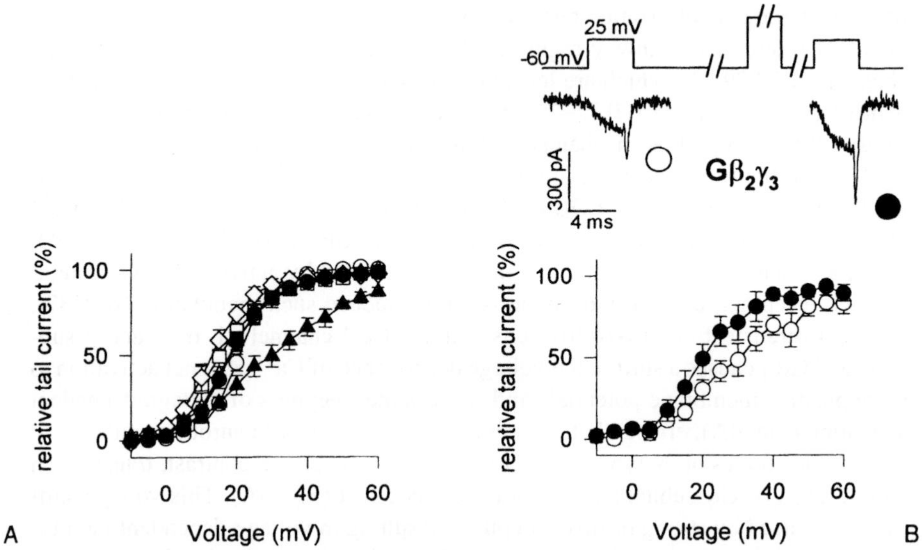

FIG. 4. Modulation of transfected P/Q-type Ca^{2+} channels by G-protein βγ subunits. **A:** Effect of $G\beta_{2\gamma3}$ on the voltage dependence of activation of Ba^{2+} currents through transfected P/Q-type Ca^{2+} channels. tsA201 cells were transfected with cDNAs encoding α_{1A} and β_{1b} alone (*filled circles*) or with $G\alpha_0$ (*open circles*), $G\alpha_{i1}$ (*open squares*), $G\alpha_{i2}$ (*open triangles*), $G\alpha_{i3}$ (*open diamonds*), or $G\beta_{2\gamma3}$ (*filled triangles*) in a 1:1:1 molar ratio in either calcium phosphate or lipofectamine (Stratagene), and were incubated for at least 72 hr. Tail currents were recorded for 4 ms at a holding potential of -60 mV after a test pulse to the indicated test potentials for 4 ms. Tail currents were normalized, and mean ± SEM values of the currents were plotted as a function of the test voltage. **B:** Effect of $G\beta_{2\gamma3}$ on facilitation of the Ba^{2+} currents by a 10-ms prepulse to +100 mV. For measuring facilitation, a 4-ms test pulse (test 1) to the indicated test potentials was applied from the holding potential of -60 mV. After 1 s, a 10-ms conditioning prepulse to +100 mV was applied, the cell was repolarized to -60 mV for 10 ms, and a second 4-ms test pulse (test 2) to the indicated potential was applied. Test pulse 1 and test pulse 2 always stepped to the same test potential. Tail currents were normalized to the largest tail current in each series of test pulses, and mean ± SEM values of the tail currents were plotted versus test pulse potential. Test 1: *open circles*; test 2: *filled circles*. An increase in I_{Ba} in test 2 compared to test 1 indicates facilitation. The current traces shown in each panel were recorded during test 1 and test 2 at +25 mV.

brane segments S5 and S6. The large, hydrophilic loops connecting domains I, II, and III, and the N-terminal and C-terminal sequences, differ strikingly in sequence among Na^+ channels, but the homologous domains and the short intracellular loop between domains III and IV are highly conserved.

The primary structures of Na^+ channel β1 subunits have been determined only recently (97). The β1 subunit cloned from rat brain is a small protein of 218 amino acids (22,821 daltons), with a substantial extracellular domain having four potential sites of N-linked glycosylation, a single α-helical membrane-spanning segment, and a very small intracellular domain. β2 subunits have a similar overall structure, but no significant amino-acid sequence identity.

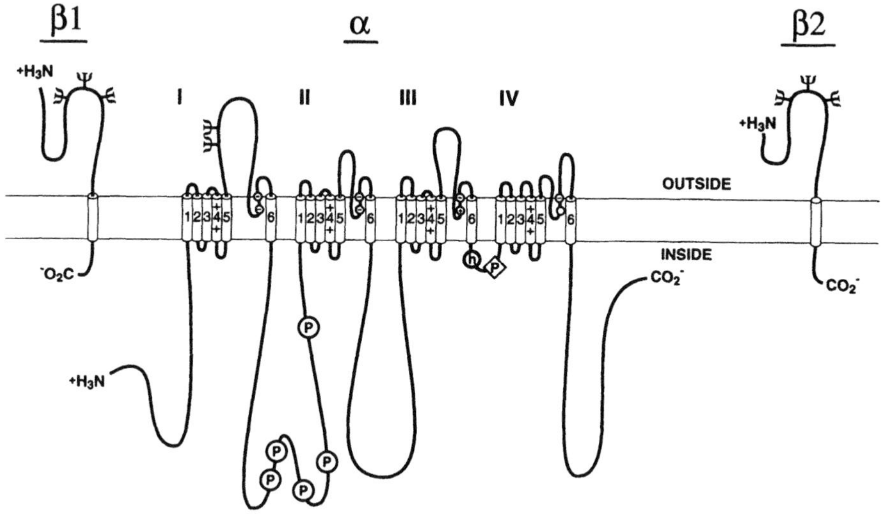

FIG. 5. Primary structures of α, β1, and β2 subunits of Na^+ channels illustrated as transmembrane folding diagrams. *Bold line:* polypeptide chains of α and β1 subunits, with the length of each segment approximately proportional to its true length in rat-brain Na^+ channel. *Cylinders* represent probable transmembrane α-helices; additional membrane-associated segments are drawn as *loops* in extended conformation, like the remainder of the sequence. Sites of experimentally demonstrated glycosylation (Ψ), cAMP-dependent phosphorylation (*circled P*), PKC phosphorylation (*P in a diamond*), amino-acid residues required for tetrodotoxin binding (*small circles with +, -, or open field* depict positively charged [Lys[1,422]], negatively charged, or neutral [Ala[1,714]] residues, respectively), and amino-acid residues that form inactivation particle (*circled h*).

Modulation by cAMP-Dependent Protein Phosphorylation

The possibility of modulation of Na^+ channel function by cAMP-dependent phosphorylation was first suggested by biochemical experiments showing that the α subunit of the Na^+ channel purified from rat brain was rapidly phosphorylated by cAMP-dependent protein kinase on at least three sites (98). Na^+-channel α subunits in intact synaptosomes are rapidly phosphorylated in response to agents that increase cAMP, and neurotoxin-activated ion flux through Na^+ channels is reduced concomitantly (99). Stimulation of rat-brain neurons in primary cell culture with agents that increase cAMP also causes rapid phosphorylation of Na^+-channel α subunits (100). Substantial phosphorylation is observed at the basal level of cAMP in the cultured neurons, and a 2-fold increase is observed upon stimulation.

The physiologic effect of phosphorylation of Na^+ channels is revealed most clearly by analysis of the effect of direct phosphorylation of Na^+ channels in excised membrane patches by purified cAMP-dependent protein kinase (Fig. 6) (101). Phosphorylation of the inside-out membrane patches from rat-brain neurons or transfected Chinese hamster ovary (CHO) cells reduces peak Na^+ currents by approximately 50%, with no change in the time course or voltage dependence of activation

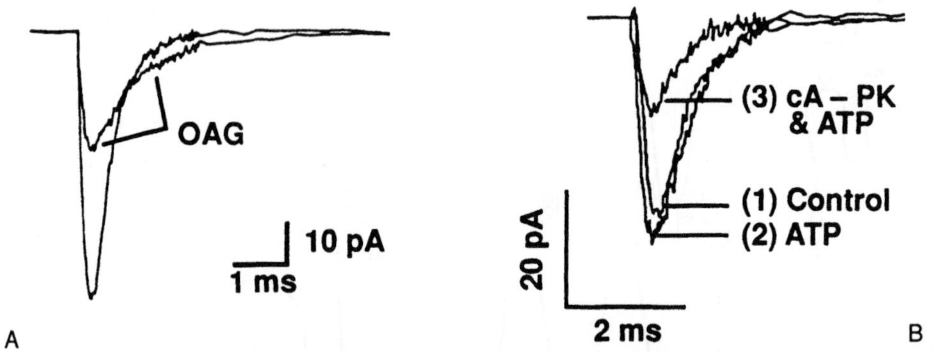

FIG. 6. Differential modulation of Na⁺ currents by protein phosphorylation. **A:** Na⁺ currents were recorded in the cell-attached patch configuration in CHO cells expressing Type IIA Na⁺-channel α subunits during depolarizations from a holding potential of -110 mV to a test potential of 0 mV. Ensemble average currents were calculated from macropatches containing up to 30 active Na⁺ channels. Current traces are illustrated under control conditions and after activation of PKC with the synthetic diacylglycerol oleylacetylglycerol (OAG). **B:** Na⁺ currents were recorded from the same cells in the excised patch-clamp configuration during depolarization from a holding potential of -130 mV to a test potential of -20 mV. Current traces are illustrated under control conditions, after addition of 1 mM, ATP, and after addition of 1 mM ATP and 2 μM cAMP-dependent protein kinase. Ensemble average currents were recorded from macropatches containing approximately 15 or more active Na⁺ channels.

or inactivation of the Na^+ current (Fig. 6B). If the basal activity of cAMP-dependent protein kinase in transfected cells is blocked by coexpression of a dominant-negative mutant form of the regulatory subunit (102), the level of Na^+ current per expressed Na^+ channel is increased, indicating that the level of channel activity is subject to tonic modulation at the basal level of activity of cAMP-dependent protein kinase in CHO cells. Phosphorylation of Na^+ channels in excised membrane patches from these kinase-negative cells by purified cAMP-dependent protein kinase causes a reduction of up to 80% in Na^+ current. Thus, the dynamic range over which Na^+-channel activity can be modulated by cAMP-dependent phosphorylation is substantial.

The modulation of Na^+-channel function in intact neuronal preparations such as brain slices has not yet been studied. However, Na^+ currents in acutely dissociated medium aspiny neurons from the striatum are reduced by agonists acting at D1 dopamine receptors, which activate adenylate cyclase, and are increased by agonists acting at D2 dopamine receptors, which inhibit adenylate cyclase (103). Neither the time course nor the voltage-dependence of the Na^+ current is altered. These results show that neuronal Na^+ channels can be modulated by neurotransmitters acting through cAMP as a second messenger. This dynamic modulation of Na^+-channel function by the wide range of neurotransmitters and neuromodulators that alter cAMP levels is expected to have profound effects on the excitability of central neurons.

The sites of phosphorylation of the Na^+ channel by cAMP-dependent protein kinase have been identified by a combination of two-dimensional phosphopeptide mapping, immunoprecipitation of phosphopeptides with site-directed antipeptide antibodies, and microsequence determination (104–106). Five sites of *in vitro* phosphorylation are clustered in the large intracellular loop connecting homologous Domains I and II (Fig. 3). These sites are all phosphorylated in intact neurons under conditions in which Na^+ channels are modulated (106), and site-directed mutation of this set of phosphorylation sites blocks modulation in *Xenopus* oocytes (107) and transfected mammalian cells (*unpublished results*). The different rates of phosphorylation of these sites *in vitro* (106) suggest that a subset of the sites may play a predominant role in channel regulation. Mutation of individual serine residues, followed by expression and functional analysis, will be required to define the role of each site in regulation by cAMP-dependent phosphorylation.

Modulation of Other Na^+ Channels by cAMP-Dependent Phosphorylation

The intracellular loop of the Na^+-channel α subunit that is phosphorylated by cAMP-dependent protein kinase is conserved among the Types I, II, and III Na^+ channels that are expressed predominantly in the central nervous system, but is highly variable in other Na^+ channels. Thus, Na^+ channels in other tissues may be differently regulated by cAMP-dependent phosphorylation. The μ1 Na^+ channel expressed in adult skeletal muscle has no consensus sites for cAMP-dependent phosphorylation in its relatively short loop between Domains I and II (94). The μ1 α subunit is phosphorylated by cAMP-dependent protein kinase *in vitro*, but its regulation by phosphorylation has not been reported. The h1 Na^+ channel α subunit expressed in adult heart has consensus phosphorylation sites in the corresponding intracellular loop (95), and is phosphorylated by cAMP-dependent protein kinase *in vitro* (108). Voltage-clamp studies of cardiac myocytes indicate that the h1 Na^+ channel is modulated by the action of norepinephrine at β-adrenergic receptors via two parallel pathways. Activation of the guanyl-nucleotide regulatory protein G_s itself shifts the voltage dependence of inactivation toward more negative membrane potentials (110). In addition, activation of adenylate cyclase by G_s causes a further negative shift in the voltage dependence of inactivation through cAMP-dependent phosphorylation (109,110). Thus, in the heart, cAMP-dependent protein phosphorylation has a negative regulatory influence on Na^+-channel activity, but the mechanism for this involves an increase in inactivation at normal resting membrane potentials, rather than a simple reduction in peak Na^+ current elicited from all holding potentials, such as occurs with brain Na^+ channels.

Modulation of Brain Na^+ Channels by Protein Kinase C

Alpha subunits of purified Na^+ channels from rat brain are also phosphorylated by PKC (111), suggesting that they may be modulated by the Ca^{2+}/diacylglycerol signaling pathway. Consistent with this suggestion, Na^+ currents in neuroblastoma cells are reduced by treatment with fatty acids that can activate PKC (112), and Na^+ currents in *Xenopus* oocytes injected with rat-brain mRNA are reduced by treatment with phorbol esters that activate PKC (113,114). Activation of PKC in rat-brain neurons or CHO cells transfected with cDNA encoding the Type IIA Na^+-channel α subunit by treatment with diacylglycerols causes two functional effects: slowing of inactivation and reduction of peak current (Fig. 4A) (115). Both of these actions are prevented by prior injection of the pseudosubstrate inhibitory peptide of PKC into the cells, indicating that they reflect phosphorylation by PKC. Moreover, both effects can be observed by directly phosphorylating Na^+ channels in excised, inside-out membrane patches with purified PKC (115). These results support the conclusion that PKC can modulate Na^+-channel function by phosphorylation of the α subunit of the Na^+-channel protein itself, as observed with purified Na^+ channels (111).

The intracellular loop connecting Domains III and IV has been implicated in Na^+-channel inactivation, and is thought to serve as the inactivation gate (116–120). This segment of the channel has a consensus sequence for phosphorylation by PKC centered at $Ser^{1,506}$. Mutagenesis of this serine residue to alanine blocks both of the modulatory effects of PKC (121). Evidently, phosphorylation of this site in the inactivation gate is required for both slowing of Na^+-channel inactivation and reduction of the peak Na^+ current by PKC.

Treatment of neurons or transfected cells with increasing concentrations of diacylglycerol reveals a biphasic modulation; low concentrations slow Na^+-channel inactivation, while higher concentrations are required to reduce peak Na^+ currents (122). These results suggest that a second site of phoshorylation is required for reduction of peak Na^+ currents. Because cAMP-dependent phosphorylation of sites in the intracellular loop between domains I and II causes a reduction of peak Na^+ currents, mutant Na^+ channels with alterations in consensus sequences for PKC phosphorylation in that region of the channel were examined for modulation by protein kinase C. Mutation of a serine residue located in a protein kinase consensus sequence toward the N-terminal end of this intracellular loop (Fig. 3) prevented the reduction in Na^+ current by PKC (122). These results implicate phosphorylation of this residue in the reduction in peak Na^+ current caused by PKC, and suggest that this effect of PKC phosphorylation has the same underlying molecular mechanism as the reduction of peak Na^+ current by cAMP-dependent protein phosphorylation.

Physiologic Significance of PKC Modulation of Brain Na^+ Channels

Na^+ channels in the cell bodies of major projection neurons in the central nervous system are responsible for action-potential initiation and therefore set the threshold for spike generation. In addition, these channels generate sustained Na^+ currents of approximately 1% to 3% of the peak Na^+ current (123–125). Sustained Na^+ currents are thought to be important determinants of action-potential firing patterns of these central neurons. The central neurons that generate these currents have a specific localization of Type I Na^+ channels in their cell bodies (126). Activation of muscarinic acetylcholine receptors in hippocampal pyramidal neurons reduces peak Na^+ currents and slows inactivation of the Na^+ current (127). Both peak and sustained Na^+ currents are comparably reduced. These effects are caused by release of diacylglycerol and activation of PKC (127). Thus, in hippocampal neurons, regulation of the Na^+ channel can be traced all the way from the binding of a neurotransmitter to a cell-surface muscarinic acetylcholine receptor through activation of PKC to phosphorylation of a specific site on the α subunit of the Na^+ channel.

Modulation of Na^+ Channels by Protein Kinase C in Other Cells

The PKC consensus site at $Ser^{1,506}$ is conserved in Na^+ channels expressed in heart and skeletal muscle (94–96). As for central neurons, activation of PKC in these cells causes slowing of Na^+-channel inactivation and a reduction of the peak Na^+ current (128,129). In cardiac cells, activation of PKC also causes a shift of steady-state inactivation to more negative membrane potential (129). All of the effects of PKC on cardiac Na^+ channels require phosphorylation of $Ser^{1,505}$ in the inactivation gate (130), as observed for brain Na^+ channels. Thus, PKC modulation of Na^+ currents is a common feature of most excitable cells.

Modulation of Brain Na^+ Channels by G Proteins

G proteins couple hormone and neurotransmitter receptors to effectors including neuronal K^+ and Ca^{2+} channels through multiple pathways that may involve membrane-delimited G-protein–ion-channel interactions or diffusible second messengers (reviewed in [1,2,78,79,131–135]). In heart, the opening of an inwardly rectifying K^+ channel by muscarinic acetylcholine receptors requires guanosine triphosphate [GTP] and a pertussis toxin (PTX)-sensitive G protein (131–135), and β-adrenergic receptors modulate Na^+ channels by parallel pathways involving cAMP-dependent protein phosphorylation and possibly direct interaction with G_s (110,131). We have found that activation of brain Na^+ channels is enhanced by activation of G proteins in both hippocampal neurons and in CHO cells expressing the α subunit of the Type IIA Na^+ channel (CNaIIA-1 cells) (136). These findings suggest that activation of G proteins in brain in response to neurotransmitters can modulate Na^+-channel function and alter the excitability of neurons.

Na^+ channels in acutely dissociated rat hippocampal neurons and in CHO cells transfected with a cDNA encoding the α subunit of rat-brain Type IIA Na^+ channel (CNaIIA-1 cells) are modulated by G-protein-coupled pathways under conditions of whole-cell voltage clamping. Activation of G proteins by 0.2 to 0.5 mM guanosine 5′-O-(3-thiotriphosphate) (GTPγS), a nonhydrolyzable GTP analogue, increased Na^+ currents recorded in both cell types. The increase in current amplitude was caused by an 8 mV to 10 mV negative shift in the voltage-dependence of both activation and inactivation. The effects of G-protein activators were blocked by treatment with pertussis toxin or guanosine 5′-O-2–thiodiphosphate (GDPβS), a nonhydrolyzable guanosine diphosphate (GDP) analogue, but not by cholera toxin. GDPβS (2 mM) alone had effects opposite to those of GTPγS, shifting Na^+-channel gating 8 mV to 10 mV toward more positive membrane potentials, suggesting that basal activation of G proteins in the absence of stimulation is sufficient to modulate Na^+channels. In CNaIIA-1 cells, thrombin, which activates pertussis toxin-sensitive G proteins in CHO cells, caused a further negative shift in the voltage dependence of Na^+-channel activation and inactivation beyond that observed with GTP alone. The results in CNaIIA-1 cells indicate that the α subunit of the Na^+ channel alone is sufficient to mediate G protein effects on gating. The modulation of Na^+ channels via a G-protein-coupled pathway acting on Na^+ channel α subunits may regulate electrical excitability through integration of different G-protein-coupled synaptic inputs.

CONCLUSION

As is evident from this brief survey of experimental results with three representative ion channels, modulation of ion-channel function through second-messenger pathways is a ubiquitous, flexible, and complex regulatory mechanism. Multiple second-messenger pathways modulate individual ion channels through phosphorylation by multiple protein kinases and binding of G-protein subunits to the target ion channel. Ion channels in the cell-surface membrane of excitable cells are therefore controlled in a second-to-second manner by intracellular regulatory pathways. This control is critical in regulating the contraction of muscle cells, secretion by endocrine cells, and electrical activity and synaptic transmission in neurons. Modulation of electrical activity and synaptic transmission in neurons by these pathways is likely to be the initial cellular event in learning and memory. Elucidation of the molecular mechanisms of ion-channel modulation, and of the interactions of multiple regulatory pathways at the level of individual ion channels, will provide a molecular template for understanding the regulation and integration of these complex regulatory processes at the cellular and multicellular levels.

REFERENCES

1. Hescheler J, Schultz G. G-proteins involved in the calcium channel signalling system. *Curr Opin Neurobiol* 1993;3:360–7.
2. Wickman KD,Clapham DE. G-protein regulation of the ion channels. *Curr Biol* 1995;5:278–85.
3. Krebs EG, Beavo JA. Phosphorylation-dephosphorylation of enzymes. *Annu Rev Biochem* 1979;48: 923–59.
4. Greengard P. Phosphorylated proteins as physiological effectors. *Science* 1978;199:146–52.
5. Cohen P. Protein phosphorylation and hormone action. *Proc R Soc Lond* 1988;234:115–44.
6. Nestler EJ, Greengard P. Protein phosphorylation in the brain. In: Protein phosphorylation in the nervous system. New York: John Wiley & Sons; 1984.
7. Nishizuka Y. The role of protein kinase C in cell surface signal transduction and tumor promotion. *Nature* 1984;308:693–8.
8. Shearman MS, Seikiguchi K, Nishizuka Y. Modulation of ion channel activity: a key function of the protein kinase C enzyme family. *Pharmacol Rev* 1989;41:211–37.

9. Levitan IB. Modulation of ion channels in neurons and other cells. *Annu Rev Neurosci* 1988;11: 119–36.
10. Kaczmarek LK. The role of protein kinase C in the regulation of ion channels and neurotransmitter release. *TINS* 1987;10:30–4.
11. Reuter H. Properties of two inward membrane currents in the heart. *Annu Rev Physiol* 1979;41:413–24.
12. Tsien RW. Calcium channels in excitable cell membranes. *Annu Rev Physiol* 1983;45:341–58.
13. Bean BP. Classes of calcium channels in vertebrate animals. *Annu Rev Physiol* 1989;51:367–84.
14. Hess P. Calcium channels in vertebrate cells. *Annu Rev Neurosci* 1990;13:337–56.
15. Tsien RW, Lipscome D, Madison DV, Bley KR, Fox AP. Multiple types of neuronal calcium channels and their selective modulation. *Trends Neurosci* 1988;11:431–8.
16. Reuter H. Calcium channel modulation by neurotransmitters, enzymes and drugs. *Nature* 1983;301: 569–74.
17. Catterall WA. Excitation-contraction coupling in vertebrate skeletal muscle: a tale of two calcium channels. *Cell* 1991;64:871–4.
18. Adams BA, Beam KG. Muscular dysgenesis in mice: a model system for studying excitation-contraction coupling. *FASEB J* 1990;4:2809–16.
19. Ríos E, Pizarro G. Voltage sensor of excitation-contraction coupling in skeletal muscle. *Physiol Rev* 1991;71:849–908.
20. Tanabe T, Mikami A, Numa S, Beam KG. Cardiac-type excitation-contraction coupling in dysgenic skeletal muscle injected with cardiac dihydropyridine receptor cDNA. *Nature* 1990;344:451–3.
21. Lu X, Xu L, Meissner G. Activation of the skeletal muscle calcium release channel by a cytoplasmic loop of the dihydropyridine receptor. *J Biol Chem* 1994;269:6511–6.
22. Ríos E, Ma J, González A. The mechanical hypothesis of excitation-contraction (EC) coupling in skeletal muscle. *J Muscle Res Cell Motil* 1991;12:127–35.
23. Ashley CC, Mulligan IP, Lea TJ. Ca^{2+} and activation mechanisms in skeletal muscle. *Q Rev Biophys* 1991;24:1–73.
24. Arreola J, Calvo J, Garcia MC, Sánchez JA. Modulation of calcium channels of twitch skeletal muscle fibres of the frog by adrenaline and cyclic adenosine monophosphate. *J Physiol* 1987;393:307–30.
25. Sculptoreanu A, Scheuer T, Catterall WA. Voltage-dependent potentiation of L-type calcium channel activity during tetanic stimulation of skeletal muscle cells due to phosphorylation by cAMP-dependent protein kinase. *Nature* 1993;364:240–3.
26. Catterall WA, Seagar MJ, Takahashi M. Molecular properties of dihydropyridine-sensitive calcium channels in skeletal muscle. *J Biol Chem* 1988;263:3535–8.
27. Campbell KP, Leung AT, Sharp AH. The biochemistry and molecular biology of the dihydropyridine-sensitive calcium channel. *Trends Neurosci* 1988;11:425–30.
28. Tanabe T, Takeshima H, Mikami A, et al. Primary structure of the receptor for calcium channel blockers from skeletal muscle. *Nature* 1987;328:313–8.
29. Ruth P, Röhrkasten A, Biel M, et al. Primary structure of the beta subunit of the DHP-sensitive calcium channel from skeletal muscle. *Science* 1989;245:1115–8.
30. Jay SD, Ellis SB, McCue AF, et al. Primary structure of the gamma subunit of the DHP-sensitive calcium channel from skeletal muscle. *Science* 1990;248:490–2.
31. Ellis SB, Williams ME, Ways NR, et al. Sequence and expression of mRNAs encoding the alpha 1 and alpha 2 subunits of a DHP-sensitive calcium channel. *Science* 1988;241:1661–4.
32. De Jongh KS, Warner C, Catterall WA. Subunits of purified calcium channels. $\alpha 2$ and δ are encoded by the same gene. *J Biol Chem* 1990;265:14738–41.
33. Jay SD, Sharp AH, Kahl SD, Vedvick TS, Harpold MM, Campbell KP. Structural characterization of the dihydropyridine-sensitive calcium channel α_2-subunit and the associated δ peptides. *J Biol Chem* 1991;266:3287–93.
34. Gurnett CA, De Waard M, Campbell KP. Dual function of the voltage-dependent Ca^{2+} channel $\alpha_2\delta$ subunit in current stimulation and subunit interaction. *Neuron* 1996;16:431–53.
35. Perez-Reyes E, Kim HS, Lacerda AE, et al. Induction of calcium currents by the expression of the alpha 1-subunit of the dihydropyridine receptor from skeletal muscle. *Nature* 1989;340:233–6.
36. Glossmann H, Striessnig J. Molecular properties of calcium channels. *Rev Physiol Biochem Pharmacol* 1990;114:1–105.
37. Curtis BM, Catterall WA. Phosphorylation of the calcium antagonist receptor of the voltage-sensitive calcium channel by cAMP-dependent protein kinase. *Proc Natl Acad Sci USA* 1985;82:2528–32.

38. Takahashi M, Seagar MJ, Jones JF, Reber BF, Catterall WA. Subunit structure of dihydropyridine-sensitive calcium channels from skeletal muscle. *Proc Natl Acad Sci USA* 1987;84:5478–82.
39. Jahn H, Nastainczyk W, Röhrkasten A, Schneider T, Hofmann F. Site-specific phosphorylation of the purified receptor for calcium-channel blockers by cAMP- and cGMP-dependent protein kinases, protein kinase C, calmodulin-dependent protein kinase II and casein kinase II. *Eur J Biochem* 1988; 178:535–42.
40. O'Callahan CM, Hosey MM. Multiple phosphorylation sites in the 165–kilodalton peptide associated with dihydropyridine-sensitive calcium channels. *Biochemistry* 1988;27:6071–7.
41. Flockerzi V, Oeken H-J, Hofmann F, Pelzer D, Cavalie A, Trautwein W. Purified dihydropyridine-binding site from skeletal muscle t-tubules is a functional calcium channel. *Nature* 1986;323:66–8.
42. Hymel L, Striessnig J, Glossman H, Schindler H. Purified skeletal muscle 1,4-dihydropyridine receptor forms phosphorylation-dependent oligomeric calcium channels in planar bilayers. *Proc Natl Acad Sci USA* 1988;85:4290–4.
43. Nunoki K, Florio V, Catterall WA. Activation of purified calcium channels by stoichiometric protein phosphorylation. *Proc Natl Acad Sci USA* 1989;86:6816–20.
44. Mundina-Weilenmann C, Chang CF, Gutierrez LM, Hosey MM. Demonstration of the phosphorylation of dihydropyridine-sensitive calcium channels in chick skeletal muscle and the resultant activation of the channels after reconstitution. *J Biol Chem* 1991;266: 4067–73.
45. De Jongh KS, Merrick DK, Catterall WA. Subunits of purified calcium channels: a 212-kDa form of alpha 1 and partial amino acid sequence of a phosphorylation site of an independent beta subunit. *Proc Natl Acad Sci USA* 1989;86:8585–9.
46. De Jongh KS, Warner C, Colvin AA, Catterall WA. Characterization of the two size forms of the $\alpha 1$ subunit of skeletal muscle L-type calcium channels. *Proc Natl Acad Sci USA* 1991;88:10778–82.
47. Lai Y, Seagar MJ, Takahashi M, Catterall WA. Cyclic AMP-dependent phosphorylation of two size forms of $\alpha 1$ subunits of L-type calcium channels in rat skeletal muscle cells. *J Biol Chem* 1990;265: 20839–48.
48. Tanabe T, Beam KG, Powell JA, Numa S. Restoration of excitation-contraction coupling and slow calcium current in dysgenic muscle by dihydropyridine receptor complementary DNA. *Nature* 1988;336:134–9.
49. Beam KG, Adams BA, Niidome T, Numa S, Tanabe T. Function of a truncated dihydropyridine receptor as both voltage sensor and calcium channel. *Nature* 1992;360:169–71.
50. Röhrkasten A, Meyer HE, Nastainczyk W, Sieber M, Hofmann F. cAMP-dependent protein kinase rapidly phosphorylates serine-687 of the skeletal muscle receptor for calcium channel blockers. *J Biol Chem* 1988;263:15325–9.
51. Rotman EI, De Jongh KS, Florio V, Lai Y, Catterall WA. Specific phosphorylation of a COOH-terminal site on the full-length form of the $\alpha 1$ subunit of the skeletal muscle calcium channel by cAMP-dependent protein kinase. *J Biol Chem* 1992;267:16100–5.
52. Rotman EI, Murphy BJ, Catterall WA. Sites of selective cAMP-dependent phosphorylation of the L-type calcium channel $\alpha 1$ subunit from intact rabbit skeletal muscle myotubes. *J Biol Chem* 1995; 270:16371–7.
53. Rotman EI, Murphy BM, and Catterall WA. Phosphorylation of a novel site in the β subunit of skeletal muscle calcium channels by cAMP-dependent protein kinase. Submitted.
54. Schwartz LM, McCleskey EW, Almers W. Dihydropyridine receptors in muscle are voltage-dependent but most are not functional calcium channels. *Nature* 1985;314:747–51.
55. Kernell D, Eerbeek O, Verhey BA. Relation between isometric force and stimulus rate in cat's hindlimb motor units of different twitch contraction time. *Exp Brain Res* 1983;50:220–7.
56. Kotsias BA, Muchnik S, Obejero-Paz CA. Co^{2+}, low Ca^{2+}, and verapamil reduce mechanical activity in rat skeletal muscles. *Am J Physiol* 1986;250:C40–6.
57. Dulhunty AF, Gage PW. Effects of extracellular calcium concentration and dihydropyridines on contraction in mammalian skeletal muscle. *J Physiol (London)* 1988;399:63–80.
58. Oz M, Frank GB. Decreases in the size of tetanic responses produced by nitrendipine or by extracellular calcium ion removal without blocking twitches or action potentials in skeletal muscle. *J Pharmacol Exp Ther* 1991;257:575–81.
59. Zygmunt AC, Maylie J. Stimulation-dependent facilitation of the high threshold calcium current in guinea-pig ventricular myocytes. *J Physiol (London)* 1990;428:653–71.
60. Pietrobon D, Hess P. Novel mechanism of voltage-dependent gating in L-type calcium channels. *Nature* 1990;346:651–5.
61. Noble S, Shimoni Y. Voltage-dependent potentiation of the slow inward current in frog atrium. *J Physiol (London)* 1981;310:77–95.

62. Lee K. Potentiation of the calcium currents of internally perfused heart cells by repetitive depolarization. *Proc Natl Acad Sci USA* 1987;84:3941–5.
63. Fedida D, Noble D, Spindler AJ. Mechanism of the use dependence of Ca^{2+} current in guinea-pig myocytes. *J Physiol (London)* 1988;405:461–77.
64. Artalejo CR, Rossie S, Perlman RL, Fox AP. Voltage-dependent phosphorylation may recruit Ca^{2+} current facilitation in chromaffin cells. *Nature (London)* 1992;358:63–6.
65. Sculptoreanu A, Rotman E, Takahashi M, Scheuer,T, Catterall WA. Voltage-dependent potentiation of the activity of cardiac L-type calcium channel $\alpha 1$ subunits due to phosphorylation by cAMP-dependent protein kinase. *Proc Natl Acad Sci USA* 1993;90:10135–9.
66. Carr DW, Hausken ZE, Fraser IDC, Stofko-Hahn RE, Scott JD. Association of the type II cAMP-dependent protein kinase with a human thyroid RII-anchoring protein. Cloning and characterization of the RII-binding domain. *J Biol Chem* 1992;267:13376–82.
67. Rosenmund C, Carr DW, Bergeson SE, Nilaver G, Scott JD, Westbrook GL. Anchoring of protein kinase A is required for modulation of AMPA/kainate receptors on hippocampal neurons. *Nature (London)* 1994;368:853–6.
68. Carr DW, Stofko-Hahn RE, Fraser ID, et al. Interaction of the regulatory subunit (RII) of cAMP-dependent protein kinase with RII-anchoring proteins occurs through an amphipathic helix binding motif. *J Biol Chem* 1991;266:14188–92.
69. Johnson BJ, Scheuer T, Catterall WA. Voltage-dependent potentiation of L-type Ca^{2+} channels in skeletal muscle cells requires anchored cAMP-dependent protein kinase. *Proc Natl Acad Sci USA* 1994;91:11492–6.
70. Westenbroek RE, Sakurai T, Elliott EM, et al. Immunochemical identification and subcellular distribution of the α_{1A} subunits of brain calcium channels. *J Neurosci* 1995;15:6403–18.
71. Hell JW, Westenbroek RE, Warner C, et al. Identification and differential subcellular localization of the neuronal class C and class D L-type calcium channel $\alpha 1$ subunits. *J Cell Biol* 1993;123:949–62.
72. Hirning LD, Fox AP, McCleskey EW, et al. Dominant role of N-type Ca^{2+} channels in evoked release of norepinephrine from sympathetic neurons. *Science* 1988;239:57–61.
73. Wheeler DB, Randall A, Tsien RW. Roles of N-type and Q-type Ca^{2+} channels in supporting hippocampal synaptic transmission. *Science* 1994;264:107–11.
74. Luebke JI, Dunlap K, Turner TJ. Multiple calcium channel types control glutamatergic synaptic transmission in the hippocampus. *Neuron* 1993;11:895–902.
75. Takahashi T, Momiyama A. Different types of calcium channels mediate central synaptic transmission. *Nature* 1993;366:156–8.
76. Bading H, Ginty DD, Greenberg, ME. Regulation of gene expression in hippocampal neurons by distinct calcium signaling pathways. *Science* 1993; 260:181–6.
77. Deisseroth K, Bito H, Tsien RW. Signaling from synapse to nucleus: postsynaptic CREB phosphorylation during multiple forms of synaptic plasticity. *Neuron* 1996; 16:89–101.
78. Hille B. G protein-coupled mechanisms and nervous signaling. *Neuron* 1992; 9:187–95.
79. Hille B. Modulation of ion-channel function by G-protein-coupled receptors. *Trends Neurosci* 1994; 17:531–6.
80. Dolphin AC. Voltage-dependent calcium channels and their modulation by neurotransmitters and G proteins. *Exp Physiol* 1995;80:1–36.
81. Marchetti C, Carbone E, Lux HD. Effects of dopamine and noradrenaline on Ca channels of cultured sensory and sympathetic neurons of chick. *Pflugers Arch* 1986;406:104–11 .
82. Bean BP. Neurotransmitter inhibition of neuronal calcium currents by changes in channel voltage dependence. *Nature* 1989;340:153–6.
83. Mintz IM, Bean BP. $GABA_B$ receptor inhibition of P-type Ca^{2+} channels in central neurons. *Neuron* 1993;10:889–98.
84. Hescheler J, Rosenthal W, Trautwein W, Schultz G. The GTP-binding protein, Go, regulates neuronal calcium channels. *Nature* 1987;325:445–6.
85. Wilk-Blaszczak MA, Singer WD, Gutowski S, Sternweis PC, Belardetti F. The G protein G_{13} mediates inhibition of voltage-dependent calcium current by bradykinin. *Neuron* 1994;13:1215–24 .
86. Zhu Y, Ikeda SR. VIP inhibits N-type Ca^{2+} channels of sympathetic neurons via a pertussis toxin-insensitive but cholera toxin-sensitive pathway. *Neuron* 1994;13:657–69.
87. Herlitze S, Garcia DE, Mackie K, Hille B, Scheuer T, Catterall WA. Modulation of Ca^{2+} channels by G protein $\beta\gamma$ subunits. *Nature* 1996;380:258–62.
88. Agnew S. Voltage-regulated sodium channel molecules. *Annu Rev Physiol* 1984;46:517–30.
89. Barchi R. Probing the molecular structure of the voltage-dependent sodium channel. *Annu Rev Neurosci* 1988;11:455–95.

90. Catterall WA Molecular properties of voltage-sensitive sodium channels. *Annu Rev Biochem* 1986; 55:953–85.
91. Noda M, Ikeda T, Kayano T, et al. Existence of distinct sodium channel messenger RNAs in rat brain. *Nature* 1986a;320:188–92.
92. Kayano T, Noda M, Flockerzi V, Takahashi H, Numa S. Primary structure of rat brain sodium channel III deduced from the cDNA sequence. *FEBS Lett* 1988;228:187–94.
93. Auld VJ, Goldin AL, Krafte DS, et al. A rat brain Na^+ channel alpha subunit with novel gating properties. *Neuron* 1988;1:449–61.
94. Trimmer JS, Cooperman SS, Tomiko SA, et al. Primary structure and functional expression of a mammalian skeletal muscle sodium channel. *Neuron* 1989;3:33–49.
95. Rogart RB, Cribbs LL, Muglia LK, Kephar, DD, Kaiser MW. Molecular cloning of a putative tetrodotoxin-resistant rat heart sodium channel isoform. *Proc Natl Acad Sci USA* 1989;86:8170–4.
96. Kallen RG, Sheng ZH, Yang J, Chen LQ, Rogart RB, Barchi RL. Primary structure and expression of a sodium channel characteristic of denervated and immature rat skeletal muscle. *Neuron* 1990;4: 233–42.
97. Isom LL, De Jongh KS, Reber BFX, et al. Primary structure and functional expression of the $\beta 1$ subunit of the rat brain sodium channel. *Science* 1992;256:839–42.
98. Costa MR, Casnellie JE, Catterall WA. Selective phosphorylation of the alpha subunit of the sodium channel by cAMP-dependent protein kinase. *J Biol Chem* 1982;257:7918–21.
99. Costa M.R, Catterall WA. Cyclic AMP-dependent phosphorylation of the alpha subunit of the sodium channel in synaptic nerve ending particles. *J Biol Chem*, 1984a;259:8210–8.
100. Rossie S, Catterall WA. Cyclic AMP-dependent phosphorylation of voltage-sensitive sodium channels in primary cultures of rat brain neurons. *J Biol Chem* 1987;262:12735–44.
101. Li M, West JW, Lai Y, Scheuer T, Catterall WA. Functional modulation of brain sodium channels by cAMP-dependent phosphorylation. *Neuron* 1992;8:1151–9.
102. Schecterson LC, McKnight, GS. Role of cyclic adenosine 3′,5′-monophosphate-dependent protein kinase in hormone-stimulated beta-endorphin secretion in AtT20 cells. *Mol Endocrinol* 1991;5: 170–8.
103. Surmeier DJ, Eberwine J, Wilson CJ, Cao Y, Stefani A, Kitai ST. Dopamine receptor subtypes colocalize in rat striatonigral neurons. *Proc Natl Acad Sci USA* 1992;89:10178–82.
104. Rossie S, Gordon D, Catterall WA. Identification of an intracellular domain of a sodium channel having multiple cyclic AMP-dependent phosphorylation sites. *J Biol Chem* 1987;262:17530–5.
105. Rossie S, Catterall WA. Phosphorylation of the alpha subunit of rat brain sodium channels by cAMP-dependent protein kinase at a new site containing Ser^{686} and Ser^{687}. *J Biol Chem* 1989;264: 14220–4.
106. Murphy BJ, Rossie S, De Jongh KS, Catterall WA. Identification of the sites of selective phosphorylation and dephosphorylation of the rat brain Na^+ channel α subunit by cAMP-dependent protein kinase and phosphoprotein phosphatases. *J Biol Chem* 1993;268:27355–62.
107. Smith RD, Goldin AL. Phosphorylation of brain sodium channels in the I-II linker modulates channel function in *Xenopus* oocytes. *J Neurosci* 1996;16:1965–74.
108. Gordon D, Merrick D, Wollner DA, Catterall WA. Biochemical properties of sodium channels in a wide range of excitable tissues studied with site-directed antibodies. *Biochemistry* 1988;27:7032–8.
109. Ono K, Kiyosue T, Arita M. Isoproterenol, DBcAMP, and forskolin inhibit cardiac sodium current. *Am J Physiol* 1989;256:C1132–7.
110. Schubert B,Vandongen AMJ, Kirsch GE, Brown AM. β-adrenergic inhibition of cardiac sodium channels by dual G-protein pathways. *Science* 1989;245:516–9.
111. Costa MR, Catterall WA. Phosphorylation of the alpha subunit of the sodium channel by protein kinase C. *Cell Mol Neurobiol* 1984b;4:291–7.
112. Linden DJ, Routeenberg A. *Cis* fatty acids, which activate protein kinase C, attenuate Na^+ and Ca^{2+} currents in mouse neuroblastoma cells. *J Physiol* 1989;419:95–119.
113. Sigel E, Baur R. Activation of protein kinase C differentially modulates neuronal Na^+, Ca^{2+}, and γ-aminobutyrate type A channels. *Proc Natl Acad Sci USA* 1988;85:6192–6.
114. Dascal N, Lotan I. Activation of protein kinase C alters voltage dependence of a Na^+ channel. *Neuron* 1991;6:165–75.
115. Numann R, Catterall WA, Scheuer T. Functional modulation of brain sodium channels by protein kinase C phosphorylation. *Science* 1991;254:115–8.
116. Vassilev PM, Scheuer T, Catterall WA. Identification of an intracellular peptide segment involved in sodium channel inactivation. *Science* 1988;241:1658–61.

117. Vassilev PM, Scheuer T, Catterall WA. Inhibition of inactivation of single sodium channels by a site-directed antibody. *Proc Natl Acad Sci USA* 1989;86:8147–51.
118. Stühmer W, Conti F, Suzuki H, et al. Structural parts involved in activation and inactivation of the sodium channel. *Nature* 1989;339:597–603.
119. Patton DE, West JW, Catterall WA, Goldin AL. Amino acid residues required for fast sodium channel inactivation. Charge neutralizations and deletions in the III-IV linker. *Proc Natl Acad Sci USA* 1992;89:10905–9.
120. West JW, Patton DE, Scheuer T, Wang Y-L, Goldin AL, Catterall WA. A cluster of hydrophobic amino acid residues required for fast Na^+ channel inactivation. *Proc Natl Acad Sci USA* 1992;89:10910–4.
121. West JW, Numann R, Murphy, BJ, Scheuer T, Catterall WA. A phosphorylation site in a conserved intracellular loop that is required for modulation of sodium channels by protein kinase C. *Science* 1991;254:866–8.
122. Numann R, West JW, Li M, Smith RD, Goldin AL, Scheuer T, Catterall WA. Biphasic modulation of sodium channels by phosphorylation at two different sites. *Soc Neurosci Abst* 1992;18:1133.
123. Stafstrom CE, Schwindt PC, Crill WE. Negative slope conductance due to a persistent subthreshold sodium current in cat neocortical neurons in vitro. *Brain Res* 1982;236:221–6.
124. Stafstrom CE, Schwindt PC, Chubb MC, Crill WE. Properties of persistent sodium conductance and calcium conductance of layer V neurons from cat sensorimotor cortex in vitro. *J Physiol Lond* 1985; 53:163–70.
125. French CR, Sak P, Buchett KJ, Gage PW. A voltage-dependent persistent sodium current in mammalian hippocampal neurons. *J Gen Physiol* 1990;95:1139–57.
126. Westenbroek RE, Merrick DK, Catterall WA. Differential subcellular localization of the R_I and R_{II} Na^+ channel subtypes in central neurons. *Neuron* 1989;3:695–704.
127. Cantrell AR, Ma JY, Scheuer T, Catterall WA. Muscarinic modulation of sodium current by activation of protein kinase C in rat hippocampal neurons. *Neuron* 1996;16:1019–25.
128. Numann R, Hauschka SD, Catterall WA, Scheuer T. Modulation of skeletal muscle sodium channels in a satellite cell line by protein kinase C. *J Neurosci* 1994;14:4226–36.
129. Qu Y, Rogers J, Tanada T, Scheuer T, Catterall WA. Modulation of cardiac Na^+channels expressed in a mammalian cell line and in ventricular myocytes by protein kinase C. *Proc Natl Acad Sci USA* 1994;91:3289–93.
130. Qu Y, Rogers J, Tanada T, Scheuer T, Catterall WA. Phosphorylation of S1505 in the cardiac Na^+ channel inactivation gate is required for modulation by protein kinase C. *J Gen Physiol* 1996;108: 375–79.
131. Brown AM. Membrane-delimited cell signaling complexes: direct ion channel regulation by G proteins. *J Membr Biol* 1993;131:93–104.
132. Breitwieser GE, Szabo G. Uncoupling of cardiac muscarinic and beta-adrenergic receptors from ion channels by a guanine nucleotide analogue. *Nature* 1985;317:538–40.
133. Pfaffinger PJ, Martin JM, Hunter DD, Nathanson NM, Hille B. GTP-binding proteins couple cardiac muscarinic receptors to a K channel. *Nature* 1985;317:536–8.
134. Yatani A, Mattera R, Codina J, et al. The G protein-gated atrial K^+ channel is stimulated by three distinct G_i alpha-subunits. *Nature* 1988;336:680–2.
135. Soejima M, Noma A. Mode of regulation of the ACh-sensitive K-channel by the muscarinic receptor in rabbit atrial cells. *Pflügers Arch* 1984;400:424–31.
136. Ma JY, Li M, Catterall WA, Scheuer T. Modulation of brain Na^+ channels by a G protein-coupled pathway. *Proc Natl Acad Sci USA* 1994;91:12351–5.

Signal Transduction in Health and Disease, Advances in Second Messenger and Phosphoprotein Research, Vol. 31, edited by J. Corbin and S. Francis.
Lippincott–Raven Publishers, Philadelphia © 1997.

15

Interruption of Specific Guanylyl Cyclase Signaling Pathways

Zeren Gao, *Peter S. T. Yuen, and David L. Garbers

*Howard Hughes Medical Institute and Department of Pharmacology, The University of Texas Southwestern Medical Center at Dallas, Dallas, Texas 75235; and *University of Tennessee, Memphis, Tennessee 38163*

There have now been described a number of point and deletion mutations within both the soluble and membrane forms of guanylyl cyclase (GC) that continue to form dimers, but when recombined with the wild-type subunit, lead to either a lack of basal activity or an eradication of ligand-stimulated enzyme activity. We studied the potential utility of such point mutants using rat insulinoma (RIN) cells. Two point mutants of the α subunit of soluble GC were shown to inhibit endogenous sodium nitroprusside (SNP)-stimulated GC activity of RIN cells. To determine whether the mutant α subunits specifically interrupt signaling by the nitric oxide (NO)-sensitive soluble GCs, other GC receptors were sought in RIN cells. Surprisingly, heat-stable enterotoxin (STa) of Escherichia coli was shown to cause dramatic (up to 800-fold) increases of cyclic guanosine monophosphate (cGMP). Guanylin, a peptide proposed as an endogenous STa, also markedly increased RIN cell cyclic GMP (cGMP). Northern hybridization and Western blotting confirmed relatively high expression of GC-C in RIN cells. Isolated rat pancreatic islets were subsequently shown to also express GC-C messenger ribonucleic acid (mRNA), and STa was shown to increase islet cGMP by about 3-fold. In contrast to the ability of the mutant α subunits to block sodium nitroprusside-stimulated accumulation of cGMP, STa-stimulated increases were not affected. Therefore, dominant negative subunits of GC can act to block cGMP signaling in a highly selective manner.

INTRODUCTION

The plasma-membrane forms of guanylyl cyclase (GC) include guanylyl cyclase-A (GC-A), the A-type natriuretic peptide (ANP) receptor; guanylyl cyclase-B (GC-B), the C-type natriuretic peptide (CNP) receptor; and guanylyl cyclase-C (GC-C), the heat-stable enterotoxin (STa) receptor (1–3). Three orphan receptors also have been discovered, named GC-D, -E, and -F. GC-D appears to be olfactory specific (4), while GC-E and GC-F are found principally in the eye (5). Although cGMP exists in virtually all cells that have been studied, and although its concentrations are known to change in response to a large number of different agents (1,2), the function of this putative second messenger cGMP in most cells is not understood. Furthermore, multiple GC receptors exist in virtually all cells, thus also requiring a determination of the function of each of the individual GC family members. Specific inhibition has represented a powerful means by which to determine the function of a given signaling pathway, but such avenues of approach have not been available for the GC receptors. Agents currently used as antagonists are apparently not selective (6,7). Interruption of translation (e.g., antisense oligonucleotides) or of transcription (e.g., gene disruption) represent two potentially powerful methods by which to define the function of proteins.

Various works now suggest that the active catalytic unit of GCs is a dimer. It has been established by various laboratories that an active soluble form of GC requires the expression of both the α and β subunits (8,9), that an active adenylyl cyclase appears to require a C1 and a C2 domain (10,11), and that truncated forms of GC-A are active in the form of homodimers (12,13). Thus, the engineering of mutant subunits that would continue to dimerize but lack enzyme activity could result in effective dominant negative proteins. Yuen et al. (14) were the first to demonstrate that a single point mutation within one subunit (the α subunit) of the soluble form of GC resulted in no enzymatic activity when recombined with native β subunit. Subsequently, similar mutations in ANP-stimulated GC-A (13), a His^{105} mutation in the β1 subunit of soluble GC (15), a soluble enzyme deletion mutant (16), or a naturally occurring α-subunit insert of soluble GC (17) were shown to block either basal enzyme activity or nitric oxide (NO)-stimulated activity of GC.

MATERIALS AND METHODS

Chemicals

Guanylin was a gift from Dr. Mark Currie, Monsanto (St. Louis, MO), STa was purchased from Sigma Chemical Co. (St. Louis, MO). Atrial natriuretic peptide (ANP) and C-type natriuretic peptide (CNP) were obtained from Peninsula Laboratories (Belmont, CA).

Cells and Tissues

The insulinoma cell line RINr1046-38 and rat islets were gifts of Dr. Christopher B. Newgard and Dr. John H. Johnson at the University of Texas Southwestern Medical Center.

Effects of Various Stimuli on cGMP Concentrations

RIN cells were split at a ratio of 1:2.5 into six-well plates and incubated in M199 medium containing 11 mM glucose for 24 hr. The cells were then incubated in a buffer containing 20 mM 4-(2-hydroxyethyl)-1-piperazine-N′-2-ethanesulfonic acid (HEPES), pH 7.2; 25.5 mM $NaHCO_3$; 0.5% bovine serum albumin (BSA); 2.5 mM $CaCl_2$; 114 mM NaCl; 4.7 mM KCl; 1.2 mM KH_2PO_4; 1.16 mM $MgSO_4$; and 0.3 mM 3-isobutyl-1-methylxanthine (IBMX) for 10 min prior to the addition of the various agents for 1 to 3 hr. At the end of the reaction, cells were extracted with perchloric acid at 1-ml/well and assayed for cGMP as described previously (3).

Fresh isolated islets (18) were cultured in six-well plates (80 islets/well) for 2 days in RPMI 1640 medium, or were used immediately. In either case, the islets were then incubated for two 30-min intervals in the buffer described earlier, containing 2 mM glucose, and were incubated for 10 min with the buffer containing 0.3 mM IBMX prior to incubation for 30 min with or without STa. After incubation, islets were treated with perchloric acid at 1 ml/well and were assayed for their cGMP concentration.

Immunoblots

Insulinoma cells from nine plates were pooled and homogenized. The pellet obtained by centrifugation was then solubilized in 1% Triton X-100. Supernatant fluids were precleared by incubating with preimmune serum at 4°C for 1 hr, followed by 150 μl 10% (vol/vol) protein G agarose, rocked at 4°C for 30 min, and spun at 4°C for 5 min. The supernatant fluid was added to 2 μl anti-GC-C serum (Z659) (19), which had been previously incubated with either the GC-C carboxy-terminal (C-terminal) peptide (NNSDHDSTYF) or with a control peptide (QLEKKKEELRVLSNHLAIEKKTET), and was then incubated at 4°C for 1 hr; this was followed by incubation with protein-G–agarose and rocking for 30 min at 4°C. The pellet obtained by centrifugation was washed three times with a solution containing 50 mM HEPES, pH 7.4; 150 mM NaCl; 1% Triton X-100; and 0.1% sodium dodecyl sulfate (SDS). It was then heated at 100°C for 3 min in a solution containing 62.5 mM Tris-HCl, 20% glycerol, 2.3% SDS, and 5% β-mercaptoethanol. Subsequently the sample was subjected to SDS-polyacrylamide gel electrophoresis (SDS-PAGE) and then transferred to nitrocellulose. The nitrocellulose blots were incubated with Z659 serum (1:5,000), followed by 1:50,000 goat antirabbit IgG conjugated to horseradish peroxidase (Biosourrce, Camarillo, CA) and detected by Enhanced Chemiluminescent (ECL) (Arlington Heights, IL).

RNA ANALYSIS

RNA from RIN cells and rat islets were purified by the method of Chomcynski and Sacchi (20). Small intestine and kidney RNA were prepared as previously described (3). RNA (30 μg) was applied to a denaturing gel, transferred to nylon, and fixed with ultraviolet (UV) light. The filters were then probed with the *Eco*R1/*Kpn*1 fragment of the GC-C extracellular domain. Hybridization was performed at 42°C for 16 hr. Filters were washed in 0.2x standard saline citrate (SSC), and 0.2% SDS at 62°C.

RESULTS

Two point mutations within the α subunit of soluble GC are capable of blocking all GC activity when the α subunit is recombined with the wild-type β subunit, and these subunits are capable of acting as dominant negative proteins in cultured cells expressing endogenous α and β subunits (14). NO has been suggested to regulate insulin secretion through its effects on soluble GC (21,22), and therefore a dominant negative subunit could prove of great utility in defining whether or not the NO effects on insulin secretion are in fact mediated by activation of the soluble form of GC.

In initial experiments, insulinoma cells at either Passage 17 or 40 were treated with 100 nM STa, 1 nM ANP, 30 nM CNP, 100 μM sodium nitroprusside (SNP), or 1 μM guanylin, and the cells' cGMP responses were measured (Fig. 1). CNP, which interacts with GC-B, failed to significantly increase cGMP, suggesting an absence of the GC-B receptor. The ANP receptor (GC-A) and a soluble form of guanylyl cyclase that responds to SNP were apparently expressed within the insulinoma cells (Fig. 1). Dramatic increases of cGMP (200-fold to 350-fold) were seen in response to STa or guanylin, and although the increases were similar at the concentrations used in this initial study, STa at higher concentrations caused up to 800-fold increases of cGMP (Fig. 2). The relative potency of STa to increase cGMP was similar to that previously observed in GC-C-overproducing cells (3) or in T84 cells (23), suggesting that GC-C mediated the STa/guanylin response.

RNA blots of insulinoma cells confirmed the presence of high levels of mRNA for GC-C (Fig. 3B), and immunoblots demonstrated high expression of the protein (Fig. 3A). Therefore, GC-C appears to be the insulinoma-cell receptor for STa and guanylin.

Given that GC-C and the soluble form of GC were both present in the insulinoma cells, we could then ask whether or not the dominant negative α subunits would block the SNP-stimulated increases in cGMP but not the STa-stimulated increases.

Stable RIN cell lines were established that overproduced either wild-type or mutant forms of the α1 subunit. Such stable cell lines were then treated with SNP or STa, and the cellular levels of cGMP were measured (Table 1). Two clonal lines expressing the mutant α1 subunit showed a marked depression in response to SNP, but the responses of cGMP to STa were not altered.

Although an ultimate goal of the studies was to then determine whether or not NO or other agents modulated insulin secretion by increasing cGMP, the results of previous work (21,22) could not be reproduced in the RIN cells that we possess, even

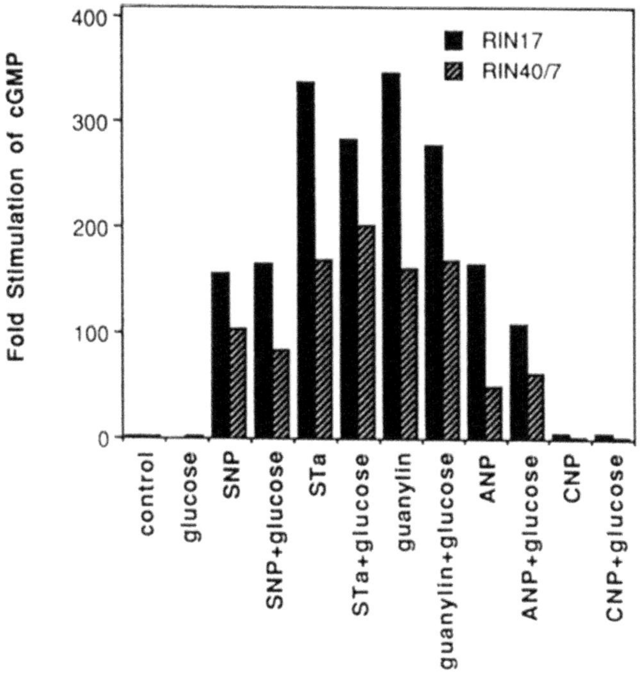

FIG. 1. Effects of glucose, sodium nitroprusside (SNP), STa, guanylin, ANP, or CNP on cGMP concentrations of insulinoma cells at Passage 17 or 40. Cells either at Passage 17 or 40 were incubated with 100 μM SNP, 100 nM STa, 1 μM guanylin, 1 nM ANP, or 30 nM CNP in the presence or absence of 5 mM glucose for 1 hr, as described in text. IBMX (0.3 mM) was included in all reaction mixtures. Each point represents the mean of triplicate, independent determinations.

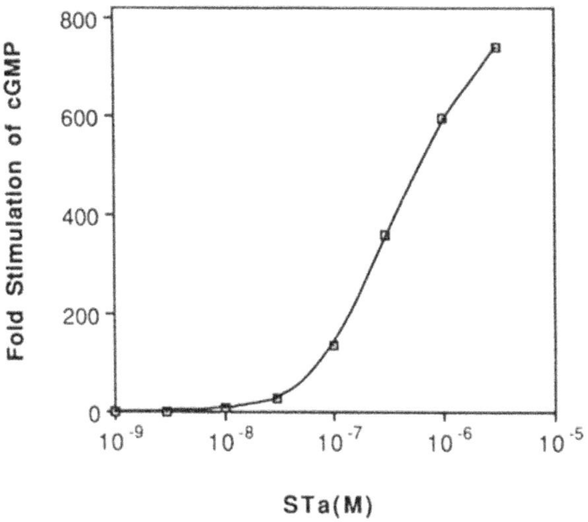

FIG. 2. Increases in cGMP in insulinoma cells as a function of STa concentration. Insulinoma cells at Passage 9 were incubated with the concentrations of STa indicated for 10 min in the presence of 0.3 mM IBMX. The experiment is typical of three experiments. The values are means of three independent measurements.

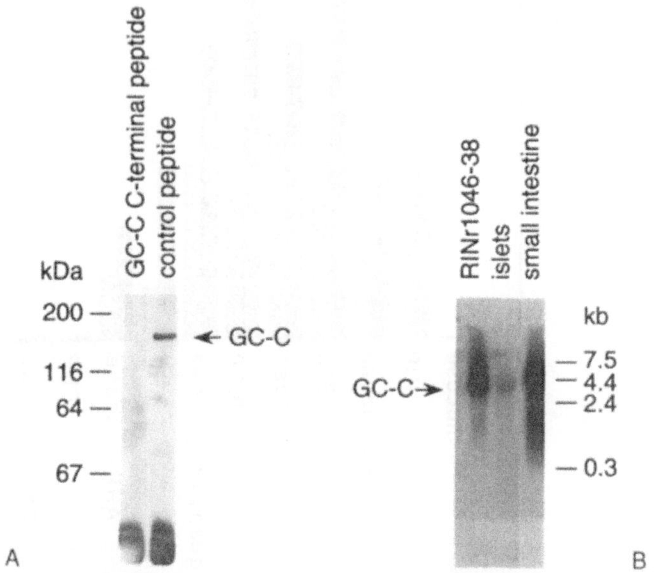

FIG. 3. Immunoblots and RNA blots for GC-C. **A:** In immunoblot analysis, insulinoma cells at Passage 36 were obtained and solubilized membrane protein was immunoprecipitated with GC-C antiserum (Z659) after incubation with either GC-C c-terminal peptide or a control peptide. The resultant pellet was then dissolved in loading buffer and subjected to SDS–PAGE, after which the gel protein was transferred to nitrocellulose. Blots were incubated with Z659 antiserum (1/5,000). **B:** In Northern blot analysis, RNA was purified from either isolated rat pancreatic islets or insulinoma cells at Passage 36. Total RNA (30 μg) was fractionated on a 1% denaturing agarose gel and blotted onto nylon. The blot was probed with the *Eco*R1/*Kpn*1 fragment of the GC-C cDNA (extracellular domain). Migration of molecular standards is shown on the side of the figure.

TABLE 1. *Effect of STa or SNP on cGMP Content of RIN Cells Stably Transfected with pCMV5 Alone, Wild-type α1 DNA, or Mutant α1 DNA: α1D513A or α1D529A*

	Fold Stimulation	
DNA-transfected	SNP (100 μM)	Sta (100nM)
pCMV5	84.5 ± 14.1	102 ± 9.7
α1	257 ± 34	90.6 ± 13
α1D513A	6.4 ± 1.5	116 ± 28.9
α1D529A	1 ± 0.06	71.8 ± 9.3

RIN-cell cGMP levels were determined as described under "Methods" in the absence or presence of 100 nM STa or 100 μM SNP. Values in parentheses represent SEM (n=3).

after transfection with a glucose transporter (Z. Gao, P. S. T. Yuen, C. B. Newgard, Y. Nagasawa, R. H. Unger, D. L. Garbers, *unpublished observations*).

ACKNOWLEDGMENTS

We thank Drs. R. H. Unger, S. Ferber, and S. Schulz for their advice, and Dr. C. McAllister, Ms. C. Green, and K. McCorkle for their technical assistance in these studies. The work was supported in part by Grant I-1233 from the Robert A. Welch Foundation.

REFERENCES

1. Drewett JG, Garbers DL. The family of guanylyl cyclase receptors and their ligands. *Endocrine Rev* 1994;15:135–62.
2. Garbers DL. Guanylyl cyclase receptors and their endocrine, paracrine, and autocrine ligands. *Cell* 1992;71:1–4.
3. Schulz S, Green CK, Yuen PST, Garbers DL. Guanylyl cyclase is a heat-stable enterotoxin receptor. *Cell* 1990;63:941–8.
4. Fulle H-J, Vassar R, Foster DC, Yang R-B, Axel R, Garbers DL. A receptor guanylyl cyclase expressed specifically in olfactory sensory neurons. *Proc Natl Acad Sci USA* 1995;92:3571–5.
5. Yang R-B, Foster DC, Garbers DL, Fulle H-J. Two membrane forms of guanylyl cyclase found in the eye. *Proc Natl Acad Sci USA* 1994;92:602–6.
6. Mayer B, Brunner F, Schmidt K. Inhibition of nitric oxide synthesis by methylene blue. *Biochem Pharm* 1993;45:367–74.
7. Mayer B, Brunner F, Schmidt K. Novel actions of methylene blue. *Europ Heart J* 1993;14, Suppl. I:22–6.
8. Harteneck C, Wedel B, Koesling D, Malkewitz J, Bohme E, Schultz G. Molecular cloning and expression of a new alpha-subunit of soluble guanylyl cyclase. Interchangeability of the alpha-subunits of the enzyme. *FEBS Lett* 1991;292:217–22.
9. Buechler WA, Nakane M, Murad F. Expression of soluble guanylate cyclase activity requires both enzyme subunits. *Biochem Biophys Res Commun* 1991;174:351–7.
10. Tang WJ, Gilman AG. Construction of a soluble adenylyl cyclase activated by Gs alpha and forskolin. *Science* 1995;268:1769–72.
11. Tang WJ, Gilman AG. Adenylyl cyclases. *Cell* 1992;70:869–72.
12. Thorpe DS, Niu S, Morkin E. Overexpression of dimeric guanylyl cyclase cores of an atrial natriuretic peptide receptor. *Biochem Biophys Res Commun* 1991;180:538–44.
13. Thompson DK, Garbers DL. Dominant negative mutations of the guanylyl cyclase-A receptor. Extracellular domain deletion and catalytic domain point mutations. *J Biol Chem* 1995;270:425–30.
14. Yuen PST, Doolittle L, Garbers DL. Dominant negative mutants of nitric oxide-sensitive guanylyl cyclase. *J Biol Chem* 1994;269:791–3.
15. Wedel B, Humbert P, Harteneck C, Foerster J, Malkewitz J, Bohem E, Schultz G, Koesling D. Mutation of His-105 in the beta 1 subunit yields a nitric oxide-insensitive form of soluble guanylyl cyclase. *Proc Natl Acad Sci USA* 1994;91:2592–6.
16. Wedel B, Harteneck C, Foerster J, Friebe A, Schultz G, Koesling D. Functional domains of soluble guanylyl cyclase. *J Biol Chem* 1995;270:24871–7.
17. Behrends S, Harteneck C, Scheltz G, Koesling D. A variant of the alpha 2 subunit of soluble guanylyl cyclase contains an insert homologous to a region within adenylyl cyclases and functions as a dominant negative protein. *J Biol Chem* 1995;270:21109–13.
18. Johnson JH, Crider BP, McCorkle K, Alford M, Unger RH. Inhibition of glucose transport into rat islet cells by immunoglobulins from patients with new-onset insulin-dependent diabetes mellitus. *N Engl J Med* 1990;322:653–9.
19. Vaandrager AB, Schulz S, deJonge HR, Garbers DL. Guanylyl cyclase C is an N-linked glycoprotein

receptor that accounts for multiple heat-stable enterotoxin-binding proteins in the intestine. *J Biol Chem* 1993;268:2174–9.
20. Chomczynski P, Sacchi N. Single-step method of RNA isolation by acid guanidinium thiocyanate-phenol-choroform extraction. *Analytical Biochem* 1987;162:156–9.
21. Schmidt HHHW, Warner, TD, Ishii K, Sheng H, Murad F. Insulin secretion from pancreatic B cells caused by L-anginine-derived nitrogen oxides. *Science* 1992;255:721–3.
22. Laychock SG, Modica ME, Cavanaugh DT. L-arginine stimulates cyclic guanosine 3′,5′-monophosphate formation in rat islets of Langerhans and RINm5F insulinoma cells; evidence for L-arginigne:nitric oxide synthase. *Endocrinology* 1991;129:3043–52.
23. Singh S, Singh G, Heim J-M, Gerzer R. Isolation and expression of a guanylate cyclase-coupled heat stable enterotoxin receptor cDNA from a human colonic cell line. *Biochem Biophys Res Commun* 1991;179:1455–63.

Signal Transduction in Health and Disease, Advances in Second Messenger and Phosphoprotein Research, Vol. 31, edited by J. Corbin and S. Francis. Lippincott–Raven Publishers, Philadelphia © 1997.

16

Structure, Function, and Regulation of Human cAMP-Dependent Protein Kinases

Kjetil Taskén, Bjørn S. Skålhegg, Kristin Austlid Taskén, Rigmor Solberg, Helle K. Knutsen, Finn Olav Levy, Mårten Sandberg, Sigurd Ørstavik, Turid Larsen, Ann Kirsti Johansen, Torkel Vang, Hans Petter Schrader, Nils T. K. Reinton, Knut Martin Torgersen, Vidar Hansson, and Tore Jahnsen

Institute of Medical Biochemistry, University of Oslo, N-0317 Oslo, Norway

The cyclic adenosine monophosphate (cAMP) signaling pathway involves hormone receptors that, upon binding of ligand, transduce their signal across the cell membrane via G-proteins that interact with membrane-bound adenylyl cyclase either to increase or reduce the production of cAMP. cAMP has been implicated in a number of cellular processes such as metabolism, gene regulation, cell growth and division, cell differentiation, and sperm motility, as well as ion-channel conductivity and neurotransmitter release (reviewed and referenced in [1,2]). With the exception of certain ion channels directly regulated by cAMP, all known effects of cAMP in eukaryotic cells are mediated by cAMP-dependent protein kinase (cAK).

STRUCTURAL FEATURES OF cAK

In the absence of cAMP, the dormant cyclic adenosine monophosphate-dependent protein kinase (cAK) holoenzyme is a tetramer consisting of two catalytic subunits (C) bound to a regulatory-subunit (R) dimer. cAMP binds cooperatively to two sites on each R protomer (reviewed in [3,4]). Upon binding of four molecules of cAMP, the enzyme dissociates into an R-subunit dimer with four bound molecules of cAMP and two free, active C subunits that phosphorylate serine and threonine residues on specific substrate proteins.

Initially, two different isozymes of cAK, termed types I and II (cAKI and cAKII, respectively), were identified on the basis of their pattern of elution from 2-diethylaminoethanol (DEAE)-cellulose columns (3). The cAKI and cAKII isozymes, eluting at salt concentrations between 25 and 50 mM and 150 and 200 mM NaCl, respectively, were shown to contain different R subunits, termed RI and RII (3). Molecular cloning techniques have, however, revealed a great heterogeneity in both the R and C subunits of these isozymes.

DISTINCT GENES ENCODE SEVEN SUBUNITS OF cAK IN HUMANS

Cloning of complementary deoxyribonucleic acids (cDNAs) for human regulatory subunits of cAK has identified two RI subunits, termed RIα (5) and RIβ (6) and two RII subunits, termed RIIα (7) and RIIβ (8), as separate gene products. The RIα and RIβ subunits reveal high similarity (81% identity at the amino-acid level), as do the RIIα and RIIβ subunits (68% identity at the amino-acid level). The distinct R subunits are conserved in higher eukaryotes, and reveal high interspecies homology (87% to 98% identity at the amino-acid level).

Two distinct C subunits of cAK were initially identified by molecular cloning, and were designated Cα (9) and Cβ (10). The cloning of the Cα and Cβ subunits from human testis by homology screening at low stringency revealed an additional cDNA, encoding a distinct C subunit designated Cγ, that has so far only been demonstrated in human testis (11). The Cγ cDNA translates a full-length protein by *in vitro* translation (12). Furthermore, transfection and expression of Cγ in kin8 cells revealed expression of a cAMP-dependent kinase activity that phosphorylated histone, but not Kemptide, and was insensitive to inhibition by protein kinase inhibitor I (PKI) (13). The recent cloning of the Cγ gene revealed an intronless open-reading frame with 5′- and 3′-untranslated regions corresponding to the Cγ cDNA (N. Reinton et al., *unpublished results*). Furthermore, comparison of the Cγ gene with the Cα cDNA revealed that the 5′- and 3′-flanking regions of the Cγ gene are colinear with the entire Cα messenger RNA (mRNA) (82% identity). In addition, the Cγ gene contains a rudimentary polyA tail and is flanked by 11-bp direct repeats, indicating that the Cγ gene originated as a Cα retroposon. Based on the degree of similarity with the Cα mRNA outside the reading frame, the Cγ gene is approximately 60 million years old. A Cα pseudogene, termed Cx, has been reported from mouse (14). Cx has several stops and frameshift mutations, and no detectable expression, and is by sequence comparison apparently not of the same origin as Cγ. In contrast, the Cγ gene is transcribed in the testis and translates a full-length enzyme with apparent activity (11,13). The Cγ gene is thus a functional retroposon. The few other reported functional retroposons also display selective expression in the testis (for references, see [14]). The Cγ subunit is homologous to both Cα and Cβ (83% and 79% identity at the amino-acid level, respectively), but similarity between Cγ and Cα/Cβ is higher at the nucleotide level than at the amino-acid level (11). This observation indicates loss of selection pressure for some period of time. However, the fact that the Cγ gene still contains an open-reading frame may indicate introduction of a novel selection pressure for the maintenance of the Cγ gene product for some specialized function in the testis.

Thus, in humans, molecular cloning has revealed a total of seven distinct genes

TABLE 1. *Chromosomal Localization of Human cAK Subunits*

Subunit	Locus Name	Chromosome	Reference
Cα	PRKACA	19p13.1	(57)
Cβ	PRKACB	1p31–32	(58, 59; R. Solberg *unpublished results*)
Cγ	PRKACG	9q13	(60)
RIα	PRKAR1A	17q23–24	(23,61)
RIα pseudogene	PRKAR1AP	1p21–31	(23)
RIβ	PRKAR1B	7p22–pter	(62)
RIIα	PRKAR2A	3p21.3–21.2	(63)
RIIβ	PRKAR2B	7q22	(62,64)

encoding subunits of cAK (5–8,11). The chromosomal mapping of the cAK-subunit genes shows that the distinct genes are localized to various chromosomes, with no evidence of clustering (Table 1). At present, no hereditary diseases have been assigned to any of the cAK genes. It is possible that the closely related isomeric forms of RI, RII, and C subunits serve to rescue function in case of mutations, whereas multiple gene-function knockouts may be lethal. Characterization of knockout mice from gene disruption experiments (15) will probably enlighten this topic.

Upstream regulatory sequences have been reported for the genes encoding RIα (16,17), RIβ (18), RIIα (19), RIIβ (20,21), Cα (22), and Cβ (22). All of these genes have GC-rich and TATA-less promoters with multiple transcription start sites, which are characteristics of housekeeping genes not normally subject to extensive regulation. However, it has now been established that many TATA-less promoters confer cell-specific expression and regulation by external stimuli (for references, see [18]). We are also in the process of characterizing the promoter responsible for regulating the transcription of the Cγ gene. Furthermore, the cloning of a processed human pseudogene for RIα with an alternate 5′-nontranslated area originating from an upstream region (exon 1a) of the RIα gene indicated an alternate transcription-initiation site, and that transcription of the RIα gene may be directed by two different promoters (23). The cloning of the human RIα gene demonstrated that the gene consists of nine coding exons preceded by two different, untranslated leader exons (17). Two alternately spliced RIα mRNAs (RIα1a, RIα1b) that originate from initiation at distinct promoters preceding the two different leader exons could be identified by reverse transcriptase–polymerase chain reaction (RT–PCR) (17). This further enhances the possibilities of regulation.

DOMAIN STRUCTURE AND ISOZYME COMPOSITION

Structure of the C Subunits

All of the C subunits (Cα, Cβ, Cγ) of cAK have catalytic core motifs that are common to all protein kinases (24) and involve a Mg-adenosine triphosphate (MgATP) binding site as well as a peptide binding site. The crystal structure of the murine Cα subunit has been reported, and was the first protein kinase crystal structure available (25).

Structure of the R Subunits

The RI and RII subunits of cAK contain an amino-terminal (N-terminal) dimerization domain and a region responsible for interaction with the C subunit, and the carboxy terminus (C-terminus) contains two tandem cAMP-binding sites, termed Site A and Site B (3,4). Of the two tandem cAMP-binding sites, only Site B is exposed in the inactive tetrameric cAK complex. Binding of cAMP to this site enhances binding of cAMP to the A site in a positively cooperative fashion, as a result of a conformational change in the molecule. The characteristics of the two cAMP-binding sites have been described in detail elsewhere (reviewed in [2–4]), as have the relative affinities and site selectivities of a wide array of chemically modified cAMP analogues (26). The crystal structure of a monomeric RI deletion mutant (Δ1–91) was recently reported (27), and provides new insight into the molecular interactions of cAMP with the A and B sites of the R subunit.

Isozyme Composition and Characteristics

It is generally assumed that the catalytic subunits of cAK associate freely with homodimers of all the R subunits. The cAKI ($RI\alpha_2C_2$ and $RI\beta_2C_2$) and cAKII ($RII\alpha_2C_2$ and $RII\beta_2C_2$) holoenzymes have been reported to have distinct biochemical properties. RIβ holoenzymes are 2-fold to 7-fold more sensitive to cyclic nucleotides than are the RIα holoenzymes (28,29). Results from kinetic studies employing human recombinant R-subunit proteins indicate that the different activation constants (K_{act}) of $RI\alpha_2C_2$ and $RI\beta_2C_2$ holoenzymes are due to a 2.8-fold difference in affinity for the C subunit, and not to differences in cAMP binding (K. Taskén, R. Kopperud, and S. O. Døskeland, *unpublished results*). Furthermore, the RIIα and RIIβ holoenzymes elute from DEAE-cellulose columns at different positions in the cAKII area, and RIIα expressed at high levels will compete with RIIβ for binding the C subunit, indicating either a higher affinity for the C subunit or a higher threshold for cAMP-induced dissociation (30).

Characterization of a cell line almost completely devoid of cAKII revealed the presence of a new isozyme of cAK consisting of RIα–RIβ heterodimers with associated phosphotransferase activity. This isozyme elutes in the position of cAKII upon DEAE-cellulose chromatography (31). Formation of RIα–RIβ heterodimeric complexes was also demonstrated *in vitro* by coimmunoprecipitation, using recombinant proteins (Fig. 1). Furthermore, some indications of RIIα–RIIβ heterodimeric complexes are also available (K. Taskén, *unpublished results*), but further work is required to demonstrate such isozymes in living cells. Heterodimerization of R subunits increases the number of potential isozymes within a cell, and adds to the possibilities for diversification of cAMP-mediated signals.

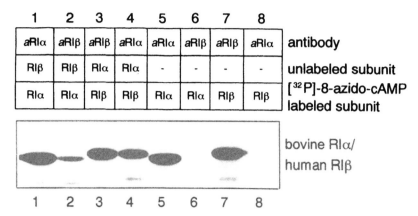

1	2	3	4	5	6	7	8	
aRIα	aRIβ	aRIβ	aRIα	aRIα	aRIβ	aRIβ	aRIα	antibody
RIβ	RIβ	RIα	RIα	-	-	-	-	unlabeled subunit
RIα	RIα	RIβ	RIβ	RIα	RIα	RIβ	RIβ	[^{32}P]-8-azido-cAMP labeled subunit

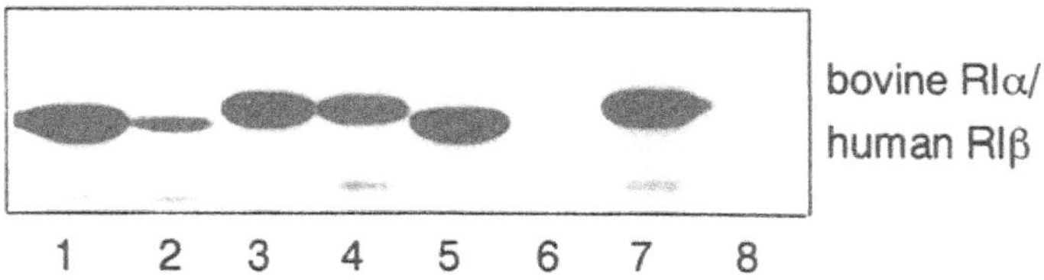

FIG. 1. *In vitro* formation of RIα–RIβ heterodimers of cAK (31). Recombinant RIα (Lanes 1 and 2, Lanes 5 and 6) or RIβ (Lanes 3 and 4, Lanes 7 and 8) were photoaffinity labeled with 8-azido-[^{32}P]cAMP and subsequently allowed to dimerize in the presence (Lanes 1 to 4) or absence (Lanes 5 to 8) of a 5-fold molar excess of the opposite unlabeled subunit. Immunoprecipitation with antibodies directed against the labeled subunit (Lanes 1, 3, 5, and 7) revealed strong bands representing RIα or RIβ. When antibodies directed against the opposite subunit were employed (Lanes 2, 4, 6, and 8), coimmunoprecipitation of labeled RIα by the RIβ antibody was demonstrated in the presence of unlabeled RIβ (Lane 2), whereas in the absence of RIβ, no coimmunoprecipitation was observed (Lane 6). Conversely, coimmunoprecipitation of labeled RIβ by the RIα antibody was observed only in the presence of unlabeled RIα (Lane 4), whereas the control experiment showed no coimmunoprecipitation (Lane 8). From (31), with permission.

CELL SPECIFIC EXPRESSION AND REGULATION OF cAK SUBUNITS

Differential Expression

The α subunits (Cα, RIα, RIIα) of cAK are expressed in almost all cells and tissues (2). The β subunits (Cβ, RIβ, RIIβ) are predominantly expressed in brain and gonadal tissues (2), although their low-level expression can be shown in a wide range of human tissues (6; K. Taskén, *unpublished observations*). Expression of the Cγ subunit has so far been demonstrated only in the human testis (11).

Gonadal tissues have a high level of α subunits as well as β subunits of cAK, and the rat testis has proved to be a good model system for studies of differential regulation of the various cAK subunits. Distinct developmental changes in the expression of cAK subunits occur (32,33). The RIα, RIβ, and Cα subunits in germ cells are induced at premeiotic and meiotic stages, whereas the RII subunits are induced only during spermatid elongation. The Cβ mRNA is present in peritubular cells and Leydig tumor cells, but not in Sertoli cells or germ cells (32).

Hormonal Regulation

Levels of expression of the different cAK subunits are subject to regulation by peptide hormones and cAMP (34–36), as well as by glucocorticoids and vitamin D

(37,38). Regulation of levels of cAK subunits by hormones acting through cAMP may serve as an autologous sensitization/desensitization mechanism of the cAMP effector system. cAMP-mediated regulation of cAK subunits acts through alterations in gene transcription (39,40) and mRNA stability (41), as well as through altered stability of the R- and C-subunit proteins after dissociation of the holoenzyme by cAMP (40,42). The protein kinase C (PKC) pathway represents another major signaling cascade in cells, and crosstalk between these two signaling systems is seen at the level of cAK (43,44).

Rat Sertoli cells serve as a good model system for studies of hormone responsiveness in general and of cAK regulation in particular. Follicle-stimulating hormone (FSH) and cAMP induce mRNA for RIα, RIIα, RIIβ, and Cα. However, the responses differ greatly in magnitude. Whereas cAMP-dependent stimulation of RIα, RIIα, and Cα mRNAs raises their levels 2-fold to 4-fold above the basal levels, the increase in RIIβ mRNA is approximately 50-fold above basal (36,39). The upregulation of RIα, RIIβ, and Cα mRNAs by cAMP is primarily due to an increased transcriptional activity (39,40), and in the case of RIIβ also seems to involve increased stability of the mRNA (41). In Sertoli cells, changes in mRNA levels are associated with qualitatively similar changes in RIα, RIIα, and RIIβ protein (35).

Different mechanisms are involved in the regulation of the RIIβ and RIα genes. Whereas transcriptional activity of the RIα gene is induced rapidly (maximal at 30 min), the induction of the RIIβ gene is much slower (maximum > 120 min) in nuclear run-on experiments (39,40). Furthermore, the RIα gene is superinduced by combined treatment with cAMP and a protein synthesis inhibitor (cycloheximide). In contrast, inhibition of protein synthesis almost completely blocks the cAMP-mediated induction of the RIIβ gene (39). Regulation of the RIIα gene is qualitatively similar to that of RIIβ, but quantitatively less pronounced.

In transfection studies, a 20-fold induction of the RIIβ gene is observed after stimulation of rat Sertoli cells by cAMP (H. K. Knutsen and K. A. Taskén, *unpublished results*). A cAMP-responsive region has been identified and is currently under investigation. This region does not have a consensus cAMP response-element (CRE), and shows only weak homology to cAMP-responsive regions identified in other slowly cAMP-responding genes.

The RIα and RIIβ genes are also subject to differential regulation by PKC (44). Thus, there is extensive evidence showing differential mechanisms of regulation of the R-subunit genes. The RIα gene, which contains a CRE, seems to be regulated by cAMP, with characteristics similar to those of the CRE-regulated c-*fos* gene. Furthermore, the two alternately spliced RIα mRNAs (RIα1a and RIα1b) are regulated differentially by cAMP (17). The basal levels of the RIα1a mRNA initiated at the upstream promoter are low, but the mRNA is strongly induced by cAMP (40-fold) to approximately the same levels as RIα1b mRNA (4-fold increase). In addition, transfection studies in Sertoli cells confirm the low basal activity of the RIα1a promoter, which confers a 15-fold increase in cAMP-responsiveness on a downstream chloramphenicol acetyl transferase (CAT) reporter. In contrast, the stronger RIα1b promoter reveals higher basal activity but a weaker response (7-fold) to cAMP (R. Solberg, unpublished results).

SUBCELLULAR LOCALIZATION AND ISOZYME-SPECIFIC EFFECTS OF cAMP-DEPENDENT PROTEIN KINASE

The heterogeneity in the R and C subunits of cAK, and the possibility that all of the C subunits may associate freely with different R subunit homo- or heterodimers, provides the potential for a large number of isozymes. Among the lines of evidence mentioned above that are compatible with the notion that specific functions can be assigned to the various isozymes are that the various isozymes have distinct biochemical properties, and that the R and C subunits are differentially regulated by hormones and reveal cell-specific expression. In addition to this, the distinct subcellular localization of cAK subunits and the demonstration of compartmentalized effects of cAMP strongly supports the concept of isozyme-specific effects. Studies employing cAMP analogue pairs that selectively activate cAKI and cAKII (4,26) also demonstrate isozyme-specific effects (see below).

Isozyme-Specific Effect of cAMP-Dependent Protein Kinase in T-Cell Proliferation

Human peripheral blood T lymphocytes contain both cAKI ($RI\alpha_2C\beta_2$) and cAKII ($RII\alpha_2C\beta_2$) in a proportion of 3:1 (45). In resting T cells the cAKI is 75% soluble, whereas 95% of the cAKII is particulate. Quiescent T lymphocytes can be activated to proliferate by cross-linking the T-cell receptor (TCR)/CD3 complex (TCR/CD3). T-cell proliferation induced through the TCR is sensitive to inhibition by cAMP. Figure 2 demonstrates that this is a cAKI-mediated effect. cAKI-specific analogue pairs (8-piperidino-cAMP [8-pip] and 8–aminohexylamino-cAMP [8-AHA]) had a synergistic effect of more than 1 log unit in inhibiting incorporation of [^{3}H]thymidine into proliferating T cells when compared to the effect of 8-AHA alone. No such synergism was observed with a cAKII-specific analogue pair (8-(4-chlorophenylthio)cAMP [8-CPT] and N^6-benzoyl-cAMP [N^6-Bnz]). Thus, inhibition of T-cell proliferation by cAMP appears to be a cAKI-mediated effect.

cAKI Redistributes to and Colocalizes with the T-Cell Receptor Complex During Anti-CD3-Induced T-Lymphocyte Activation and Capping

The subcellular localization of cAKI and cAKII was examined during T-cell activation (46). Figure 3 shows that in quiescent T cells the TCR/CD3 complex is widely distributed on the cell surface (Fig. 3A). In this situation, RIα is almost homogenously distributed within the cell (Fig. 3B), whereas RIIα is localized to one distinct spot (Fig. 3C) in close proximity to the nucleus. This is in agreement with our previous observation showing that cAKI is primarily soluble whereas cAKII is particulate in resting T lymphocytes (45), and with observations of centrosomal anchoring in lymphoblasts (47). When capping is induced (Fig. 3D–F), the TCR/CD3

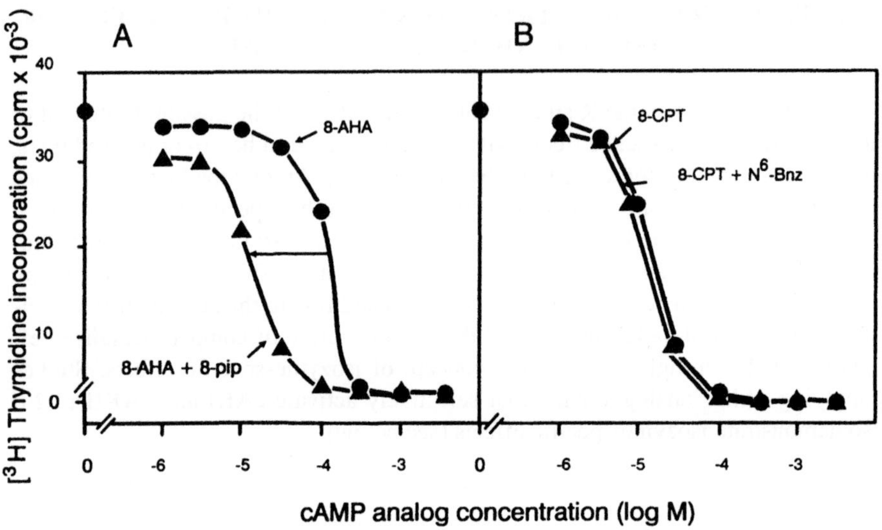

FIG. 2. Synergistic activation of cAKI using site-selective cAMP analogues in intact T lymphocytes (45). Site-selective cAMP analogues that would complement each other in activating either cAKI or cAKII were tested for their ability to synergize in the inhibition of T-lymphocyte proliferation. Inhibition of anti-CD3 stimulated [^{3}H]thymidine incorporation by 8-AHA in the absence (●) or presence (▲) of a subinhibitory concentration (7% of IC_{50}, 90 μM) of 8-pip (**A**) or by 8-CPT in the absence (●) or presence (▲) of a subinhibitory concentration (7% of IC_{50}, 30 μM) of N^6-Bnz (**B**). Note that the addition of a small concentration of 8-pip, that had no effect by itself, produced a distinct left-shift of the inhibition curve by 8-AHA (*arrow*). 8-pip and 8-AHA complements each other in the preferential activation of cAKI by binding to the cAMP binding sites A and B of RI, respectively. In contrast, no left-shift of the inhibition curve by 8-CPT was observed when N^6-Bnz was added to complement 8-CPT in the preferential activation of cAKII. From Skålhegg, BS et al. *J Biol Chem* (1992) 267:15707–15714 "Cyclic AMP-dependent protein kinase type I mediates the inhibitory effect of 3′,5′ cyclic adenosine monophosphate on cell replication in human T lymphocytes", with permission.

complex is capped at one pole of the cells (Fig. 3D). Double immunofluorescence staining of the same cell with an anti-RIα antibody clearly shows that RIα redistributes to and colocalizes with the TCR/CD3 complex during activation and capping. In contrast, no effect of TCR/CD3-capping is observed in the subcellular distribution of RIIα. Immunoprecipitations of TCR/CD3 complex from capped and uncapped cells further demonstrate that 70% to 75% of cAKI phosphotransferase activity, cAMP binding activity, and immunoreactive RIα and C are redistributed to and coimmunoprecipitated with the TCR/CD3 complex only after capping and activation of T lymphocytes. Cross-linking of MHC Class I antigens, not involving T-cell activation, does not induce any redistribution or cocapping of cAKI (46).

Colocalization of cAKI with the TCR/CD3 complex during capping and activation of T cells strongly supports our observation that inhibition of T-cell proliferation by cAMP is mediated by cAKI. The localization of cAKI in close proximity to the TCR/CD3 complex probably serves to establish an inhibitory pathway that uncouples the TCR/CD3 complex from its intracellular signaling system. In addition, T-cell activation by stimulation of the CD28 cell-surface molecule in conjunction

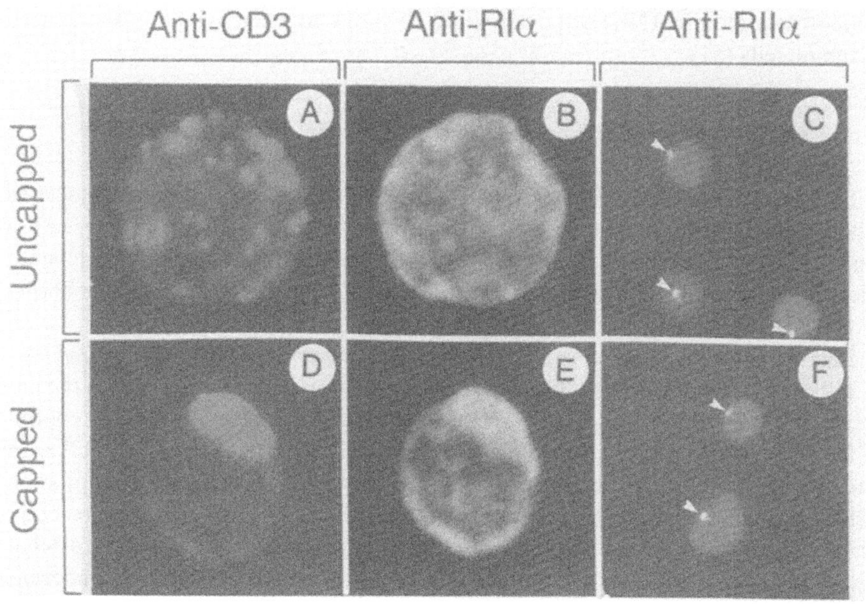

FIG. 3. Subcellular localization of cAKI and cAKII in quiescent and activated T cells (46). Localization of the TCR-CD3 complex, RIα, and RIIα in uncapped **(A–C)** or TCR-CD3 capped **(D–F)** T cells was examined after immunofluorescence labeling in a confocal immunofluorescence microscope. **A:** TCR/CD3 complex in uncapped T cells visualized with anti-CD3 antibody and rhodamine-isothiocyanate (RITC)-conjugated antibody to IgG_1 in the second layer. (Note: the TCR/CD3 is scattered on the cell surface on uncapped T cells and a few small patches are seen.) **B:** The same cell as shown in (A), after overnight incubation of permeabilized cells with anti-RIα and fluorescence-isothiocyanate (FITC)-conjugated antibody to IgG_{2A} in the second layer. (Note: the homogeneous distribution of RIα in uncapped T cells.) **C:** Fluorescent isothiocyanate (FITC) fluorescence of permeabilized cells incubated with anti-RIIα (*arrowheads*) and counterstained with 7-amino-actinomycin D to visualize the nucleus. (Note: RIIα is localized to a distinct spot in close proximity to the nucleus.) **D:** RITC fluorescence of anti-CD3 capped T cells. (Note: the TCR/CD3 complex is capped.) **E:** The same cell as shown in (D) is incubated with anti-RIα and FITC-labeled antibody to IgG in the second layer. (Note: distinct capping of RIα.) **F:** FITC fluorescence in capped T cells after incubation with anti-RIIα (*arrowheads*) and counterstaining of the nucleus with 7-amino-actinomycin D. (Note: the subcellular distribution of RIIα is not influenced by TCR/CD3 activation and capping.) (From Skålhegg BS, Taskén K, Hansson V, Huitfeldt HS, Jahnsen T, Lea T. Location of cAMP-dependent protein kinase type I with the TCR/CD3 complex. *Science* 1994;263:84–87, with permission.)

with the PKC activator TPA is not inhibited by cAMP, whereas activation by CD28 in conjunction with TCR/CD3 is sensitive to inhibition by cAMP (48). This is compatible with the notion that the target for cAKI is a component of the TCR/CD3 complex or associated proteins that reside distal from TCR stimulation but proximal for PKC activation.

B-lymphoid cells activated by cross-linking of Ig-receptors are also sensitive to cAMP-mediated inhibition of cell proliferation (49). The observation that cAKI, but not cAKII, colocalizes with the Ig-receptor complex during anti-Ig-induced activation and capping (50) suggests that cAKI cocapping with mitogenic receptors is a general phenomenon in lymphoid cells. Furthermore, selective activation of cAKI inhibits natural killer (NK) cell cytotoxicity toward allogeneic lymphoblasts as well as tumor cells (51).

cAKII: Subcellular Localization and Specific Anchoring Proteins

In many tissues, cAKII isozymes ($RII\alpha_2C_2$, $RII\beta_2C_2$) are primarily associated with the particulate fraction. However, the ratio of soluble to particulate cAKII varies, and is probably dependent on the level of expression of RII and the level of specific proteins that bind and anchor RII. A number of different A kinase anchoring proteins (AKAPs) have been identified and cloned (reviewed in [52,53]).

Both RIIα and RIIβ have been reported to localize to the Golgi-centrosomal area of different cell types (47,54). Centrosomal localization is in agreement with the observations in T cells shown in Fig. 3, and suggests involvement of cAKII in control of the cell cycle and formation of the spindle apparatus during mitosis, as well as in microtubule organization in interphase. Colocalization and coimmunoprecipitation of RIIα of cAKII with $p34^{cdc2}$ kinase have been reported (55), and RIIβ has recently been shown to serve as a substrate for cdc2 kinase *in vitro* (56). A centrosomal AKAP has been identified (47), and the identity of this protein is currently under investigation.

AKAPs show specific, high-affinity interaction with dimers of either RIIα or RIIβ. Anchoring proteins showing cell-specific expression and distinct subcellular localization have been reported (53). Thus, such anchoring proteins may, owing to their distinct localization and cell-specific expression, serve to target cAKII toward specific substrates at defined subcellular loci.

SUMMARY

A large number of hormones, neurotransmitters, and other signaling substances that bind to G-protein-coupled cell-surface receptors have their signals converge at one sole second messenger, cAMP. The question of how specificity can be maintained in a signal-transduction system in which many extracellular signals leading to a vast array of intracellular responses are all mediated through one second-messenger system has been the subject of thorough investigation and a great deal of speculation. An increasing number of cAK isozymes, consisting of homo- or heterodimers of R subunits (RIα, RIβ, RIIα, RIIβ) with associated catalytic subunits (Cα, Cβ, Cγ), may, at least in part, explain this specificity.

The various cAK isozymes display distinct biochemical properties, and the heterogeneous subunits of cAK reveal cell-specific expression and differential regulation at the level of gene transcription, mRNA stability, and protein stability in response to a wide range of hormones and other signaling substances. The existence of a number of anchoring proteins specific to either RIIα or RIIβ, and which localize cAKII isozymes toward distinct substrates at defined subcellular loci, strongly supports the idea that specific functions can be assigned to the various cAK isozymes.

The demonstration that selective activation of cAKI is necessary and sufficient for cAMP-mediated inhibition of T-cell proliferation, and the observation that T-cell activation is associated with redistribution and colocalization of cAKI to the TCR, is also compatible with the notion of isozyme-specific effects.

REFERENCES

1. McKnight GS. Cyclic AMP second messenger systems. *Curr Opin Cell Biol* 1991;3:213–7.
2. Scott JD. Cyclic nucleotide-dependent protein kinases. *Pharmacol Ther* 1991;50:123–45.
3. Beebe SJ, Corbin JD. Cyclic nucleotide-dependent protein kinases. *Enzymes* 1986;17:43–111.
4. Døskeland SO, Maronde E, Gjertsen BT. The genetic subtypes of cAMP-dependent protein kinase—functionally different or redundant? *Biochim Biophys Acta* 1993;1178:249–58.
5. Sandberg M, Taskén K, Øyen O, Hansson V, Jahnsen T. Molecular cloning, cDNA structure and deduced amino acid sequence for a type I regulatory subunit of cAMP-dependent protein kinase from human testis. *Biochem Biophys Res Commun* 1987;149:939–45.
6. Solberg R, Taskén K, Keiserud A, Jahnsen T. Molecular cloning, cDNA structure and tissue-specific expression of the human regulatory subunit RIβ of cAMP-dependent protein kinases. *Biochem Biophys Res Commun* 1991;176:166–72.
7. Øyen O, Myklebust F, Scott JD, Hansson V, Jahnsen T. Human testis cDNA for the regulatory subunit RII_{α} of cAMP-dependent protein kinase encodes an alternate amino-terminal region. *FEBS Lett* 1989;246:57–64.
8. Levy FO, Øyen O, Sandberg M, et al. Molecular cloning, complementary deoxyribonucleic acid structure and predicted full-length amino acid sequence of the hormone-inducible regulatory subunit of 3′,5′-cyclic adenosine monophosphate-dependent protein kinase from human testis. *Mol Endocrinol* 1988;2:1364–73.
9. Uhler MD, Carmichael DF, Lee DC, Chrivia JC, Krebs EG, McKnight GS. Isolation of cDNA clones coding for the catalytic subunit of mouse cAMP-dependent protein kinase. *Proc Natl Acad Sci USA* 1986;83:1300–4.
10. Uhler MD, Chrivia JC, McKnight GS. Evidence for a second isoform of the catalytic subunit of cAMP-dependent protein kinase. *J Biol Chem* 1986;261:15360–3.
11. Beebe SJ, Øyen O, Sandberg M, Frøysa A, Hansson V, Jahnsen T. Molecular cloning of a tissue-specific protein kinase (Cγ) from human testis, representing a third isoform for the catalytic subunit of cAMP-dependent protein kinase. *Mol Endocrinol* 1990;4:465–75.
12. Foss KB, Landmark B, Skalhegg BS, et al. Characterization of in-vitro-translated human regulatory and catalytic subunits of cAMP-dependent protein kinases. *Eur J Biochem* 1994;220:217–23.
13. Beebe SJ, Salomonsky P, Jahnsen T, Li Y. The Cγ subunit is a unique isozyme of the cAMP-dependent protein kinase. *J Biol Chem* 1992;267:25505–12.
14. Cummings DE, Edelhoff S, Disteche CM, McKnight GS. Cloning of a mouse protein kinase A catalytic subunit pseudogene and chromosomal mapping of C subunit isoforms. *Mamm Genome* 1994;5:701–6.
15. Brandon EP, Gerhold KA, Qi M, McKnight GS, Idzerda RL. Derivation of novel embryonic stem cell lines and targeting of cyclic AMP-dependent protein kinase genes. *Rec Prog Horm Res* 1995;50:403–8.
16. Nowak I, Seipel K, Schwarz M, Jans DA, Hemmings BA. Isolation of a cDNA and characterization of the 5′ flanking region of the gene encoding the type I regulatory subunit of the cAMP-dependent protein kinase. *Eur J Biochem* 1987;167:27–33.
17. Solberg R, Sandberg M, Natarajan V, et al. The human gene for the regulatory subunit RIα of cAMP-dependent protein kinase—two distinct promoters provide differential regulation of alternately spliced mRNAs. *Endocrinology* 1997;138:169–181.
18. Clegg CH, Koeiman NR, Jenkins NA, Gilbert DJ, Copeland NG, Neubauer MG. Structural features of the murine gene encoding the RIβ subunit of cAMP-dependent protein kinase. *Mol Cell Neurosci* 1994;5:153–64.
19. Foss KB, Solberg R, Simard J, et al. Molecular cloning and promoter studies of the human gene for the regulatory subunit RIIα of cAMP-dependent protein kinase. *Biochim Biophys Acta* 1996;1350:98–108.
20. Kurten RC, Levy LO, Shey J, Durica JM, Richards JS. Identification and characterization of the GC-

rich and cyclic adenosine 3′,5′-monophosphate (cAMP)-inducible promoter of the type IIβ cAMP-dependent protein kinase regulatory subunit gene. *Mol Endocrinol* 1992;6:536–50.

21. Singh IS, Luo ZJ, Eng A, Erlichman J. Molecular cloning and characterization of the promoter region of the mouse regulatory subunit RIIβ of type II cAMP-dependent protein kinase. *Biochem Biophys Res Commun* 1991;178:221–6.
22. Chrivia JC, Uhler MD, McKnight GS. Characterization of genomic clones coding for the Cα and Cβ subunits of mouse cAMP-dependent protein kinase. *J Biol Chem* 1988;263:5739–44.
23. Solberg R, Sandberg M, Spurkland A, Jahnsen T. Isolation and characterization of a human pseudogene for the regulatory subunit RIα of cAMP-dependent protein kinases and its sublocalization on chromosome 1. *Genomics* 1993;15:591–7.
24. Taylor SS, Knighton DR, Zheng J, Ten Eyck LF, Sowadski JM. Structural framework for the protein kinase family. *Annu Rev Cell Biol* 1992;8:429–62.
25. Knighton DR, Zheng JH, Ten Eyck LF, et al. Crystal structure of the catalytic subunit of cyclic adenosine monophosphate-dependent protein kinase. *Science* 1991;253:407–14.
26. Øgreid D, Ekanger R, Suva RH, Miller JP, Døskeland SO. Comparison of the two classes of binding sites (A and B) of type I and type II cyclic-AMP-dependent protein kinases by using cyclic nucleotide analogs. *Eur J Biochem* 1989;181:19–31.
27. Su Y, Dostmann WR, Herberg FW, et al. Regulatory subunit of protein kinase A: structure of deletion mutant with cAMP binding domains. *Science* 1995;269:807–13.
28. Cadd GG, Uhler MD, McKnight GS. Holoenzymes of cAMP-dependent protein kinase containing the neural form of type I regulatory subunit have an increased sensitivity to cyclic nucleotides. *J Biol Chem* 1990;265:19502–6.
29. Solberg R, Taskén K, Wen W, et al. Human regulatory subunit RIβ of cAMP-dependent protein kinases: expression, holoenzyme formation, and microinjection into living cells. *Exp Cell Res* 1994; 214:595–605.
30. Otten AD, Parenteau LA, Døskeland SO, McKnight GS. Hormonal activation of gene transcription in ras-transformed NIH3T3 cells overexpressing RIIα and RIIβ subunits of the cAMP-dependent protein kinase. *J Biol Chem* 1991;266:23074–82.
31. Taskén K, Skålhegg BS, Solberg R, et al. Novel isozymes of cAMP-dependent protein kinase exist in human cells due to formation RIα-RIβ heterodimeric complexes. *J Biol Chem* 1993;268:21276–83.
32. Øyen O, Myklebust F, Scott JD, et al. Subunits of cyclic adenosine 3′,5′-monophosphate-dependent protein kinase show differential and distinct expression patterns during germ cell differentiation: alternative polyadenylation in germ cells gives rise to unique smaller-sized mRNA species. *Biol Reprod* 1990;43:46–54.
33. Landmark BF, Øyen O, Skålhegg BS, Fauske B, Jahnsen T, Hansson V. Cellular localization and age-dependent changes of the regulatory subunits of cAMP-dependent protein kinase in rat testis. *J Reprod Fertil* 1993;99:323–34.
34. Jahnsen T, Lohmann SM, Walter U, Hedin L, Richards JS. Purification and characterization of hormone-regulated isoforms of the regulatory subunit of type II cAMP-dependent protein kinase from rat ovaries. *J Biol Chem* 1985;260:15980–7.
35. Landmark BF, Fauske B, Eskild W, et al. Identification, characterization, and hormonal regulation of 3′,5′-cyclic adenosine monophosphate-dependent protein kinases in rat Sertoli cells. *Endocrinology* 1991;129:2345–54.
36. Øyen O, Sandberg M, Eskild W, et al. Differential regulation of messenger ribonucleic acids for specific subunits of cyclic adenosine 3′,5′-monophosphate (cAMP)-dependent protein kinase by cAMP in rat Sertoli cells. *Endocrinology* 1988;122:2658–66.
37. Levy FO, Ree AH, Eikvar L, Govindan MV, Jahnsen T, Hansson V. Glucocorticoid receptors and glucocorticoid effects in rat Sertoli cells. *Endocrinology* 1989;124:430–6.
38. Berg JP, Ree AH, Sandvik JA, et al. 1,25-dihydroxyvitamin D3 alters the effect of cAMP in thyroid cells by increasing the regulatory subunit type II beta of the cAMP-dependent protein kinase. *J Biol Chem* 1994;269:32233–8.
39. Taskén KA, Knutsen HK, Attramadal H, et al. Different mechanisms are involved in cAMP-mediated induction of mRNAs for subunits of cAMP-dependent protein kinases. *Mol Endocrinol* 1991;5:21–8.
40. Taskén K, Andersson KB, Skålhegg BS, et al. Reciprocal regulation of mRNA and protein for subunits of cAMP-dependent protein kinase (RIα and Cα) by cAMP in a neoplastic B cell line (Reh). *J Biol Chem* 1993;268:23483–9.
41. Knutsen HK, Taskén KA, Eskild W, Jahnsen T, Hansson V. Adenosine 3′,5′-monophosphate-dependent stabilization of messenger ribonucleic acids (mRNAs) for protein kinase-A (PKA) subunits in rat Sertoli cells: rapid degradation of mRNAs for PKA subunits is dependent on ongoing RNA and

protein synthesis. *Endocrinology* 1991;129:2496–502.
42. Houge G, Vintermyr OK, Døskeland SO. The expression of cAMP-dependent protein kinase subunits in primary rat hepatocyte cultures. Cyclic AMP down-regulates its own effector system by decreasing the amount of catalytic subunit and increasing the mRNAs for the inhibitory (R) subunits of cAMP-dependent protein kinase. *Mol Endocrinol* 1990;4:481–8.
43. Taskén K, Kvale D, Hansson V, Jahnsen T. Protein kinase C activation selectively increases mRNA levels for one of the regulatory subunits (RIα) of cAMP-dependent protein kinases in HT-29 cells. *Biochem Biophys Res Commun* 1990;172:409–14.
44. Taskén KA, Knutsen HK, Eikvar L, et al. Protein kinase C activation by 12-O-tetradecanoylphorbol 13-acetate modulates messenger ribonucleic acid levels for two of the regulatory subunits of 3′,5′-cyclic adenosine monophosphate-dependent protein kinases (RIIβ and RIα) via multiple and distinct mechanisms. *Endocrinology* 1992;130:1271–80.
45. Skålhegg BS, Landmark BF, Døskeland SO, Hansson V, Lea T, Jahnsen T. Cyclic AMP-dependent protein kinase type I mediates the inhibitory effects of 3′,5′-cyclic adenosine monophosphate on cell replication in human T lymphocytes. *J Biol Chem* 1992;267:15707–14.
46. Skålhegg BS, Taskén K, Hansson V, Huitfeldt HS, Jahnsen T, Lea T. Location of cAMP-dependent protein kinase type I with the TCR/CD3 complex. *Science* 1994;263:84–7.
47. Keryer G, Rios RM, Landmark BF, Skålhegg BS, Lohmann SM, Bornens M. A high-affinity binding protein for the regulatory subunit of cAMP-dependent protein kinase II in the centrosome of human cells. *Exp Cell Res* 1993;204:230–40.
48. Skålhegg BS, Rasmussen AM, Taskén K, Hansson V, Jahnsen T, Lea T. Cyclic AMP sensitive signalling by the CD28 marker requires concomitant stimulation by the T-cell antigen receptor (TCR/CD3) complex. *Scand J Immunol* 1994;40:201–8.
49. Blomhoff HK, Smeland EB, Beiske K, et al. Cyclic AMP-mediated suppression of normal and neoplastic B cell proliferation is associated with regulation of *myc* and Ha-*ras* protooncogenes. *J Cell Physiol* 1987;131:426–33.
50. Levy FO, Rasmussen AM, Taskén K, et al. Cyclic AMP-dependent protein kinase (cAK) in human B cells: co-localization of type I cAK ($RI\alpha_2C_2$) with the antigen receptor during anti-immunoglobulin-induced B cell activation. *Eur J Immunol* 1996;26:1290–6.
51. Torgersen KM, Vaage JT, Levy FO, Hansson V, Rolstad B, Taskén K. Selective activation of cAMP-dependent protein kinase type I inhibits cytotoxicity mediated by rat IL-2 activated Natural Killer cells. *J Biol Chem* 1997;272:5495–5500.
52. Scott JD, Carr DW. Subcellular localization of the type II cAMP-dependent protein kinase. *News Physiol Sci* 1992;7:143–8.
53. Scott JD, McCartney S. Localization of A-kinase through anchoring proteins. *Mol Endocrinol* 1994; 8:5–11.
54. Rios RM, Celati C, Lohmann SM, Bornens M, Keryer G. Identification of a high affinity binding protein for the regulatory subunit RIIβ of cAMP-dependent protein kinase in Golgi enriched membranes of human lymphoblasts. *EMBO J* 1992;11:1723–31.
55. Tournier S, Raynaud F, Gerbaud P, Lohmann SM, Doree M, Evain-Brion D. Association of type II cAMP-dependent protein kinase with p34cdc2 protein kinase in human fibroblasts. *J Biol Chem* 1991;266:19018–22.
56. Keryer G, Luo Z, Cavadore JC, Erlichman J, Bornens M. Phosphorylation of the regulatory subunit of type IIβ cAMP-dependent protein kinase by cyclin B/$p34^{cdc2}$ kinase impairs its binding to microtubule-associated protein 2. *Proc Natl Acad Sci USA* 1993;90:5418–22.
57. Taskén K, Solberg R, Zhao Y, Hansson V, Jahnsen T, Siciliano MJ. The gene encoding the catalytic subunit Cα of cAMP-dependent protein kinase (locus PRKACA) localizes to human chromosome region 19p13.1. *Genomics* 1996;36:535–8.
58. Simard J, Bérubé D, Sandberg M, et al. Assignment of the gene encoding the catalytic subunit Cβ of cAMP-dependent protein kinase to the p36 band on chromosome 1. *Hum Genet* 1992;88:653–7.
59. Van Roy N, Laureys G, Versteeg R, Opdenakker G, Speleman F. High-resolution fluorescence mapping of 46 DNA markers to the short arm of human chromosome 1. *Genomics* 1993;18:71–8.
60. Foss KB, Simard J, Berube D, et al. Localization of the catalytic subunit Cγ of the cAMP-dependent protein kinase gene (PRKACG) to human chromosome region 9q13. *Cytogenet Cell Genet* 1992; 60:22–5.
61. Boshart M, Weih F, Nichols M, Schütz G. The tissue-specific extinguisher locus TSE1 encodes a regulatory subunit of cAMP-dependent protein kinase. *Cell* 1991;66:849–59.

62. Solberg R, Sistonen P, Träskelin AL, et al. Mapping of the regulatory subunits RIβ and RIIβ of cAMP-dependent protein kinase genes on human chromosome 7. *Genomics* 1992;14:63–9.
63. Taskén K, Naylor SL, Hansson V, Jahnsen T, Solberg R. The gene for the cAMP-dependent protein kinase subunit RIIα (locus PRKAR2A) localize to human chromosome region 3p21.2–3. *Submitted* 1996;
64. Wainwright B, Lench N, Davies K, et al. A human regulatory subunit of type II cAMP-dependent protein kinase localized by its linkage relationship to several cloned chromosome 7q markers. *Cytogenet Cell Genet* 1987;45:925–32.

Signal Transduction in Health and Disease, Advances in Second Messenger and Phosphoprotein Research, Vol. 31, edited by J. Corbin and S. Francis.
Lippincott–Raven Publishers, Philadelphia © 1997.

17

Structural Order of the Slow and Fast Intrasubunit cGMP-Binding Sites of Type Iα cGMP-Dependent Protein Kinase

*Robin B. Reed, †Mårten Sandberg, †Tore Jahnsen, ‡Suzanne M. Lohmann, *Sharron H. Francis, and *Jackie D. Corbin

**Department of Molecular Physiology and Biophysics, Vanderbilt University School of Medicine, Nashville, Tennessee 37232; †Institute of Medical Biochemistry, University of Oslo, N-0317 Oslo, Norway, and ‡Department of Clinical Biochemistry, Medical University Clinic, Wurzburg 8700, Germany*

The two isoforms (α and β) of Type I cyclic guanosine monophosphate (cGMP)-dependent protein kinase (cGK) are products of alternative messenger ribonucleic acid (mRNA) splicing (1,2), which results in identical proteins with the exception of approximately the first 100 amino-acid residues at the amino terminus (N-terminus). The N-termini for Type Iα cGK and Type Iβ cGK are quite different in their amino-acid sequences, with only 36% identity, but this region in both proteins shares common functional roles, since the dimerization and the autoinhibitory/pseudosubstrate domains are contained in these segments. The identical sequence shared by Types Iα cGK and Iβ cGK begins at Ser^{89} in Type Iα cGK and Ser^{104} in Type Iβ cGK, near the start of cGMP-binding Site A, and extends through both cGMP-binding sites and catalytic domain to the carboxyl-terminal (C-terminal) end of each protein (2–4).

Despite having identical amino-acid sequences in both their cGMP-binding sites and catalytic regions, Type Iα cGK and Type Iβ cGK have markedly different cyclic-nucleotide-binding properties and kinase-activation characteristics. The two isozymes have slightly different kinase-activation constants for cGMP (Fig. 1, top), and both exhibit a 100-fold selectivity in binding cGMP versus cyclic adenosine monophosphate (cAMP). However, Type Iα and Type Iβ cGK have quite different dissociation rates for cGMP (Fig. 1, bottom), and the cGMP analogue specificities of the two isoforms vary by as much as 200-fold (3,5,6). These functional differences in the cGMP-binding sites of the two isozymes must be conferrred by the respective N-termini (7), since the remainder of the amino-acid sequence in the two isozymes is identical.

Like cAMP-dependent protein kinase (cAK), cGK contains two cyclic nucleotide-

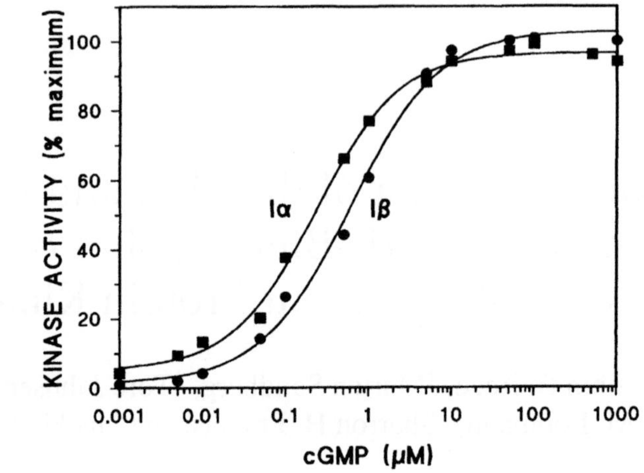

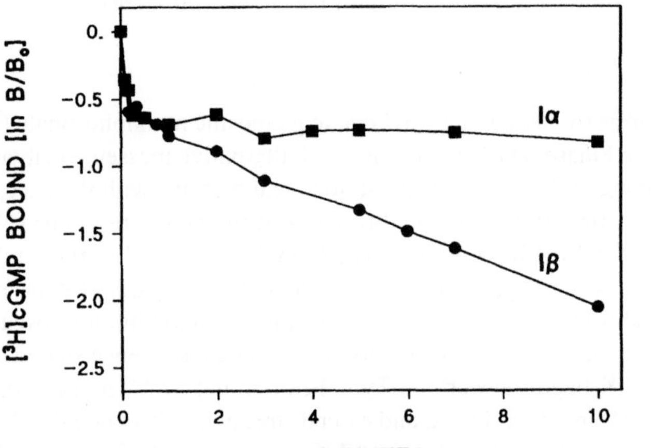

FIG. 1. Top: Kinase activation of Type I cGK isozymes by cGMP. Protein kinase activity of native Type Iα cGK (■) and native Type Iβ cGK (●) was measured in the presence of increasing concentrations of cGMP in a 30-min incubation, as described in Materials and Methods. Bottom: [^{3}H] cGMP dissociation of Type I cGK isozymes. [^{3}H]cGMP-dissociation curves for native Type Iα cGK (■) and native Type Iβ cGK (●) were generated as described in Materials and Methods. B_o is the total amount of bound [^{3}H]cGMP at time zero. B is the amount of bound [^{3}H]cGMP remaining at the time points sampled after addition of a 100-fold molar excess of unlabeled cGMP. Each panel is representative of four experiments.

binding sites (Fig. 2) that are distinguished by having fast or slow cGMP-dissociation characteristics. On the amino-acid sequence level, the two homologous cGMP-binding sites of cGK show similarity to the two cAMP-binding sites of cAK. However, the more N-terminal cyclic nucleotide-binding site of cGK (Site A) is more similar to Site A of cAK, while Site B of cGK is more similar to Site B of cAK. This

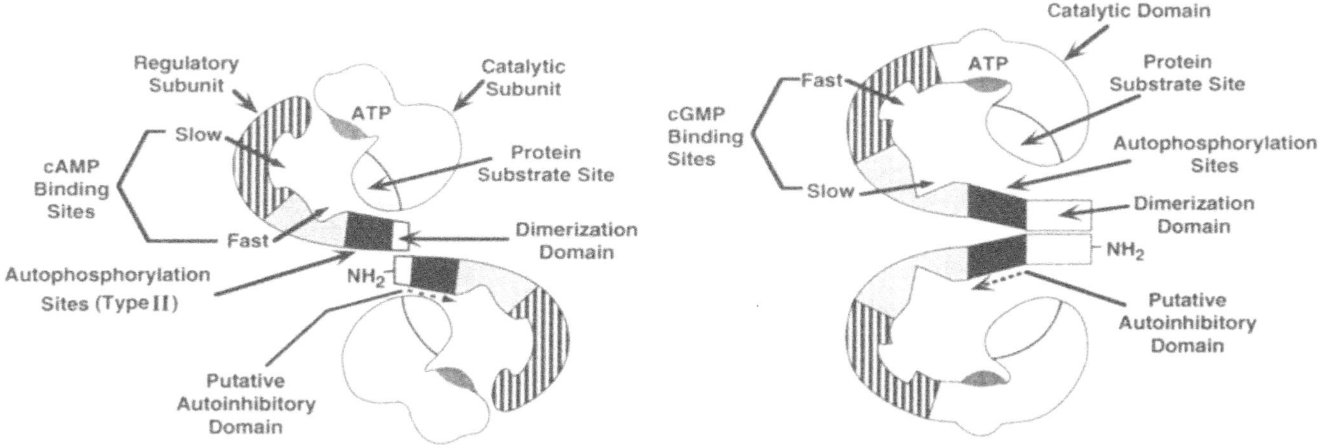

FIG. 2. Proposed functional domains of cAK and cGK. The cAMP-dependent protein kinase tetramer (left) and the Type Iβ cGMP-dependent protein kinase homodimer (right) are represented. Locations of the cyclic nucleotide-binding sites are indicated as either fast or slow to reflect cyclic nucleotide-dissociation characteristics. The fast-dissociation site in cAK is Site A, whereas in Type Iβ cGK the slow-dissociation site is Site A. Site B in cAK is the slow-dissociation site, whereas Site B in Type Iβ cGK is the fast-dissociation site. Type Iα cGK and Type Iβ cGK differ in their amino-acid sequences in the dimerization and inhibitory domains only.

homology between cAK and cGK led to the proposal that the identity of the kinetically defined cGMP-binding sites within the structure of cGK is the same as in cAK (i.e., the N-terminal site is the fast-dissociation site, whereas the more C-terminal site is the slow-dissociation site) (8). However, conversion of a specific threonine to alanine in either of the two cGMP-binding sites of Type Iβ cGK strongly diminishes cGMP binding to the respective site; biochemical analysis of these mutants led to the surprising discovery that the structural order of the two cGMP-binding sites of this isozyme of cGK is transposed from the structural order of the two cAMP-binding sites of cAK (9).

Since the different N-termini of the Type I cGK isozymes confer quite different cyclic nucleotide-analogue-binding specificities and cGMP-dissociation kinetics on these enzymes, it was thought that the different sequences might also endow the cGMP-binding sites in the two enzymes with very different specific kinetic parameters, such as opposite slow- and fast-dissociation kinetics, despite their amino-acid sequence identity. This hypothesis was tested by creating threonine-to-alanine mutations in the cGMP-binding sites of Type Iα cGK, analogous to those made in Type Iβ cGK (9), in order to determine the structural order of the two cGMP-binding sites in the Type Iα isozyme.

MATERIALS AND METHODS

Subcloning of Human Type Iα cGK cDNA

One partial complementary DNA (cDNA) clone for the wild-type (WT) Iα human cGK (hcGKIα) was subcloned into the WT, T317A, T193A, and T317A/T193A pVL1392 hcGKIβ plasmids (9). The 490-bp *Eco*RI/*Nco*I fragment of hcGKIα (containing 20 bp of 5′ nontranslated region) was excised from pCRII hcGKIα and ligated into analogous restriction sites in pVL1392 hcGKIβ WT and threonine-to-alanine mutant plasmids to replace the N-terminus of Type Iβ cGK with the N-terminus of Type Iα cGK. pVL1392 vectors were propagated in *Escherichia coli* (DH5α). All plasmids were sequenced manually (United States Biochemical, Cleveland, OH) or on an Applied Biosystems, Inc. [AB1] (Foster City, CA), Model 377A DNA Sequencer by the dideoxy chain-termination method (10) prior to cotransfection. Type Iα cGK cDNA is predicted to encode cGK proteins that are 15 amino acids shorter in polypeptide length than those encoded by Type Iβ cDNA; thus, T193A of the Type Iβ protein becomes T178A in the Type Iα protein, and T317A in the Type Iβ protein becomes T302A in the Type Iα protein.

Expression of Wild-Type and Mutant cGK

All tissue-culture procedures were performed with Sf9 insect cells (*Spodoptera frugiperda*, Invitrogen [Carlsbad, CA]) maintained at 27°C in TNM-FH media (Grace's *Antheraea* medium plus TC yeastolate plus lactalbumin hydrolyzate) plus

10% fetal bovine serum (FBS) (JRH Biosciences, Leneka, KS). pVL1392-hcGKIα transfer vectors (4 μg) were cotransfected with linear WT *Autographia Californica* nuclear polyhedrosis virus (AcMNPV) (1 μg), using cationic liposomes (Invitrogen) as a carrier. Recombinant viruses were harvested 3 days after cotransfection and purified through two serial agarose overlay plaque assays (11). Purified recombinant viruses were amplified by infection of Sf9 cells for 7 days in 25-cm^2 flasks (Falcon), Becton-Dickinson, Inc (Lincoln Park, NJ). Infected cells were harvested by gentle rapping and pelleted by centrifugation (1000 × *g*, 10 min, 20°C). High-titer viruses in the supernatant were harvested as extracellular viral particles and stored at 4°C as stock for further experiments. The cell pellet containing cGK was resuspended in cold, sterile KPEM (10 mM potassium phosphate; pH 6.8, 1 mM ethylene diamine tetraacetic acid [EDTA], 25 mM 2-mercaptoethanol) plus the protease inhibitors pepstatin A (1 μg/ml) and leupeptin (0.5 μg/ml) (Sigma Chemical Co., St. Louis, MO) and homogenized on ice by a 10-second burst in an Ultra-Turrax microhomogenizer (Janke and Kunkel, Cincinnati, OH). Crude extracts were tested as described below for cGK activity, [^{3}H]cGMP-binding activity, and immunoreactivity to rabbit anti-bovine-lung Type Iα cGK antibodies to confirm expression of WT and mutant cGK.

Protein Kinase Activation

The kinase activities of native, wild-type, and mutant cGK were determined by the method of Wolfe et al. (3), using a synthetic heptapeptide (RKRSRAE) (12) (Peninsula Laboratories, Belmont, CA) as substrate. Twenty microliters of crude extract in KPEM were added to 5 μl of cyclic nucleotide or H_2O plus 25 μl of reaction mixture (20 mM Tris, pH 7.4; 20 mM magnesium acetate; 200 μM adenosine triphosphate [ATP]; 100 μM isobutylmethylxanthine [IBMX]; 136 μg/ml substrate; 0.9 μM protein kinase inhibitor peptide 5-24 (Peninsula Laboratories); and 20,000 cpm/μl [γ-^{32}P]ATP). Assays were conducted for 5 min or 30 min at 30°C, and the amount of ^{32}Pi transferred to Whatman P-81 cation-exchange paper was calculated. Hill plots were used to determine K_a values according to Shabb et al. (13), with the log [γ/(1 − γ)] plotted versus the log of the cyclic nucleotide concentration (in μM), where γ is the fractional activation of the enzyme. K_a was calculated as the antilog of the *x*-intercept of the log–log plot.

cGMP-Binding and -Dissociation

The [^{3}H]cGMP-binding assay used in the study was the $(NH_4)_2SO_4$ Millipore filtration assay (Millipore Corp., Bedford, MA) (14), modified as described by Wolfe et al. (3), except that 10 μM cGMP and 1 μM [^{3}H]cGMP (Amersham [Arlington Heights, IL]; specific activity 15 to 30 Ci/mmol) were used in the reaction mixture (50 mM KH_2PO_4, pH 6.8; 1 mM EDTA; 0.5 mg/ml Sigma histone IIA). Equal volumes of crude extract in KPEM and [^{3}H]cGMP-binding mixture were incubated at 30°C for 30 min to saturate the binding sites. Incubations were then cooled to 4°C, stopped by the addition of 2 ml cold saturated aqueous $(NH_4)_2SO_4$, filtered through a 0.45-μm nitrocellulose Millipore filter, and washed with 6 ml of the $(NH_4)_2SO_4$ solution. Filters were placed in counting vials containing 1 ml of 2% sodium dodecyl sulfate (SDS) and vortexed for 10 s prior to the addition of 5 ml Beckman Readysafe aqueous scintillant (Beckman Instruments, Mountain View, CA). For [^{3}H]cGMP-dissociation assays, the addition of a 100-fold molar excess of unlabeled cGMP at time zero (B_o) initiated the dissociation (exchange) of bound [^{3}H]cGMP. Thirty-microliter aliquots were sampled at the times indicated (B), and were filtered and washed as described above. The half-life of the bound-cGMP was determined by the method of Rannels and Corbin (15).

Sodium Dodecyl Sulfate–Polyacrylamide Gel Electrophoresis of cGK

An aliquot of the crude extract (2 μg protein) containing cGK from infected Sf9 cells was boiled for 5 min in the presence of 10% SDS, 2 M 2-mercaptoethanol, and 1 mg/ml bromophenol blue and subjected to 8% SDS–polyacrylamide gel electrophoresis (PAGE) (16). Proteins were visualized with an affinity-purified rabbit anti-bovine-lung type Iα cGK IgG antibody in immunoblot experiments. For immunoblotting, proteins were transferred to 0.45 μm polyvinylidene difluoride (PVDF) (Millipore) membranes for 1 hr at 100 V. Membranes were blocked for at least 1 hr at 4°C in 5% nonfat milk (Carnation) in Tris-buffered saline plus 3% Tween-20 (TBST), and were then incubated for 1 hr at 20°C of the primary antibody (1:5,000). The antibody was the affinity-purified rabbit anti-bovine-lung Type Iα cGK IgG antibody. Membranes were rinsed in copious amounts of TBST and then incubated with secondary antibody (1:5,000 horseradish peroxidase [HRP]–GAR), detected by chemiluminescence (Amersham) and autoradiography.

Protein Quantification

Protein quantities were determined by the method of Bradford (17), using protein assay dye Bio-Rad (Melville, NY) and bovine serum albumin (BSA) fraction V (Sigma) as standard. This method routinely overestimates the amount of protein by 37% (3).

RESULTS

A baculovirus expression system was used to express WT human Type Iα cGK and cGMP-binding site threonine-to-alanine mutants of the enzyme (T178A, T302A, and T178A/T302A) in Sf9 cells. Expression of recombinant cGKs was achieved by infection of Sf 9 cells with plaque-purified recombinant baculoviruses. The appearance of active, full-length cGKs was determined by measuring cGMP-binding activity and protein kinase activity in the absence and presence of cGMP, and by performing immunoblots in the crude extracts using anti-Type Iα cGK antibodies. In immunoblot experiments, WT Iα cGK and threonine-to-alanine mutant

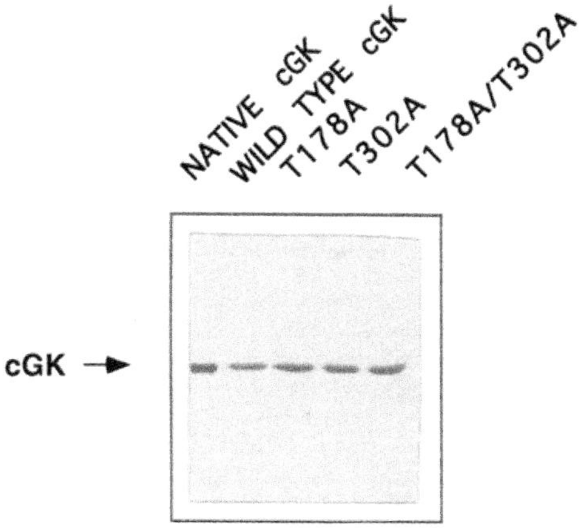

FIG. 3. Comparison of native, WT, and mutant Type Iα cGK. Crude extracts were electrophoresed on 8% SDS-polyacrylamide minigels and immunoblotted with anti-Type Iα cGK antibodies as described in Materials and Methods.

enzymes comigrated with purified, native, bovine Type Iα cGK at 76 kDa (Fig. 3). Comparison of kinase-activation constants or cGMP-dissociation behavior for Type Iβ cGK in crude extracts or purified enzyme indicated no significant differences. Therefore, the results of subsequent analyses of Type Iα cGK in crude extracts, under assay conditions identical to those in the initial assay, were considered to be valid.

Type Iα cGK (WT and threonine-to-alanine mutants) was assayed in crude extracts with increasing concentrations of cAMP (Fig. 4) or cGMP (Fig. 5) in order to analyze kinase catalytic activity, and the results are summarized in Table 1. The K_a values for cAMP (Fig. 4) for all enzymes were increased 2- to 5-fold, indicating that the respective threonine residues are not negative determinants against cAMP binding (i.e., the binding of cAMP to these sites is not impeded by a threonine in this position). The results also indicated that the overall structure and function of the binding sites in the mutants were preserved. However, the K_a for cGMP was increased by approximately 35-fold by mutation of Thr^{178} to alanine, whereas mutation of Thr^{302} to alanine led to a much smaller increase in the K_a for cGMP (approximately 3-fold). The double mutation (T178A/T302A) drastically increased the K_a for cGMP (approximately 700-fold). Of the two single mutants, T178A caused a much greater shift of the K_a values for cGMP than did T302A, indicating that the loss in cGMP-binding affinity was greater when the more N-terminal site was modified. These results are consistent with the interpretation that in Type Iα cGK, the N-terminal cGMP-binding site (Site A) binds cGMP with higher affinity than does the tandem cGMP-binding site (Site B). This structural order is identical to that determined for Type Iβ cGK (9), but is transposed from the structural order found in cAK.

Two different times of assay—5 min and 30 min—were used for determining the K_a values for the native, WT, and mutated Type Iα cGK enzymes, in order to examine whether or not autophosphorylation of Type Iα cGK influenced the results. The extent of autophosphorylation of the enzyme was determined to be 0.2 mol/mol sub-

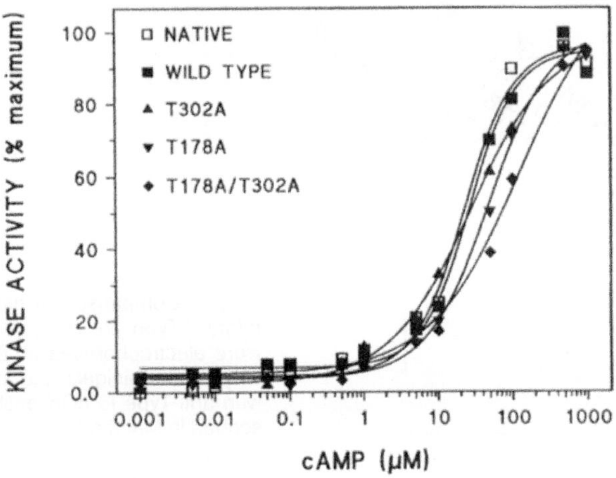

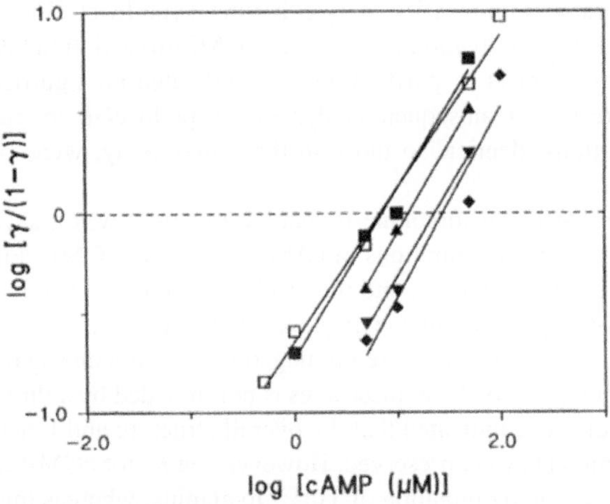

FIG. 4. cAMP activation of native, WT, and mutant Type Iα cGK. Protein kinase activity of native (□), WT (■), T302A (▲), T178A (▼), and T178A/ T302A (◆) cGK was measured in the presence of increasing concentrations of cAMP in a 30-min incubation, as described in Materials and Methods. Hill plots (bottom) were generated to calculate K_a values as the antilogs of the *x*-intercepts. Each panel is representative of four experiments.

unit in 5 min and 1.3 mol/mol subunit in 30 min. Assay times of 5 min or 30 min did not affect the order or magnitude of affinity changes for the binding-site mutants, providing further evidence that the structural orders of the cGMP-binding sites in Type Iα and Type Iβ cGK are identical (data not shown).

The dissociation of [^{3}H]cGMP from native, WT, and each threonine-to-alanine mutant Type Iα cGK enzyme was determined in the presence of a 100-fold molar

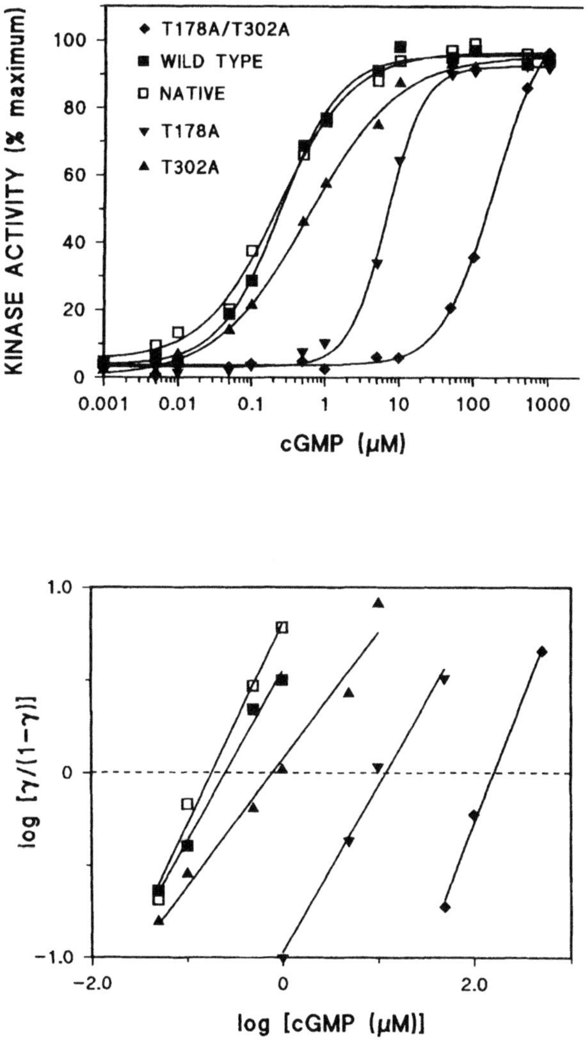

FIG. 5. cGMP activation of native, WT, and mutant Type Iα cGK. Protein kinase activity of native (□), WT (■), T302A (▲), T178A (▼), and T178A/ T302A (◆) cGK was measured in the presence of increasing concentrations of cGMP in a 30-min incubation, as described in Materials and Methods. Hill plots (bottom) were generated to calculate K_a values as the antilogs of the *x*-intercepts. Each panel is representative of four experiments.

excess of unlabeled cGMP (Fig. 6) in order to provide further evidence that the structural order of the two sites in Type Iα cGK is the same as that found in Type Iβ cGK. The results are summarized in Table 1. Native and WT cGK exhibited similar biphasic dissociation curves, consistent with cGMP dissociation from two sites with distinctly different affinities. The initial steep slope represents dissociation of

TABLE 1. *Cyclic Nucleotide Kinetics of Type Iα cGMP-Dependent Protein Kinase*

	K_a cGMP (μM)	K_a cAMP (μM)	[³H]cGMP Dissociation $t_{1/2}$ (min) Site A	Site B
Native	0.15±0.04	6.5±1.0	37	0.10
Wild Type	0.23±0.02	5.4±0.7	32	0.21
T178A	9.0±1.0	25±2.9	0.22	0.04
T302A	0.76±0.08	12±1.2	12.3	0.22
T178A/T302A	170±23	35±5.5	NC	NC

Protein kinase activation constants (K_a) were determined in assays of 30-min incubation from the *x*-intercept values of Hill plots like those given in Figs. 4 and 5, and are the averages from four experiments, with the SEM shown. [³H]cGMP-dissociation rates ($t_{1/2}$) were calculated from results like those given in Fig. 6, and are the averages from four experiments, with the SEM (not shown) not exceeding 25% of any given value. NC=not calculated.

[³H]cGMP from the fast site, while the more shallow slope measures [³H]cGMP dissociation from the slow site. In the T302A mutant, the slow component of the [³H]cGMP dissociation curve ($t_{1/2}$= 12 min) was slightly faster than that of WT or native Type Iα cGK ($t_{1/2}$= 35 min), but the fast component nearly disappeared. Conversely, T178A displayed a dramatic alteration of the slow component of [³H]cGMP dissociation ($t_{1/2}$= 0.22 min), with approximately 50-fold faster dissociation, which was consistent with mutation of the higher-affinity site. The dissociation of [³H]cGMP from the double mutant (T178A/T302A) was extremely fast, indicating that the cGMP affinities of both sites were greatly reduced. These combined results indi-

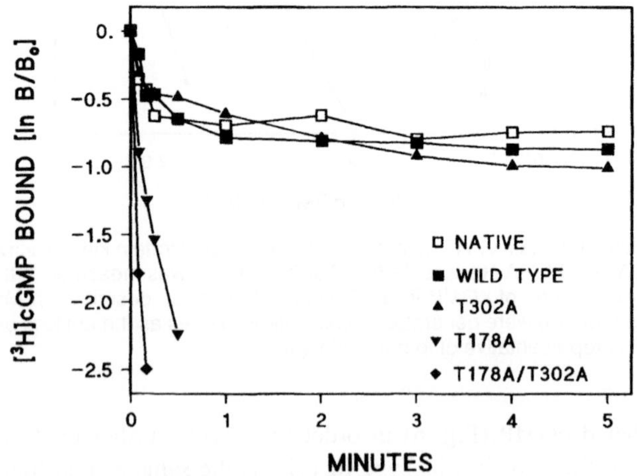

FIG. 6. Dissociation of [³H]cGMP from native, WT, and mutant Type Iα cGK. [³H]cGMP-dissociation curves for native (□), WT (■), T302A (▲), T178A (▼), and T178A/ T302A (◆)cGK were measured as described in Materials and Methods. B_o is the total amount of bound [³H]cGMP at time zero. B is the amount of bound [³H]cGMP remaining at the time points sampled after addition of a 100-fold molar excess of unlabeled cGMP. The figure is representative of four experiments.

cated that substitution of an alanine for the conserved threonine in either cGMP-binding site of cGK markedly reduces the affinity of that site for cGMP, but only slightly reduces the affinity for cAMP, and provided further support for the proposal that in Type Iα cGK, as in Type Iβ cGK, the slow site is N-terminal to the fast site, and that this structural order is transposed from that observed for the cAMP-binding sites of cAK.

DISCUSSION

The structural order for the slow and fast cGMP-binding sites of Type Iβ cGK was determined to be transposed from the order of these sites in cAK (9). The original incorrect assignment for cGK was based on similarities in the amino-acid sequences of these sites in the linear structures of cAK and cGK (8). That is, the amino-acid sequence of the more N-terminal cGMP-binding site in cGK has greater similarity to the more N-terminal cAMP-binding site in cAK than to the more C-terminal cGMP-binding site in cGK, and the sequence of the more C-terminal site in cGK has greater similarity to the more C-terminal site in cAK than to the more N-terminal site in cGK.

The results of this study indicate that the cGMP-binding sites in Type Iα cGK are arranged identically to the sites in Type Iβ cGK; the N-terminal Site A is the slow site and the tandem Site B is the fast site, so that the order of kinetically defined binding sites in both Type I cGKs is reversed from that of these sites in cAK. Therefore, the different N-terminal amino-acid sequences in Type Iα and Type Iβ cGK (dimerization and/or autoinhibitory domains), despite influencing the kinetics and analogue specificities of the binding sites, do not play different roles in determining which is the higher affinity site for the two isozymes, although they clearly modulate the binding sites differently in the two isozymes. It will be of interest to determine the structural order of the two cyclic nucleotide-binding sites for Type II cGK, since the sequence similarities between Type I and Type II cGK cGMP-binding sites are rather weak (18). Since earlier conjecture about the structural order of the cGMP-binding sites in the Type I cGKs proved to be erroneous, any supposition about the structural order of these sites in Type II cGK that is based on similarities with the respective sites in the Type I cGKs or cAK is inappropriate.

These results firmly establish the structural order of both fast and slow cGMP-binding sites for both Type Iα and Type Iβ cGK isozymes. The fact that cGK contains a regulatory domain fused with the catalytic domain, unlike cAK, in which the C subunit dissociates from R subunit upon cAMP binding and activation, may account for the different structural order of the binding sites in cAKs and Type I cGKs. A possible method for testing this hypothesis would be to analyze truncation mutants of WT and threonine-to-alanine mutant cGKs that lack the catalytic domain, and to examine the dissociation characteristics retained by each site. However, it is possible that structural features in the distinct N-termini in cGKs and cAKs could impose different properties on the cyclic-nucleotide-binding sites in these two enzyme families.

Our results suggest that the N-terminal segment of some cyclic nucleotide-dependent protein kinases definitely modulates cyclic nucleotide binding, but that most of the cyclic nucleotide-affinity determinants are probably contained in the immediate binding pocket itself. Apparently, outside structural influences are not needed to provide the highly disparate high or low affinities of the different binding sites. Residues that produce high-affinity interactions with cAMP in the N-terminal binding site of cAK could be conserved in the N-terminal site of cGK, and analogously, such residues in the C-terminal site could also be conserved. However, other residues outside the binding pocket could also contribute to cGMP-binding affinity and thus produce the structural order of these sites in Type I cGKs. Evolutionary processes may have modified these peripheral residues to produce a low-affinity Site A in cAK and a high-affinity Site A in cGK, and, conversely, a high-affinity Site B in cAK and a low-affinity Site B in cGK. Perhaps the structural order of two tandem binding sites is somewhat inconsequential for the concentration of cyclic nucleotide required to activate these kinases in intact cells, since occupation of both sites of either enzyme is required for full activation (19). This question cannot be fully answered until the molecular mechanism for kinase activation by cyclic nucleotide binding to each site is resolved. Generally, the kinases are present at high concentrations (> 0.1 μM) in cells, and cyclic nucleotide levels are only slightly lower than the levels of the kinases, so that a substantial portion of the cyclic nucleotide-binding sites would contain bound cGMP. Evolutionary pressures could have designed the kinases to respond stoichiometrically to changes in physiologic concentrations of cyclic nucleotides, regardless of the structural order of the two cGMP-binding sites.

ACKNOWLEDGMENTS

The authors are grateful to Dr. Kennard Grimes for assistance in purification of native Type Iα cGK, to Dr. Jeffrey A. Smith for measurements of autophosphorylation levels, and for National Institutes of Health Grant DK40029 (J. D. C.), NIH Trainee Grant #5T32DK07563-05 (R. B. R.), and Deutsche Forschungsgemeinschaft Grant SFB 355 (S. M. L.).

REFERENCES

1. Francis SH, Woodford TA, Wolfe L, Corbin JD. Types Iα and Iβ isozymes of cGMP-dependent protein kinase: alternative mRNA splicing may produce different inhibitory domains. *Second Messengers Phosphoprot* 1988–89;12:301–10.
2. Sandberg M, Natarajan V, Ronander I, et al. Molecular cloning and predicted full-length amino acid sequence of the type Iβ isozyme of cGMP-dependent protein kinase from human placenta. *FEBS Lett* 1989;255:321–29.
3. Wolfe L, Corbin JD, Francis SH. Characterization of a novel isozyme of cGMP-dependent protein kinase from bovine aorta. *J Biol Chem* 1989;264:7734–41.
4. Wernet W, Flockerzi V, Hofmann F. The cDNA of the two isoforms of bovine cGMP-dependent protein kinase. *FEBS Lett* 1989;251:191–6.
5. Corbin JD, Ogreid D, Miller JP, Suva RH, Jastorff B, Doskeland SO. Studies of cGMP-analog speci-

ficity and function of the two intrasubunit-binding sites of cGMP-dependent protein kinase. *J Biol Chem* 1986;261:1208–14.
6. Sekhar KR, Hatchett RJ, Shabb JB, et al. Relaxation of pig coronary arteries by new and potent cGMP analogs that selectively activate type Iα, compared with type Iβ, cGMP-dependent protein kinase. *Mol Pharmacol* 1992;42:103–8.
7. Landgraf W, Hofmann F. The amino terminus regulates binding to and activation of cGMP-dependent protein kinase. *Eur J Biochem* 1989;181:643–50.
8. Takio K, Wade RD, Smith SB, Krebs EG, Walsh KA, Titani K. Guanosine cyclic 3',5'-phosphate dependent protein kinase, a chimeric protein homologous with two separate protein families. *Biochemistry* 1984;23:4207–18.
9. Reed RB, Sandberg M, Jahnsen T, Lohmann SM, Francis SH, Corbin JD. Fast and slow cyclic nucleotide-dissociation sites in cAMP-dependent protein kinase are transposed in type Iβ cGMP-dependent protein kinase. *J Biol Chem* 1996;271:17570–5.
10. Sanger F, Niklen S, Coulson AR. DNA sequencing with chain-terminating inhibitors. *Proc Natl Acad Sci USA* 1977;74:5463–7.
11. O'Reilly DR, Miller LK, Luckow VA. *Baculovirus expression vectors: a laboratory manual.* New York: W.H. Freeman; 1992.
12. Glass DB, Krebs EG. Phosphorylation by guanosine 3':5'-monophosphate-dependent protein kinase of synthetic peptide analogs of a site phosphorylated in histone H2B. *J Biol Chem* 1982;257: 1196–1200.
13. Shabb JB, Ng L, Corbin JD. One amino acid change produces a high affinity cGMP-binding site in cAMP-dependent protein kinase. *J Biol Chem* 1990;265:16031–4.
14. Doskeland SO, Ogreid D. Ammonium sulfate precipitation assay for the study of cyclic nucleotide binding to proteins. *Methods Enzymol* 1988;159:147–50.
15. Rannels SR, Corbin JD. Using analogs to study selectivity and cooperativity of cyclic nucleotide-binding sites. *Methods Enzymol* 1983;99:168–75.
16. Laemmli, UK. Cleavage of structural proteins during the assembly of the head of bacteriophage T4. *Nature* 1970;227:680–5.
17. Bradford MM. A rapid and sensitive method for the quantitation of microgram quantities of protein utilizing the principle of proteind-dye binding. *Anal Biochem* 1976;72:248–54.
18. Uhler, MD. Cloning and expression of a novel cyclic GMP-dependent protein kinase from mouse brain. *J Biol Chem* 1993;268:13586–91.
19. Corbin JD, Doskeland SO. Studies of two different interchain cGMP-binding sites of cGMP-dependent protein kinase. *J Biol Chem* 1983;258:11391–7.

Signal Transduction in Health and Disease, Advances in Second Messenger and Phosphoprotein Research, Vol. 31, edited by J. Corbin and S. Francis.
Lippincott–Raven Publishers, Philadelphia © 1997.

18

The Pseudosubstrate Sequences Alone Are Not Sufficient for Potent Autoinhibition of cAMP- and cGMP-Dependent Protein Kinases as Determined by Synthetic Peptide Analysis

Celeste E. Poteet-Smith, Jackie D. Corbin, and Sharron H. Francis

Department of Molecular Physiology and Biophysics, Vanderbilt University School of Medicine, Nashville, Tennessee 37232

The autoinhibitory domains of cyclic adenosine monophosphate (cAMP)-dependent protein kinases (cAK) and cyclic guanosine monophosphate (cGMP)-dependent protein kinases (cGK) contain a pseudosubstrate sequence (^{94}RRG*A*I for Type Iα cAK and ^{74}KRQ*A*I for Type Iβ cGK) that resembles the consensus substrate site for both kinases (RRXS^{P}X) and is believed to competitively inhibit substrate phosphorylation. Synthetic peptides corresponding to these sequences in the Type Iα cAK regulatory subunit (RI-subunit) and in Type Iβ cGK were used to determine the adequacy of the pseudosubstrate sequences for potent inhibition of the cAK catalytic subunit (C-subunit) and Type Iβ cGK. The IC_{50} of 12 to 18 μM for RItide(82–99), -(92–99), or -(92–122) toward the C-subunit was four to five orders of magnitude greater than that of the wild-type (WT) RI subunit or a mutant lacking the amino-terminal (N-terminal) 93 residues (IC_{50} ~0.4 nM) (1), but the RItide(94–99) IC_{50} (800 μM) was much higher. The Type Iβ cGK pseudosubstrate peptides were considerably weaker inhibitors toward Type Iβ cGK than were the RI-subunit peptides toward the C-subunit. Type Iβ(71–82), -(58–82), -(74–82) and -(71–92) inhibited cGMP-activated Type Iβ cGK with IC_{50} values of 300 to 2700 μM. Results of peptide analogues substituted with alanine indicated that the P^{-3}/P^{-2} basic residues and P^{+1} isoleucine (P = pseudophosphorylation site) make major contributions to the potency of cAK inhibition, and these same residues are required, albeit to a lesser degree, for optimal peptide inhibition of Type Iβ cGK. The RI-subunit P^{-5} arginine and P^{-4} arginine, and the Type Iβ cGK P^{-5} arginine, are also required for optimal peptide potency. The results suggest that potent autoinhibition of cAK and cGK involves both the pseudosubstrate sequences and elements outside these pseudosubstrate sequences, particularly those located on the carboxyl-terminal (C-terminal) side of the pseudosubstrate sequences.

In the absence of activating ligands, many protein kinases exhibit a low basal activity that is maintained through a process known as autoinhibition (2–7). Autoinhibition commonly involves the interaction of a substrate-like sequence in the protein kinase with its catalytic site; this sequence competes with substrate for access to the catalytic site. Although the importance of substrate-like sequences in autoinhibition is widely accepted, the contribution of other structural components to the autoinhibitory mechanism has not been studied extensively. In the absence of cyclic adenosine monophosphate (cAMP), cAMP-dependent protein kinase (cAK) is an inactive tetrameric complex (R_2C_2). cAMP-binding to the regulatory subunit (R-subunit) decreases the affinity of the R-subunit for the catalytic subunit (C-subunit), thus relieving the potent inhibition of the C-subunit by the R-subunit (8–10). Cyclic guanosine monophosphate (cGMP)-dependent protein kinase (cGK) is homologous with cAK, but cGK is a homodimer in which each polypeptide chain contains a contiguous regulatory and catalytic domain (11,12). However, when activated by cGMP-binding, cGK, unlike cAK, does not dissociate into subunits. The regulatory components of both cAK and cGK share the same basic domain structure: a short amino-terminal (N-terminal) dimerization domain followed by an autoinhibitory domain and then by two homologous cyclic-nucleotide-binding domains.

Although some differences have been noted in the substrate specificities of cAK and cGK, these enzymes catalyze the phosphorylation of many of the same protein and peptide substrates *in vitro*, and consequently have the same consensus phosphorylation sequence (Arg-Arg-Xaa-Ser$^{(P)}$-Xaa) (13–16). The autoinhibitory domains of cAKs and cGKs contain a sequence that mimics the consensus phosphorylation sequence (Table 1). In the cAK Type II R-subunits (RII-subunits), this sequence is a substrate for the C-subunit. The cAK Type Iα R-subunit (RI-subunit) and Type Iβ cGK (IβcGK) contain similar sequences in which the phosphorylatable serine is substituted with a nonphosphorylatable alanine or glycine residue (Table 1); this is known as a pseudosubstrate sequence. However, the homologous sequences in the regulatory domains of Type Iα cGK (IαcGK), Type II cGK, and in *Drosophila* Type

TABLE 1. *Pseudosubstrate Sequences of cAMP- and cGMP-dependent Protein Kinases*

cAMP-dependent protein kinases:	
RIα-subunit	R R **R^{94} R G $\underline{A}$ I** S
RII-subunit	F D **R^{92} R X $\underline{S}^{P}$ V** C
cGMP-dependent protein kinases:	
IαcGK	R T T **R^{59} A Q $\underline{G}$ I** S
IβcGK	P R T **K^{74} R Q $\underline{A}$ I** S
IIcGK	R R G **A^{130}K A $\underline{G}$ V** S
DrIIcGK	Q R Q **R^{130}A L $\underline{G}$ I** S

The pseudosubstrate sequences are in boldface, the pseudophosphorylation site is underlined, and the autophosphorylation site is designated by a superscript P.

II cGK (DrIIcGK) contain only a single basic residue, and are not ideal consensus sequences. Nevertheless, these sequences are believed to interact with the respective catalytic sites to competitively inhibit substrate phosphorylation (17).

Considerably more is known about the requirements for autoinhibition of cAK than for cGK, owing in large part to studies of the high-affinity, heat-stable, protein kinase inhibitor (PKI) of cAK. Studies using PKI peptide analogues have demonstrated the importance of the P^{-3} arginine, P^{-2} arginine, and P^{+1} isoleucine residues in the pseudosubstrate sequence, as well as the importance of residues N-terminal to the pseudosubstrate sequence, in potent cAK inhibition (18–20). Residues N- or C-terminal to the (pseudo)phosphorylation site (P residue) are designated as minus or plus residues, respectively (21). The co-crystal of the C-subunit and a peptide derived from the inhibitory segment of PKI, PKI(5–24), confirmed that each of these residues makes important contacts with the C-subunit, thereby contributing to the potency of inhibition (21,22). The pseudosubstrate P^{-3} and P^{-2} arginine residues form hydrogen bonds with four glutamic acid residues in the C-subunit, and the P^{+1} isoleucine residue sits in a hydrophobic pocket near the surface of the enzyme. These interactions effectively lock the pseudosubstrate sequence into the active site. The PKI P^{-6} arginine hydrogen bonds with a glutamic acid residue in the C-subunit and orients the amphipathic helix at the N-terminal end of PKI. The amphipathic helix confers high-affinity binding of PKI to the C-subunit by interacting with a hydrophobic groove on the surface of the enzyme (21). A structural counterpart does not exist in the corresponding region of R-subunit, however, and the segment N-terminal to the pseudosubstrate sequence in R-subunits is not required for potent inhibition of C-subunit (1,2,23). The pseudosubstrate P^{-3} and P^{-2} arginine residues and the P^{+1} hydrophobic residue in the R-subunit are also important determinants for potent cAK autoinhibition (1,24,25).

As compared with those for cAK, the requirements for autoinhibition of cGK are not well understood. Although Type Iα cGK and Type II cGK are autoinhibited in the absence of cGMP, neither contains a prototypical pseudosubstrate sequence in the autoinhibitory domain. Researchers are only now beginning to gain some understanding of which residues within the putative cGK pseudosubstrate sequence are required for autoinhibition. Recently, it has been demonstrated by limited proteolysis of IβcGK that the P^{-2} arginine residue contributes significantly to potent autoinhibition of catalysis, but the role of the P^{-3} lysine residue could not be determined (17). These same studies also indicated that removal of the entire pseudosubstrate sequence of IβcGK (through P^{+9}) or Iα cGK (through P^{+15}) did not fully activate the enzyme. This suggested that residues C-terminal to these sequences supply important contacts for autoinhibition of Type I cGKs.

In contrast to studies of inhibition of C-subunit by PKI, similar insights into inhibition of cGK cannot be gained since PKI is a very weak inhibitor of cGK (26). In addition, cGK is inhibited only poorly by PKI peptide analogues, especially those that encompass the PKI amphipathic helix (residues P^{-16} to P^{-9}) (27). Although some differences exist at the active sites of cAK and cGK (i.e., steric/structural constraints), it is likely that there are many similarities in the mechanisms of autoinhibition for these homologous enzymes. The study described in this chapter shows that the residues within and immediately adjacent to the pseudosubstrate sequences in cAK and cGK exhibit similarities in their contributions to autoinhibition. This chapter also addresses the functional inadequacies of the isolated pseudosubstrate sequences of cAK and cGK to act as independent, potent inhibitors of the cAK C-subunit and IβcGK. It is proposed that residues C-terminal to the pseudosubstrate sequences interact with the C-subunit or the IβcGK catalytic domain to further stabilize the interaction of the pseudosubstrate sequence with the active sites, thus facilitating autoinhibition.

EXPERIMENTAL PROCEDURES

Materials

All peptides corresponding to the pseudosubstrate regions of RI-subunit and Type IβcGK were ≥ 95% pure preparations synthesized by Peptides International, Louisville, KY. Kemptide and Glasstide peptide substrates were from Peninsula Laboratories (Belmont, CA) as were protein kinase C^{19-36}, delta sleep-inducing peptide, bradykinin, and myosin light-chain kinase$^{480-501}$. The Ca^{2+}/calmodulin-dependent protein kinase$^{281-309}$ peptide, Syntide-2 peptide, and Ca^{2+}/calmodulin-dependent protein kinase II (CaMKII) were generous gifts from Dr. Roger Colbran of Vanderbilt University. [γ-^{32}P] adenosine triphosphate ([γ-^{32}P]ATP) was from Dupont (Boston, MA)/NEN. Bovine serum albumin (BSA; essentially fatty acid free, catalog No. A-6003), cyclic nucleotides, and other chemical reagents were purchased from Sigma Chemical Co., St. Louis, MO. C-subunit was prepared from bovine heart according to the method of Sugden et al. (28), and IβcGK was purified from bovine aorta according to the method of Francis et al. (29).

Peptide Sequencing and Amino Acid Analysis

The purified peptides were sequenced by sequential Edman degradation. The concentration of peptides in solution was quantitated by amino-acid analysis. These services were performed by the Peptide Sequencing and Amino Acid Analysis Shared Resource at Vanderbilt University.

Inhibition of C-Subunit by Synthetic RI-Subunit Peptides

Immediately prior to use, C-subunit was diluted in dilution buffer (50 mM potassium phosphate, pH 6.8; 0.1 mM dithiothreitol [DTT]; 1 mg/ml BSA). The RI-subunit peptides were diluted in KPEM (10 mM potassium phosphate, pH 6.8; 1 mM ethylene diamine tetraacetic acid [EDTA]; 2 mM 2-mercaptoethanol). A final concentration of 21 pM C-subunit was preincubated with varying concentrations of RI-subunit peptides in kinase buffer (20 mM Tris, pH 7.4; 20 mM magnesium acetate; 0.1 mM ATP; 0.1 mM isobutylmethylxanthine [IBMX]; and 10 mg/ml BSA) for 15 to 30 min at 30°C. The kinase assay was initiated by the addition of Kemptide peptide substrate [LRRASLG] (81 μM final concentration) and [γ-^{32}P]ATP (1,000 cpm/pmol), bringing the final ATP concentration to 171 μM (total reaction volume = 35 μl). Following a 60- to 120-min incubation at 30°C, the reaction was terminated by spotting 20 μl aliquots of reaction mixture onto P-81 phosphocellulose papers and immediately placing the papers into 75 mM phosphoric acid. The papers were washed a minimum of six times in 75 mM phosphoric acid for ≥ 1 hr, then dried, and the radioactivity on the papers counted by Cerenkov radiation to determine the amount of Kemptide phosphorylation.

Inhibition of Type Iβ cGMP-Dependent Protein Kinase by Synthetic Type Iβ cGMP-Dependent Protein Kinase Peptides

IβcGK was diluted immediately before use in dilution buffer. A final concentration of 2 nM of the enzyme was incubated with varying concentrations of synthetic peptides corresponding to the pseudosubstrate sequence of either IβcGK or RI-subunit (24 μl in KPEM) in kinase buffer containing 0.17 mM [γ-^{32}P]ATP (50 to 100 cpm/pmol), 14 μM cGMP, and 96 μM heptapeptide substrate [RKRSRAE] (total reaction volume = 35 μl). Following a 15- to 30-min incubation at 30°C, the reaction was terminated by spotting 12 μl of reaction mixture onto P-81 phosphocellulose paper and immediately placing the papers into 75 mM phosphoric acid. The papers were processed as described above to determine the amount of heptapeptide phosphorylation.

Ca^{2+}/Calmodulin-Dependent Protein Kinase II Assay

CaMKII (1 μM) was autophosphorylated on Thr^{286} as previously described (30), except that 2 mM $CaCl_2$ was used instead of 1 mM, and 1 mM DTT was included. Assays were performed essentially as described previously (31) in 50 mM 4-(2-hydroxyethyl)-1-poperazine-N′-2-ethanesulfonic acid (HEPES), pH 7.5; 1 mg/ml BSA; 1 mM DTT; 10 mM magnesium acetate; 0.4 mM [γ-^{32}P]ATP (500 cpm/pmol); 1 mM ethylene glucol-bis-(β-aminoethyl ether)-N,N,N′,N′-tetraacetic acid (EGTA); 20 μM Syntide-2 peptide substrate (PLARTLSVAGLPGKK); 1 nM autophosphorylated CaMKII; and various concentrations of RItide$^{82-99}$ peptide. Following a 10 min incubation at 30°C, the reaction was terminated by spotting 15 μl (25 μl total reaction volume) onto P81 phosphocellulose paper, which was then processed as described above to determine the amount of Syntide-2 phosphorylation.

RESULTS

Inhibition of C-Subunit by Synthetic Peptides Corresponding to the Pseudosubstrate Sequence of RI-Subunit

Synthetic peptides were used to assess the contribution that residues within and adjacent to the RI-subunit pseudosubstrate sequence (Arg[94]-Arg-Gly-*Ala*-Ile) make toward inhibition of C-subunit. The inhibitory constants for the RI-subunit peptides examined in this study are summarized in Table 2, and the inhibition curves are illustrated in Fig. 1. Peptide 1 [RItide(82–99)] inhibited C-subunit with an IC_{50} of 12

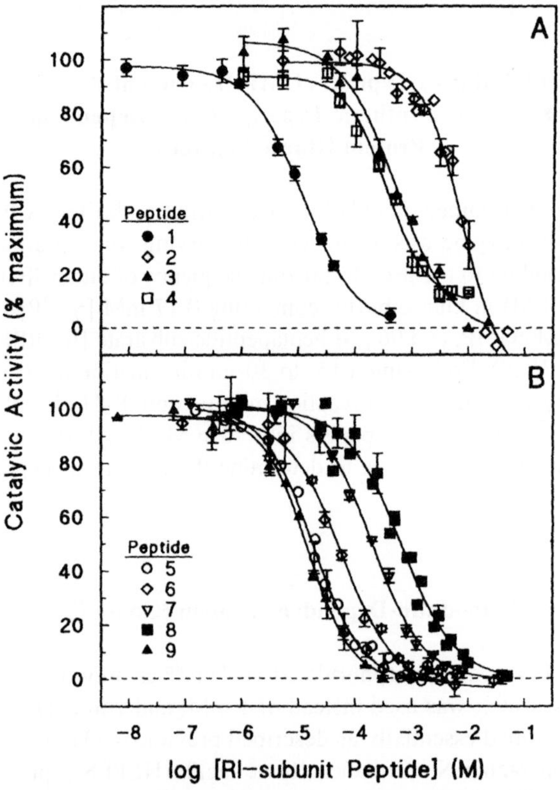

FIG. 1. Inhibition of C-subunit activity by RI-subunit pseudosubstrate sequence peptides. C-subunit (21 pM) was preincubated for 15 to 30 min at 30°C in 20 mM Tris, pH 7.4, 20 mM magnesium acetate, 0.1 mM ATP, 0.1 mM isobutylmethylxanthine (IBMX); and 10 mg/ml BSA in the presence of varying concentrations of the following RI-subunit peptides: **A**: Peptide 1 (●), n=8; Peptide 2 (◆), n=3; Peptide 3 (△), n=2; Peptide 4 (□), n=2. **B**: Peptide 5 (○), n=3; Peptide 6 (◆), n=2; Peptide 7 (▽), n=2; Peptide 8 (■), n=3; Peptide 9 (△), n=2. The reactions were initiated by the addition of Kemptide substrate (81 μM final concentration) and [γ-^{32}P]ATP (1,000 cpm/pmol) to a final concentration of 171 μM, and the incubation was continued for an additional 60 to 120 min at 30°C. The reactions were terminated and the amount of Kemptide phosphorylation was determined.

TABLE 2. *Effect of Substitutions and Length on the Inhibitory Potencies of RI-Subunit Peptides toward cAK Catalytic Subunit.*

Peptide Number	Peptide Sequence																		IC_{50} (μM)
											P^{-5}	P^{-4}	P^{-3}	P^{-2}		**P**	P^{+1}		
	82	**83**	**84**	**85**	**86**	**87**	**88**	**89**	**90**	**91**	**92**	**93**	**94**	**95**	**96**	**97**	**98**	**99**	
(1)	Pro	Pro	Pro	Pro	Asn	Pro	Val	Val	Lys	Gly	Arg	Arg	Arg	Arg	Gly	Ala	Ile	Ser	12
(2)	Pro	Pro	Pro	Pro	Asn	Pro	Val	Val	Lys	Gly	Arg	Arg	**Ala**	Arg	Gly	Ala	Ile	Ser	6230
(3)	Pro	Pro	Pro	Pro	Asn	Pro	Val	Val	Lys	Gly	Arg	Arg	Arg	**Ala**	Gly	Ala	Ile	Ser	517
(4)	Pro	Pro	Pro	Pro	Asn	Pro	Val	Val	Lys	Gly	Arg	Arg	Arg	Arg	Gly	Ala	**Ala**	Ser	350
(5)											Arg	Arg	Arg	Arg	Gly	Ala	Ile	Ser	18
(6)											**Ala**	Arg	Arg	Arg	Gly	Ala	Ile	Ser	48
(7)											**Ala**	**Ala**	Arg	Arg	Gly	Ala	Ile	Ser	203
(8)													Arg	Arg	Gly	Ala	Ile	Ser	810
(9)											Arg	Arg	Arg	Arg	Gly	Ala	Ile	Ser —	14
					Ala	Glu	Val	Tyr	Thr	Glu	Glu	Asp	Ala	Ala	Ser	Tyr	Val	Arg —	
					Lys	Val	Ile	Pro	Lys	Asp	Tyr	Lys	Thr^{122}						

The parent peptide is RItide(82–99) (Peptide 1). Substituted amino-acid residues are highlighted and underlined. Inhibitory constants were obtained from Hill plots.

μM, as compared with the IC_{50} value of ~0.4 nM obtained for the recombinant wild-type (WT) RI-subunit and a mutant lacking the N-terminal 93 residues (1). Alanine was used as the substituted residue in the RItide peptide analogues. Substitution of the P^{-3} arginine (Peptide 2) decreased the inhibitory potency of the parent peptide by ~520-fold, substitution of the P^{-2} arginine (Peptide 3) reduced the potency by about 43-fold, and substitution of isoleucine at P^{+1} (Peptide 4) reduced the potency by 29-fold. Peptide 5 (RItide[(92–99]), which lacked the N-terminal 10 residues, had a potency similar to that of Peptide 1. Substitution of the P^{-5} arginine (Peptide 6) or both the P^{-5} and P^{-4} arginine residues (Peptide 7) reduced the inhibitory potency of Peptide 5 by 2.7- and 11-fold, respectively. However, Peptide 8, which lacked the P^{-5} and P^{-4} arginine residues, was considerably less potent (45-fold) than was the P^{-5}/P^{-4} alanine-substituted Peptide 7. The possibility that residues essential for potent C-subunit inhibition might reside in the segment located immediately C-terminal to the pseudosubstrate region was examined by using Peptide 9 (RItide[92–122]), in which RI-subunit residues 100–122 were included at the C-terminal end of Peptide 5. The inhibitory potency of Peptide 9 was indistinguishable from that of Peptide 5.

Several tests were performed to ensure that the principal peptides, Peptides 1 and 5 (RItide[82–99] and RItide[92–99]), were specific inhibitors of C-subunit. Despite the fact that peptides corresponding to protein kinase $C^{(19-36)}$ (RFA*RKGAL*RQKN-VHEVKN) and calcium calmodulin-dependent protein kinase(281–309) (MHRQ-ETVDCLKKFNA*RRKLKGAI*LTT-MLA) contain sequences similar to the RI-subunit pseudosubstrate sequence (italicized), neither peptide inhibited C-subunit even at concentrations as high as 50 μM (data not shown). In addition, C-subunit was not inhibited by 50 μM of a synthetic peptide based on a sequence from myosin light-chain kinase(480–501) (AKKLSKDRMKKYMARRKWQKTG), or by 50 μM of the delta sleep-inducing (WAGGDASGE) or bradykinin (RPPGFSPFR) peptides (data not shown). Peptide 1 (RItide[82–99]) inhibited IβcGK (IC_{50} = 30 μM), albeit at a higher concentration than that for C-subunit (IC_{50} = 12 μM). However, Peptide 1 inhibited less than 50% of the kinase activity of autophosphorylated CaMKII when used at concentrations as high as 350 μM (Fig. 2).

Inhibition of Iβ cGMP-Dependent Protein Kinase by Synthetic Peptides Corresponding to the Pseudosubstrate Sequence of Iβ cGMP-Dependent Protein Kinase

Synthetic peptides based on the pseudosubstrate sequence of IβcGK were also used to assess the contribution that residues within and adjacent to the pseudosubstrate sequence (Lys^{74}-Arg-Gln-*Ala*-Ile) make toward inhibition of the kinase activity of this enzyme (Table 3). Peptides 10 and 16 were first tested as substrates for IβcGK. Although these peptides do not contain prototypical consensus phosphorylation sites for cGK, Ser^{79} is autophosphorylated in the intact enzyme and is responsible for activation of this enzyme (32). When substituted for the heptapeptide substrate in a kinase assay (see Experimental Procedures), neither peptide was

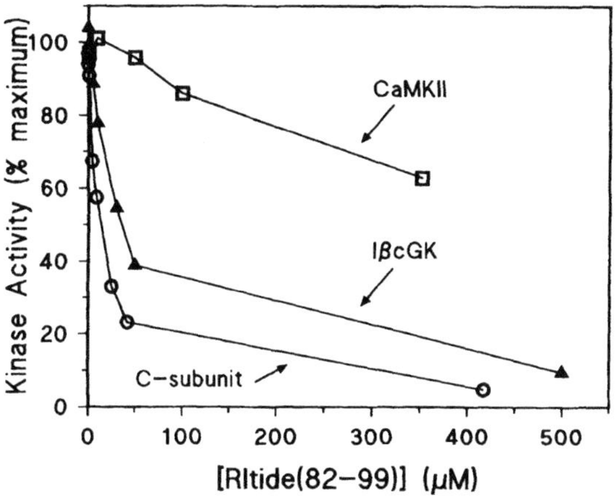

FIG. 2. Specificity of RItide(82–99) (peptide 1) for cAMP-dependent protein kinase. cAMP-dependent protein kinase C-subunit (21 pM), (○), cGMP-dependent protein kinase Type Iβ (IβcGK) (214 pM) (▲), or autophosphorylated calcium/calmodulin-dependent protein kinase (CaMKII) (1 nm), (□) were assayed against varying concentrations of RItide(82–99) peptide [PPPPNVPVKGRRRRGAIS] as described in Experimental Procedures.

detectably phosphorylated by βcGK, even when peptide concentrations as high as 10 mM were used (data not shown). The inhibition curves for the IβcGK peptides are illustrated in Fig. 3, and the IC_{50} values are summarized in Table 3. The parent peptide, Peptide 10 (Iβ[71–82]), inhibited IβcGK with an IC_{50} of 0.5 mM, a value that was 42-fold greater than the values obtained for inhibition of C-subunit by

TABLE 3. *Effect of Substitutions and Length on the Inhibitory Potencies of IβcGK Peptides toward IβcGK*

Peptide Number	Peptide Sequence	IC_{50} (mM)
	P-5 (72), P-3 (74), P-2 (75), P (77), P+1 (78)	
	71 72 73 74 75 76 77 78 79 80 81 82	
(10)	Pro Arg Thr Lys Arg Gln Ala Ile Ser Ala Glu Pro	0.50
(11)	Pro **Ala** Thr Lys Arg Gln Ala Ile Ser Ala Glu Pro	2.67
(12)	Pro Arg Thr **Ala** Arg Gln Ala Ile Ser Ala Glu Pro	1.4
(13)	Pro Arg Thr **Arg** Arg Gln Ala Ile Ser Ala Glu Pro	0.035
(14)	Pro Arg Thr Lys **Ala** Gln Ala Ile Ser Ala Glu Pro	2.1
(15)	Pro Arg Thr Lys Arg Gln Ala **Ala** Ser Ala Glu Pro	0.79
(16)	Lys Arg Gln Ala Ile Ser Ala Glu Pro	2.74
(17)	Gln[58] Ala Gln Lys Gln Ser Ala Ser Thr Leu Gln Gly Glu — Pro Arg Thr Lys Arg Gln Ala Ile Ser Ala Glu Pro	1.0
(18)	Pro Arg Thr Lys Arg Gln Ala Ile Ser Ala Glu Pro — Thr Ala Phe Asp Ile Gln Asp Leu Ser His[92]	0.30

The parent peptide is Type IβcGK (71–82) (Peptide 10) and the substituted amino-acid residues are highlighted and underlined. Inhibitory constants were obtained from Hill plots.

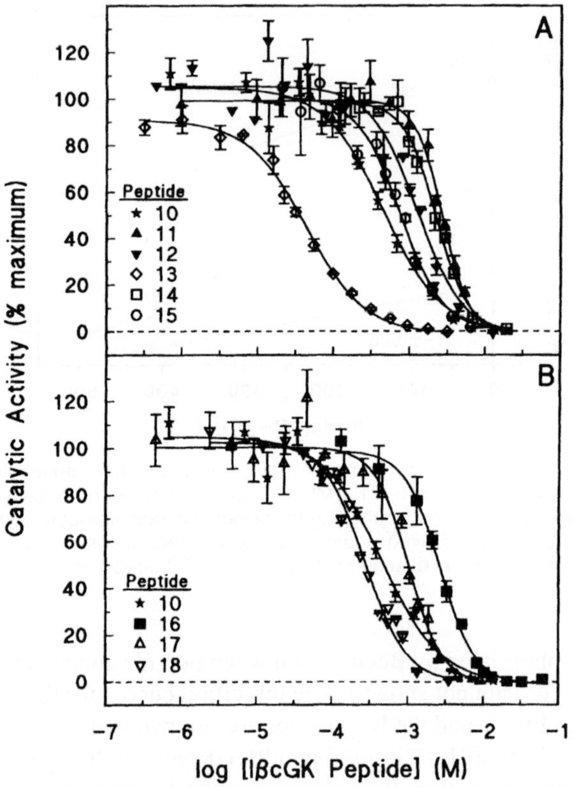

FIG. 3. Inhibition of IβcGK activity by IβcGK pseudosubstrate sequence peptides. IβcGK (2 nM) was incubated with 20 mM Tris, pH 7.4, 20 mM magnesium acetate, 0.17 mM [γ-^{32}P]ATP (50 to 100 cpm/pmol), 14 μM cGMP, 0.1 mM IBMX, 10 mg/ml BSA, and 96 μM heptapeptide substrate (RKRSRAE) in the presence of varying concentrations of the following IβcGK peptides: (**A**) Peptide 10 (★), n=4; Peptide 11 (△), n=3; Peptide 12 (▽), n=3; Peptide 13 (♦), n=3; Peptide 14, (□), n=3; Peptide 15 (○), n=3. (**B**) Peptide 10 (★), n=4; Peptide 16 (■), n=2; Peptide 17 (△), n=3; Peptide 18 (▽), n=3. Following a 15 to 30-min incubation at 30°C, the reaction was terminated and the amount of heptapeptide substrate phosphorylation was determined.

Peptide 1 (12 μM) and Peptide 5 (18 μM). Substitution of the basic residues in this sequence (P^{-5} arginine, P^{-3} lysine, or P^{-2} arginine) with alanine (Peptides 11, 12, and 14) decreased the inhibitory potency of these peptides by 5-, 3-, or 4-fold, respectively. However, replacement of the P^{-3} lysine residue with arginine (Peptide 13) improved the inhibitory potency of this peptide by ~14-fold compared to that of Peptide 10. Substitution of the P^{+1} isoleucine residue (Peptide 15) caused only a minimal decrease in the potency of this peptide (~1.5-fold) compared to Peptide 1.

Peptide 16, which lacked the first three residues of the N-terminus of Peptide 10 had a decreased inhibitory potency (6-fold). Peptide 17, which included IβcGK residues 58 to 82, was analyzed to determine if residues N-terminal to the P^{-6}

residue might be required for potent inhibition of IβcGK. The inhibitory potency of Peptide 17 was ~2-fold lower than that of Peptide 10. Since Peptide 17 contains both the primary (Ser^{63}) and secondary (Ser^{79}) sites of autophosphorylation for IβcGK, it was also tested as a substrate by substituting it for heptapeptide substrate in the kinase assay (17,32). Peptide 17 (10 mM) was not phosphorylated by IβcGK as determined by electrophoresis on an 18% sodium dodecylsulfate (SDS)–polyacrylamide gel and autoradiography (data not shown). Peptide 18, encompassing IβcGK residues 71–92, was examined to determine the effect on inhibitory potency of 10 additional IβcGK residues on the C-terminus of Peptide 10. The inhibitory potency of Peptide 18 was ~2-fold greater than that of Peptide 10.

DISCUSSION

The most potent RI-subunit peptides examined in this study (Peptides 1, 5, and 9) inhibit cAK C-subunit with similar potencies, but are about four to five orders of magnitude less potent toward C-subunit than is either the WT RI-subunit or RI-subunit in which the N-terminal 93 residues are deleted (Δ1–93) (1). Despite the fact that the RI-subunit peptides inhibit C-subunit with much lower potency than do full-length and certain truncated RI-subunits, a similar pattern of crucial amino acids is found for each (1,24,25). Since the N-terminal 93 residues of the RI-subunit are not required for autoinhibition of the C-subunit (1), the high IC_{50} values obtained for the RI-subunit peptides suggest that element(s) on the C-terminal side of the pseudosubstrate sequence, perhaps even elements within the cyclic-nucleotide-binding sites, must be required for establishing or stabilizing potent interactions between crucial pseudosubstrate residues and the C-subunit. However, when located in a peptide, residues 100–122, which reside on the immediate C-terminal side of the RI-subunit pseudosubstrate sequence, are not sufficient to provide this high-affinity interaction with RI-subunit Peptide 9. This is particularly interesting in light of the fact that this segment contains six residues that are highly conserved in cAK R-subunits, and a seventh residue (Lys^{118}) that is conserved in both cAK and cGK regulatory domains (33–35).

The P^{-3} and P^{-2} arginine residues are critical determinants for potent inhibition of C-subunit by RI-subunit peptides. These results are in agreement with previous findings using full-length (25) and truncated RI-subunits (1), RII-subunits (24), PKI peptide analogues (18,19), and the crystal structure of the C-subunit -PKI(5–24) complex (21). The data presented here suggest that the requirement for a basic residue is much greater at the P^{-3} position than at the P^{-2} position. This result is consistent with results obtained using PKI peptide analogues (19), but contrary to findings made with full-length (25) or truncated RI-subunits (1), or with the peptide substrate Kemptide (36). The reason for this difference is unclear, and cannot be explained as an inherent difference in the modes of interaction of peptides and proteins with the active site, since the Kemptide peptide arginine residues interact with

C-subunit in a manner similar to those in the RI-subunit protein. The P^{+1} isoleucine is also crucial for potent inhibition of C-subunit by RI-subunit peptides. The importance of a large P^{+1} hydrophobic residue for potent inhibition of C-subunit by PKI has been established with peptide analogues (20), as well as from the crystal structure of the C-subunit -PKI(5–24) complex (21). More recently, our laboratory has demonstrated that a large P^{+1} hydrophobic residue is also required for potent inhibition of C-subunit by RI-subunit (1).

Although neither Arg^{92} (P^{-5}) nor Arg^{93} (P^{-4}) is detectably involved in autoinhibition of the cAK C-subunit by RI-subunit mutants (1), substitution or removal of Arg^{92} and Arg^{93} from Peptide 5 dramatically reduces the inhibitory potency of the resulting Peptides 7 and 8. The inherent weakness of the peptide inhibitors may allow detection of minor interactions between the pseudosubstrate residues and the C-subunit that were not detected with the more potent truncated RI-subunits. A basic residue(s) occurs at the P^{-4}, P^{-5}, or P^{-6} positions in many cAK substrates and suggests that, in some instances, these basic residues could provide contacts with the active site (14,37–41). Indeed, each of the P^{-3}, P^{-4}, and P^{-6} basic residues in phosphorylase kinase (β subunit) are required for the cAK C-subunit to phosphorylate a synthetic peptide substrate derived from this sequence with a high affinity (37).

The IβcGK parent Peptide 10 is ~30-fold less potent toward IβcGK than is the comparable RI-subunit Peptide 5 toward cAK. However, when the pseudosubstrate IβcGK P^{-3} lysine residue is substituted with arginine in Peptide 13, this peptide is a more potent inhibitor of IβcGK than is the parent peptide, and in fact exhibits an IC_{50} (35 μM) comparable to that of the RI-subunit parent Peptide 1 toward the C-subunit (12 μM). Therefore, although the P^{-3} lysine provides a modest contribution to the potency of IβcGK inhibition by synthetic peptides derived from the pseudosubstrate sequence, it is not the most optimal residue in this position for contact with the catalytic site. The interaction of the cGK pseudosubstrate sequence and the catalytic site may be innately weak compared with the analogous interaction in cAK, since the autoinhibition of cGK is facilitated by the juxtaposition of the regulatory and catalytic domains in the same polypeptide chain. The physical association of regulatory and catalytic domains in cGK persistently maintains these functional elements in close proximity, and may diminish the importance of interactions at a high-affinity pseudosubstrate sequence. In fact, since the the two domains are physically associated, an optimized pseudosubstrate sequence could pose an impediment to cGK activation by cGMP. In order for optimum cGMP regulation of cGK to occur in intact cells, a weak interaction of the pseudosubstrate sequence with the active site might have been favored in the process of natural selection of cGK regulation. This same scheme may apply in other kinases whose regulatory and catalytic activities are colocalized in the same polypeptide.

As assessed from the peptide studies described here, the P^{-5} arginine, P^{-3} lysine, and P^{-2} arginine residues contribute only moderately to the potency of IβcGK inhibition. Nevertheless, each of these residues is required for inhibition by the peptides as compared with the IβcGK parent peptide. The P^{+1} isoleucine residue also appears to contribute to the potency of IβcGK inhibition by peptides, but to a lesser extent than do the P^{-5}, P^{-3}, and P^{-2} basic residues. Substitutions in the IβcGK peptides have a less profound effect on their inhibitory potency than similar substitutions made in the RI-subunit pseudosubstrate peptides.

In addition to the pseudosubstrate sequence, potent IβcGK autoinhibition must require additional interactions involving residues outside of the pseudosubstrate sequence. It is suggested that, as a consequence of the physical association of the regulatory and catalytic domains in cGK, there are numerous low-affinity interactions between the IβcGK autoinhibitory domain and the active site. This is supported by the observation that each of the P^{-5}, P^{-3}, P^{-2}, and P^{+1} residues in the IβcGK peptides contributes to a similar extent, albeit weakly, to inhibition of IβcGK. The addition of 10 IβcGK residues (83–92) to the C-terminal end of Peptide 10 slightly improves the inhibitory potency of the resulting peptide (Peptide 18), suggesting that residue(s) in this region may also interact with the active site region, but that these interactions are also weak.

Results with a truncated enzyme produced by limited proteolysis of IβcGK (the N-terminus begins at Arg^{75}-Gln-*Ala*-Ile-Ser-Ala . . .) suggest that the P^{-2} Arg^{75} residue can provide for potent autoinhibiton of IβcGK catalysis (17). In the intact enzyme, the Arg^{72} (P^{-5}) or Lys74 (P^{-3}) basic residues may also contribute significantly to the potency with which this region interacts with the catalytic site, but these potential contributions would not be detectable in the proteolyzed IβcGK since the catalytic activity is almost fully inhibited (17). It is notable that the Type Iα and Type II cGKs each have only a single basic residue in the putative pseudosubstrate site (Table 1). These enzymes are all autoinhibited in the absence of cGMP, which suggests that one basic residue in either the P^{-2} or P^{-3} position may provide sufficient interaction for maintaining the cGKs in an inactive state. However, substitution of either Arg^{72} (P^{-3}) or Lys^{74} (P^{-5}) residues with Ala in IβcGK Peptides 12 or 11 reduces the inhibitory potency compared to parent Peptide 10. Since some autoinhibition of IαcGK and IβcGK is retained after removal of the pseudosubstrate sequences by limited proteolysis, it is possible that contacts C-terminal to the pseudosubstrate site may contribute to the overall binding affinity of the pseudosubstrate region. It is also likely that detection of minor contributions of individual residues would be difficult to measure by changes in the basal activity of truncated forms of cGK.

The protein kinase catalytic core is highly conserved among nearly 400 members of the protein kinase superfamily (42,43). There are several invariant and highly conserved residues in the catalytic core that participate either directly or indirectly in binding MgATP and in catalysis (42,43). In contrast, the catalytic core residues involved in protein substrate binding are not well conserved among different protein kinases in the superfamily. Since cGK and cAK are homologous enzymes that exhibit similar substrate specificities and have similar pseudosubstrate sequences, it is likely that the residues involved in substrate binding and autoinhibition will be similar for both of these kinases. The decreased inhibitory potency observed with both RI-subunit and IβcGK peptides with alanine substitutions in place of the P^{-3} or P^{-2}

basic residues or the P^{+1} hydrophobic residue can be interpreted on the basis of interactions that occur in the co-crystal of the PKI(5–24) peptide and C-subunit (21). The crystal structure clearly establishes the importance of the PKI P^{-3} and P^{-2} arginine residues and the P^{+1} hydrophobic residue to potent inhibition of C-subunit (21). The PKI P^{-3} pseudosubstrate arginine residue forms hydrogen bonds with Glu^{127} and Glu^{331}, and the P^{-2} pseudosubstrate arginine hydrogen bonds with Glu^{230} and Glu^{170}, thus locking the pseudosubstrate sequence into the active site. The PKI P^{+1} isoleucine residue is situated in a hydrophobic pocket comprised of Leu^{198}, Pro^{202}, and Leu^{205}, near the surface of the C-subunit, and is believed to participate in properly orientating the P-site residue. Another high-affinity interaction is provided by the P^{-6} arginine residue, which forms two hydrogen bonds with Glu^{203}, but the analogous basic residue is absent in the RI-subunit or IβcGK. With the exception of Glu^{331}, which is outside of the conserved catalytic core, each of the glutamic acid residues involved in forming hydrogen bonds with the PKI P^{-6}, P^{-3}, and P^{-2} arginine residues is invariant in cGK Types Iα and Iβ. The C-subunit hydrophobic residues that interact with the PKI P^{+1} hydrophobic residue are also highly conserved as phenylalanine, proline, and valine in both cGK Types Iα and Iβ (43–45). Since PKI does not contain P^{-4} or P^{-5} basic residues, the crystal structure does not provide clues as to how interactions between these residues and the active site might occur with the RI-subunit or IβcGK.

In summary, results of these peptide studies suggest that the P^{-5}, P^{-3}, P^{-2}, and P^{+1} pseudosubstrate sequence residues contribute to autoinhibition of both IαcAK and IβcGK. This study also reveals additional important RI-subunit residues (P^{-5} and P^{-4} arginine residues) that were not detected using truncated RI-subunits (1). The parent peptide of the IβcGK pseudosubstrate sequence is profoundly less potent than the pseudosubstrate sequence of the RI-subunit, primarily because the IβcGK P^{-3} site contains lysine instead of arginine. The impotence of the cAK RI-subunit and IβcGK pseudosubstrate sequence peptides suggests that there must be other interactions between the autoinhibitory and catalytic domains that contribute to the interaction of these sequences with the active sites of C-subunit and IβcGK. The physical linkage of the regulatory and catalytic components of cGK, and the profound weakness of the synthetic peptides that are derived from the pseudosubstrate sequences, supports such an interaction. Since the segments N-terminal to the pseudosubstrate sequences of the RI-subunit and IβcGK are not required for potent autoinhibition of C-subunit and IβcGK (1,17), it is suggested that contributing autoinhibitory domain residues/elements must reside C-terminal to the pseudosubstrate sequences of both enzymes.

cAK is somewhat unique among the protein kinases that are regulated by pseudosubstrate sequences, in that the pseudosubstrate sequence and catalytic domain are located on separate subunits, whereas these domains are contiguous in many other protein kinases including cGK, myosin light-chain kinase, protein kinase C (PKC), and CaMKII (4). Since cAK and cGK are homologous enzymes, it is suggested that association of the cAK R-subunit and C-subunit involves interactions similar to those involved in cGK autoinhibition. Compared to that of cAK, the structural arrangement of cGK is more typical of members of the protein kinase family; consequently, consideration of autoinhibitory mechanisms in cGK may be more generally applicable, such that cGK may be a more suitable prototype for studying overall mechanisms of autoinhibition.

ACKNOWLEDGMENTS

This work was supported by NIH DK 40029 and CA 68485 and NIH Training Grant 5T32DK07563.

REFERENCES

1. Poteet-Smith CE, Shabb JB, Francis SH, Corbin JD. Identification of critical determinants for autoinhibition in the pseudosubstrate region of type Iα cAMP-dependent protein kinase. *J Biol Chem* 1997; 272:379–88.
2. Corbin JD, Sugden PH, West L, Flockhart DA, Lincoln TM, McCarthy D. Studies on the properties and mode of action of the purified regulatory subunit of bovine heart adenosine 3':5'-monophosphate-dependent protein kinase. *J Biol Chem* 1978;253:3997–4003.
3. Flockhart DA, Corbin JD. Regulatory mechanisms in the control of protein kinases. *CRC Crit Rev Biochem* 1982;12:133–86.
4. Kemp BE, Pearson RB. Intrasteric regulation of protein kinases and phosphatases. *Biochim Biophys Acta* 1991;1094:67–76.
5. Kemp BE, Parker MW, Hu S, Tiganis T, House C. Substrate and pseudosubstrate interactions with protein kinases: determinants of specificity. *Trends Biochem Sci* 1994;19:440–4.
6. Woodford TA, Taylor SJ, Corbin JD. The biological functions of protein phosphorylation-dephosphorylation. In: Bittar, EE, ed. *Fundamentals in medical cell biology*. JAI Press, 1992;453–507.
7. Soderling TR. Protein kinases. Regulation by autoinhibitory domains. *J Biol Chem* 1990;265: 1823–6.
8. Builder SE, Beavo JA, Krebs EG. Mechanism of activation and inactivation of the cAMP-dependent Protein kinases. In: Rosen OM et al, *Protein phosphorylation*. Cold Spring Harbor, NY: Cold Spring Harbor Laboratories, 1981;33–44.
9. Granot J, Mildvan AS, Hiyama K, Kondo H, Kaiser ET. Magnetic resonance studies of the effect of the regulatory subunit on metal and substrate binding to the catalytic subunit of bovine heart protein kinase. *J Biol Chem* 1980;255:4569–73.
10. Rangel-Aldao R, Rosen OM. Effect of cAMP and ATP on the reassociation of phosphorylated and nonphosphorylated subunits of the cAMP-dependent protein kinase from bovine cardiac muscle. *J Biol Chem* 1977;252:7140–5.
11. Gill GN, Holdy KE, Walton GM, Kanstein CB. Purification and characterization of cGMP-dependent protein kinase. *Proc Natl Acad Sci USA* 1976;73:3918–22.
12. Lincoln TM, Dills WL Jr, Corbin JD. Purification and subunit composition of guanosine 3':5'-monophosphate-dependent protein kinase from bovine lung. *J Biol Chem* 1977;252:4269–75.
13. Kennelly PJ, Krebs EG. Consensus sequences as substrate specificity determinants for protein kinases and protein phosphatases. *J Biol Chem* 1991;266:15555–8.
14. Zetterqvist O, Ragnarsson U, Engstrom L. Substrate specificity of cyclic AMP-dependent protein kinase. In: Kemp BE, ed. *Peptides and protein phosphorylation*. Orlando, FL: CRC Press, 1990; 171–88.
15. Glass DB. Substrate specificity of the cGMP-dependent protein kinase. In: Kemp BE, ed. *Peptides and protein phosphorylation*. Boca Raton, FL: CRC Press, 1990;210–38.
16. Francis SH, Corbin JD. Progress in understanding the mechanism and function of cyclic GMP-dependent protein kinase. *Adv Pharm* 1994;26:115–70.
17. Francis SH, Smith JA, Colbran JL, et al. Arginine-75 in the pseudosubstrate sequence of type I beta

cGMP-dependent protein kinase provides for potent autoinhibition, but the primary autophosphorylation at serine-63 is well outside this sequence. *J Biol Chem* 1996;271:20748–55.
18. Scott JD, Fischer EH, Demaille JG, Krebs EG. Identification of an inhibitory region of the heat-stable protein inhibitor of the cAMP-dependent protein kinase. *Proc Natl Acad Sci USA* 1985;82: 4379–83.
19. Cheng HC, Kemp BE, Pearson RB, et al. A potent synthetic peptide inhibitor of the cAMP-dependent protein kinase. *J Biol Chem* 1986;261:989–92.
20. Glass DB, Cheng HC, Mende-Mueller L, Reed J, Walsh DA. Primary structural determinants essential for potent inhibition of cAMP-dependent protein kinase by inhibitory peptides corresponding to the active portion of the heat-stable inhibitor protein. *J Biol Chem* 1989;264:8802–10.
21. Knighton DR, Zheng JH, ten Eyck LF, Xuong NH, Taylor SS, Sowadski JM. Structure of a peptide inhibitor bound to the catalytic subunit of cyclic adenosine monophosphate-dependent protein kinase. *Science* 1991;253:414–20.
22. Knighton DR, Zheng JH, ten Eyck LF, et al. Crystal structure of the catalytic subunit of cyclic adenosine monophosphate-dependent protein kinase. *Science* 1991;253:407–14.
23. Weber W, Hilz H. Stoichiometry of cAMP binding and limited proteolysis of protein kinase regulatory subunits RI and RII. *Biochem Biophys Res Commun* 1979;90:1074–81.
24. Wang YH, Scott JD, McKnight GS, Krebs EG. A constitutively active holoenzyme form of the cAMP-dependent protein kinase. *Proc Natl Acad Sci USA* 1991;88:2446–50.
25. Buechler YJ, Herberg FW, Taylor SS. Regulation-defective mutants of type I cAMP-dependent protein kinase. Consequences of replacing arginine 94 and arginine 95. *J Biol Chem* 1993;268: 16495–503.
26. Walsh DA, Angelos KL, Van Patten SM, Glass DB, Garetto LP. The inhibitor protein of the cAMP-Dependent Protein Kinase. In: Kemp BE, ed. *Peptides and protein phosphorylation*. Boca Raton, FL: CRC Press, 1990;43–84.
27. Glass DB, Cheng HC, Kemp BE, Walsh DA. Differential and common recognition of the catalytic sites of the cGMP-dependent and cAMP-dependent protein kinases by inhibitory peptides derived from the heat-stable inhibitor protein. *J Biol Chem* 1986;261:12166–71.
28. Sugden PH, Holladay LA, Reimann EM, Corbin JD. Purification and characterization of the catalytic subunit of adenosine 3':5'-cyclic monophosphate-dependent protein kinase from bovine liver. *Biochem J* 1976;159:409–22.
29. Francis SH, Wolfe L, Corbin JD. Purification of type I alpha and type I beta isozymes and proteolyzed type I beta monomeric enzyme of cGMP-dependent protein kinase from bovine aorta. *Methods Enzymol* 1991;200:332–41.
30. Colbran RJ, Smith MK, Schworer CM, Fong YL, Soderling TR. Regulatory domain of calcium/calmodulin-dependent protein kinase II. Mechanism of inhibition and regulation by phosphorylation. *J Biol Chem* 1989;264:4800–4.
31. Colbran RJ, Fong YL, Schworer CM, Soderling TR. Regulatory interactions of the calmodulin-binding, inhibitory, and autophosphorylation domains of Ca2+/calmodulin-dependent protein kinase II. *J Biol Chem* 1988;263:18145–51.
32. Smith JA, Francis SH, Walsh KA, Kumar S, Corbin JD. Autophosphorylation of type Iβ cGMP-dependent protein kinase increases basal catalytic activity and enhances allosteric activation by cGMP or cAMP. *J Biol Chem* 1996;271:20756–62.
33. Uhler MD. Cloning and expression of a novel cyclic GMP-dependent protein kinase from mouse brain. *J Biol Chem* 1993;268:13586–91.
34. Takio K, Smith SB, Krebs EG, Walsh KA, Titani K. Amino acid sequence of the regulatory subunit of bovine type II adenosine cyclic 3',5'-phosphate dependent protein kinase. *Biochemistry* 1984; 23:4200–6.
35. Titani K, Sasagawa T, Ericsson LH, et al. Amino acid sequence of the regulatory subunit of bovine type I adenosine cyclic 3',5'-phosphate dependent protein kinase. *Biochemistry* 1984;23:4193–9.
36. Kemp BE, Graves DJ, Benjamini E, Krebs EG. Role of multiple basic residues in determining the substrate specificity of cyclic AMP-dependent protein kinase. *J Biol Chem* 1977;252:4888–94.
37. Zetterqvist O, Ragnarsson U. The structural requirements of substrates of cyclic AMP-dependent protein kinase. *FEBS Lett* 1982;139:287–90.
38. Hemmings HC Jr, Williams KR, Konigsberg WH, Greengard P. DARPP-32, a dopamine-and adenosine 3':5'-monophosphate-regulated neuronal phosphoprotein I. Amino acid sequence around the phosphorylated threonine. *J Biol Chem* 1984;259:14486–90.
39. Aitken A, Bilham T, Cohen P. Complete primary structure of protein phosphatase inhibitor-1 from rabbit skeletal muscle. *Eur J Biochem* 1982;126:235–46.

40. Murphy BJ, Rossie S, De Jongh KS, Catterall WA. Identification of the sites of selective phosphorylation and dephosphorylation of the rat brain Na^+ channel α subunit by cAMP-dependent protein kinase and phosphoprotein phosphatases. *J Biol Chem* 1993;268:27355–62.
41. Butt E, Abel K, Krieger M, et al. cAMP- and cGMP-dependent protein kinase phosphorylation sites of the focal adhesion vasodilator-stimulated phosphoprotein (VASP) *in vitro* and in intact human platelets. *J Biol Chem* 1994;269:14509–17.
42. Hanks SK, Quinn AM, Hunter T. The protein kinase family: conserved features and deduced phylogeny of the catalytic domains. *Science* 1988;241:42–52.
43. Hanks SK, Hunter T. Protein kinases 6. The eukaryotic protein kinase superfamily: kinase (catalytic) domain structure and classification. *FASEB J* 1995;9:576–96.
44. Takio K, Wade RD, Smith SB, Krebs EG, Walsh KA, Titani K. Guanosine cyclic 3',5'-phosphate dependent protein kinase, a chimeric protein homologous with two separate protein families. *Biochemistry* 1984;23:4207–18.
45. Sandberg M, Natarajan V, Ronander I, et al. Molecular cloning and predicted full-length amino acid sequence of the type I beta isozyme of cGMP-dependent protein kinase from human placenta. Tissue distribution and developmental changes in rat. *FEBS Lett* 1989;255:321–9.

Signal Transduction in Health and Disease, Advances in Second Messenger and Phosphoprotein Research, Vol. 31, edited by J. Corbin and S. Francis.
Lippincott–Raven Publishers, Philadelphia © 1997.

19

Recent Advances in the Study of Ca^{2+}/CaM-Activated Phosphodiesterases

Expression and Physiological Functions

Allan Z. Zhao, Chen Yan, William K. Sonnenburg, and Joseph A. Beavo

Department of Pharmacology, University of Washington School of Medicine, Seattle, Washington 98195

INTRODUCTION

The second messengers cyclic adenosine monophosphate (cAMP) and cyclic guanosine monophosphate (cGMP) regulate such important cellular functions as gene transcription, neurotransmitter synthesis and release, ion-channel gating, and energy metabolism (1–6). The intracellular concentration of cyclic nucleotides is regulated by the relative activity of both adenylyl and guanylyl cyclases and phosphodiesterases (PDEs) (7). Like that of many other signal-transduction components, the activities of individual PDEs are regulated by different mechanisms under different physiologic conditions. Large numbers of biochemical, pharmacologic, and molecular genetic studies have revealed seven major families of PDEs. These are the Ca^{2+}/calmodulin-activated PDE1 family; the cGMP-stimulated PDE2 family; the cGMP-inhibited PDE3 family; the high-affinity/rolipram-sensitive cAMP-specific PDE4 family; the cGMP-specific, cGMP-binding PDE5 family; the light-stimulated photoreceptor PDE6 family; and the high-affinity, cAMP-specific/rolipram-insensitive PDE7 family (for a comprehensive review of the PDE nomenclature, see [8]). Many of these PDE families contain several gene members. Furthermore, recent studies, using molecular-biologic techniques, have identified, for many of these PDE genes, multiple, alternatively spliced isozyme variants that escaped detection in previous biochemical and pharmacologic studies. These PDE isozymes participate in regulating such diverse physiologic processes as platelet aggregation, smooth-muscle contraction, phototransduction, pulmonary airway and vascular resistance, and long-term potentiation in the central nervous system. In this review, we will focus on the members of the PDE1 family, including their expression patterns, structures, and physiologic functions.

THE PHOSPHODIESTERASE-1 FAMILY IS ENCODED BY THREE DIFFERENT GENES

Of all the phosphodiesterases, the PDE1 family is probably the most thoroughly studied. Owing to their property of being stimulated by calmodulin (CaM) in the presence of Ca^{2+}, the PDE1s are often referred to as the CaM–PDE family. It is now known that at least three different genes encode the Ca^{2+}/CaM-activated PDE family; they are named as PDE1A, PDE1B, and PDE1C. Two splice variants of PDE1A and at least four splice variants of PDE1C have been isolated and characterized (9, 10; Yan, Zhao, and Beavo, *unpublished observations*). However, alternatively spliced variants of PDE1B have not yet been defined. Figure 1 is a schematic representation of the PDE1A and PDE1C alternative-splice variants.

PDE1A1 and PDE1A2 are likely to be products of an alternatively spliced gene. Most of the coding sequence of these two isozymes is identical, except within the divergent amino-terminus (N-terminus) (9,10). Likewise, the PDE1C isozymes are most likely encoded by a single alternatively spliced gene, since the the nucleotide sequence throughout most of the coding region of the four PDE1C variants is identical (18; Yan, Zhao, and Beavo, *unpublished observation*). The original classification of CaM–PDE isozymes was based on their apparent molecular weights by sodium dodecylsulfate (SDS)–gel electrophoresis: this comprised the 58-kDa and 59-kDa isozymes (isolated from lung and heart, respectively), the 61-kDa isozyme (from brain), the 63-kDa isozyme (from brain), the 68- to 70-kDa isozyme (from testis), and the 75-kDa isozyme (from brain) (11–15). Comparison of complementary deoxyribonucleic acid (cDNA) sequences and the earlier protein sequence data indicates that PDE1A1 encodes the 59-kDa CaM–PDE (and probably the 58-kDa isozyme as well), that PDE1A2 encodes the 61-kDa CaM-PDE (15–17), and that PDE1B1 represents the 63-kDa CaM–PDE (18). It is possible that one of the PDE1C splice variants encodes the testicular 68–70-kDa CaM–PDE isozyme. However, this needs to be verified by further biochemical analysis. It is still unclear whether any of the PDE1C splice variants encodes the 75-kDa CaM–PDE originally detected in bovine brain extracts (14,15). PDE1C2, with a molecular weight of 87 kDa, is most highly expressed in olfactory sensory neurons, and is likely to encode the high-affinity CaM–PDE activity detected in olfactory cilia in early studies (19). Currently, the physiologic reasons for the expression of these CaM–PDEs in multiple forms remain elusive.

KINETIC PROPERTIES OF CALMODULIN–PHOSPHODIESTERASES

As we will discuss below, more than one CaM–PDE is often expressed in the same tissue, though not necessarily in the same cell types. Therefore, the kinetic parameters of nearly all CaM–PDEs determined with biochemically purified enzymes from various tissues often reflect the pitfall of containing more than one CaM-PDE in the enzyme preparations. This undoubtedly has contributed to the large variation

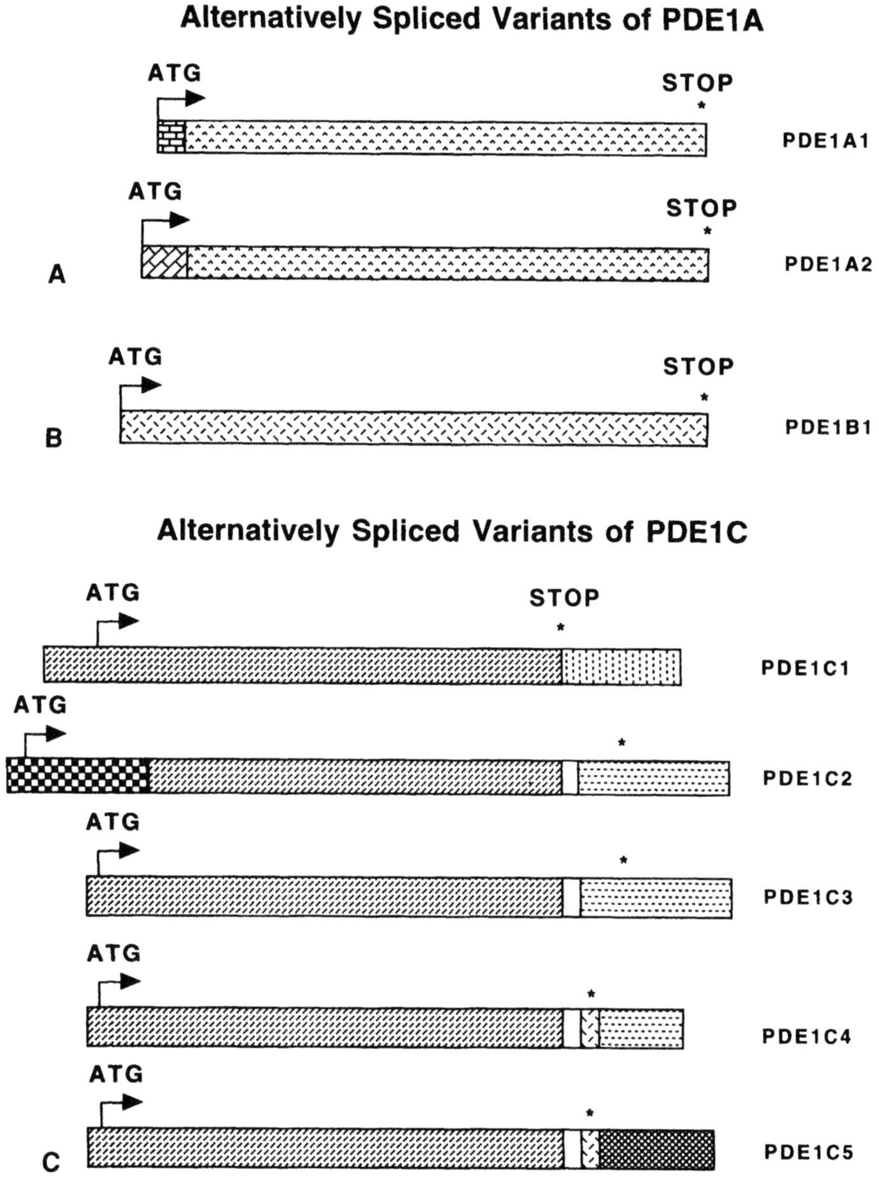

FIG. 1. A: Schematic representation of PDE1A, PDE1B, and PDE1C alternative splice variants. The major difference between the two splice variants of PDE1A lies at the N-terminus (9,10). **B:** Except for PDE1B1, no other alternatively spliced PDE1B variant has yet been defined. **C:** Five alternatively spliced variants of PDE1C with complete coding sequences have been isolated (19,23; Yan, Zhao, and Beavo, *unpublished observation*). These splice variants have been proven to exist by RNase protection assay and/or *in situ* hybridization analysis (19,23; Yan, Zhao, and Beavo, *unpublished observation*). All the diagrams are drawn to scale.

TABLE 1. *Kinetic Properties of CaM-PDEs*

	K_m (μM)		V_{max} Ratio (cAMP/cGMP)
	cAMP	cGMP	
PDE1C2	1.2±0.1	1.1±0.2	1.2±0.1
PDE1B1	24.3±2.9	2.7±0.2	0.9±0.1
PDE1A2	112.7±7.9	5.0±0.6	2.9±0.1

The K_m and V_{max} for each PDE were calculated from Eadie–Hofstee plots based on the best-fit lines. Results are expressed as mean ±SD of at least three independent experiments performed in triplicate. Table 1. Kinetic constants for CaM–PDEs expressed in COS-7 cells. (From ref. 19, with permission.)

in the kinetic data reported by different research groups for a particular CaM-PDE isozyme (20). Recent molecular cloning and expression studies have allowed us to define the kinetic characteristics of most known CaM-PDEs, through the use of recombinant enzyme for a single isoform. Table 1 shows some of these data, obtained from a transient expression system in COS-7 cells (19; Yan, Zhao, and Beavo, *unpublished observations*). In accord with previous data based on biochemically purified protein preparations, PDE1A and PDE1B have a high affinity toward cGMP as a substrate, and a relatively low affinity toward cAMP as a substrate (Table 1, [19, 20]). All of the PDE1C isozymes, however, have very high affinity toward cAMP and cGMP, with K_m values for both cyclic nucleotides of 1.0 μM (Table 1; and Yan, Zhao, and Beavo, *unpublished observations*). It remains to be seen whether tissue-specific protein modification affects the enzyme activities differently than that of the enzyme expressed in COS-7 cells. However, there is so far no reason to expect that the kinetic properties of the recombinant enzyme expressed in COS cells will be different from those of the native enzyme.

STRUCTURAL ASPECTS OF CALMODULIN–PHOSPHODIESTERASES

It is difficult at this stage to speculate which regions or amino-acid residues are responsible for some of the unique kinetic properties of CaM–PDEs. For example, it is not yet known what domains or amino-acid residues of PDE1C are critical for its high affinity toward both cAMP and cGMP. Figure 2 shows the amino-acid sequence alignments of several known CaM–PDE cDNAs isolated from bovine and rat cDNA libraries. Like all other PDEs, all the CaM-PDEs have a highly conserved core region of ~250 amino acids that is thought to be the catalytic domain (21). In addition, a recent study of PDE1A2 enzyme in our laboratory defined a previously unidentified CaM-binding domain and an inhibitory domain that maintains the enzyme in a less active state at low Ca^{2+} concentrations (10). These newly defined domains are highly conserved among all the PDE1A, PDE1B, and PDE1C isozymes, suggesting that there may be common mechanisms involved in regulating their activities.

```
Bovpde1a1  .......... .......... .......... .......... ..........
Bovpde1a2  .......... .......... .......... .......... ..MGSTATET
Bovpde1b1  .......... .......... .......... .......... ..........
Ratpde1b1  .......... .......... .......... .......... ..........
Ratpde1c2  MTDTSHKKEG FKKCRSATFS IDGYSFTIVA NEAGDKNARP LARFSRSKSQ 50

Bovpde1a1  .......... MDDHVTIRRK H......... .......... ......LQRP
Bovpde1a2  EELENTTFKY LIGEQTEKMW Q......... .......... ......RLKG
Bovpde1b1  .......... ......MELS PRSPPEMLES DCPSPLELKS APSKKMWIKL
Ratpde1b1  .......... ......MELS PRSPPEMLES DCPSPLELKS APSKKMWIKL
Ratpde1c2  NCLWNSLIDG LTGNVKEKPR PTIVQDTRPP EEILADELPQ LDSPEALVKT 100

Bovpde1a1  IFRLRCLVKQ LEKGDVNVID LKKNIEYAAS VLEAVYIDET RRLLDTDDEL
Bovpde1a2  I..LRCLVKQ LEKGDVNVID LKKNIEYAAS VLEAVYIDET RRLLDTDDEL
Bovpde1b1  RSLLRYMVKQ LENGEVNIEE LKKNLEYTAS LLEAVYIDET RQILDTEDEL
Ratpde1b1  RSLLRYMVKQ LENGEVNIEE LKKNLEYTAS LLEAVYIDET RQILDTEDEL
Ratpde1c2  SFRLRSLVKQ LERGEASVVD LKKNLEYAAT VLESVYIDET RRLLDTEDEL 150

                                                  *
Bovpde1a1  SDIQSDSVPS EVRDWLASTF TRKMGMMKKK SEEKPRFRSI VHVVQAGIFV
Bovpde1a2  SDIQSDSVPS EVRDWLASTF TRKMGMMKKK SEEKPRFRSI VHVVQAGIFV
Bovpde1b1  QELRSDAVPS EVRDWLASTF TQQTRAKG.P SEEKPKFRSI VHAVQAGIFV
Ratpde1b1  RELRSDAVPS EVRDWLASTF TQQTRAKGRR AEEKPKFRSI VHAVQAGIFV
Ratpde1c2  SDIQSDAVPS EVRDWLASTF TRQMGMMLRR SDEKPRFKSI VHAVQAGIFV 200

Bovpde1a1  ERMYRKSYHM VGLAYPEAVI VTLKDVDKWS FDVFALNEAS GEHSLKFMIY
Bovpde1a2  ERMYRKSYHM VGLAYPEAVI VTLKDVDKWS FDVFALNEAS GEHSLKFMIY
Bovpde1b1  ERMFRRTYTS VGPTYSTAVL NCLKNVDLWC FDVFSLNRAA DDHALRTIVF
Ratpde1b1  ERMFRRTYTA VGPTYSTAVH NCLKNLDVWC FDVFSLNRAA DDHALRTIVF
Ratpde1c2  ERMYRRTSNM VGLSYPPAVI DALKDVDTWS FDVFSLNEAS GDHALKFIFY 250

Bovpde1a1  ELFTRYDLIN RFKIPVSCLI AFAEALEVGY SKYKNPYHNL IHAADVTQTV
Bovpde1a2  ELFTRYDLIN RFKIPVSCLI AFAEALEVGY SKYKNPYHNL IHAADVTQTV
Bovpde1b1  ELLTRHNLIS RFKIPTVFLM TFLDALETGY GKYKNPYHNQ IHAADVTQTV
Ratpde1b1  ELLTRHSLIS RFKIPTVFLM SFLEALETGY GKYKNPYHNQ IHAADVTQTV
Ratpde1c2  ELLTRYDLIS RFKIPISALV SFVEALEVGY SKHKNPYHNL MHAADVTQTV 300

Bovpde1a1  HYIMLHTGIM HWLTELEILA MVFAAAIHDY EHTGTTNNFH IQTRSDVAIL
Bovpde1a2  HYIMLHTGIM HWLTELEILA MVFAAAIHDY EHTGTTNNFH IQTRSDVAIL
Bovpde1b1  HCFLLRTGMV HCLSEIEVLA IIFAAAIHDY EHTGTTNSFH IQTKSECAIL
Ratpde1b1  HCFLLRTGMV HCLSEIEVLA IIFAAAIHDY EHTGTTNSFH IQTKSECAIL
Ratpde1c2  HYLLYKTGVA NWLTELEIFA IIFSAAIHDY EHTGTTNNFH IQTRSDPAIL 350
```

FIG. 2. Amino-acid alignments of several known CaM–PDE isozymes. The boxed region contains the core of ~250 amino acid residues conserved among all PDEs, which is thought to be the catalytic domain. The single overlined region shows the CaM-binding domain defined in PDE1A1 by sequence comparison and biochemical studies, which is only partly conserved in all the other CaM–PDEs. The double-underlined region indicates the CaM-binding domain defined in a recent study (10). The single-underlined region represents a domain that maintains the CaM–PDEs in a less active state in the absence of Ca^{2+} (10). The asterisk denotes the serine residue that has been shown to be phosphorylated by protein kinase A (PKA) and to affect the sensitivity of PDE1A isozymes to Ca^{2+}/CaM (48). This serine residue is also conserved among all the other known CaM-PDEs.

```
Bovpde1a1  YNDRSVLENH HVSAAYRLMQ E.EEMNVLIN LSKDDWRDLR NLVIEMVLST
Bovpde1a2  YNDRSVLENH HVSAAYRLMQ E.EEMNVLIN LSKDDWRDLR NLVIEMVLST
Bovpde1b1  YNDRSVLENH HISSVFRMMQ D.DEMNIFIN LTKDEFVELR ALVIEMVLAT
Ratpde1b1  YNDRSVLENH HISSVFRMMQ D.DEMNIFIN LTKDEFVELR ALVIEMVLAT
Ratpde1c2  YNDRSVLENH HLSAAYRLLQ EDEEMNILVN LSKDDWREFR TLVIEMVMAT 400

Bovpde1a1  DMSGHFQQIK NIRNSLQQPE GLDKAKTMSL ILHAADISHP AKSWKLHHRW
Bovpde1a2  DMSGHFQQIK NIRNSLQQPE GLDKAKTMSL ILHAADISHP AKSWKLHHRW
Bovpde1b1  DMSCHFQQVK SMKTALQQLE RIDKSKALSL LLHAADISHP TKQWSVHSRW
Ratpde1b1  DMSCHFQQVK TMKTALQQLE RIDKSKALSL LLHAADISHP TKQWSVHSRW
Ratpde1c2  DMSCHFQQIK AMKTALQQPE AIEKPKALSL MLHTADISHP AKAWDLHHRW 450

Bovpde1a1  TMALMEEFFL QGDKEAELGL PFSPLCDRKS TMVAQSQIGF IDFIVEPTFS
Bovpde1a2  TMALMEEFFL QGDKEAELGL PFSPLCDRKS TMVAQSQIGF IDFIVEPTFS
Bovpde1b1  TKALMEEFFR QGDKEAELGL PFSPLCDRTS TLVAQSQIGF IDFIVEPTFS
Ratpde1b1  TKALMEEFFR QGDKEAELGL PFSPLCDRTS TLVAQSQIGF IDFIVEPTFS
Ratpde1c2  TMSLLEEFFR QGDREAELGL PFSPLCDRKS TMVAQSQVGF IDFIVEPTFT 500

Bovpde1a1  LLTDSTEKII IPLIEEDSKT KTPS...... ....YGASRR SNMKGTTNDG
Bovpde1a2  LLTDSTEKII IPLIEEDSKT KTPS...... ....YGASRR SNMKGTTNDG
Bovpde1b1  VLTDVAEKSV QPTGDDDSKS KNQPSFQW.. .......... .......RQP
Ratpde1b1  VLTDVAEKSV QPLTDDDSKS KSQPSFQW.. .......... .......RQP
Ratpde1c2  VLTDMTEKIV SPLIDETSQT GGTGQRRSSL NSINSSDAKR SGVKSSGSEG 550

Bovpde1a1  TYSPDYSLAS VDLKSFKNSL VDIIQQNKER WKELAAQGEP DPHKNSDL..
Bovpde1a2  TYSPDYSLAS VDLKSFKNSL VDIIQQNKER WKELAAQGEP DPHKNSDL..
Bovpde1b1  SLDVEVGDPN PDVVSFRSTW TKYIQENKQK WKERAASGIT NQMSIDELSP
Ratpde1b1  SLDVDVGDPN PDVVSFRSTW TKYIQENKQK WKERAASGIT NQMSIDELSP
Ratpde1c2  SAPINNSVIP VDYKSFKATW TEVVQINRER WRAKVPKEEK AKKEAEEKAR 600

Bovpde1a1  VNAEEKHAET HS
Bovpde1a2  VNAEEKHAET HS
Bovpde1b1  CEEEAPASPA EDEHNQNGNL D
Ratpde1b1  CEEEAPSSPA EDEHNQNGNL D
Ratpde1c2  LAAEEKQKEM EAKSQAEQGT TSKAEKKTSG ETKGQVNGTR TSKGDNPRGK 650

Ratpde1c2  NSKGDKAGEK QQNGDLKDGK NKADKKDHSN TGNESKKADG TKKRSHGSPA 700

Ratpde1c2  PSTSSTSRLT LPVIKPPLRH FKRPAYASSS YAPSVPKKTD DHPVRYKMLD 750

Ratpde1c2  QRIKIKKIQN ISHHWNKK 768
```

FIG. 2. *Continued*

TISSUE- AND CELL-SPECIFIC EXPRESSION OF CALMODULIN–PHOSPHODIESTERASES

Through the techniques of Northern blot analysis, ribonuclease (RNase) protection assay, and *in situ* hybridization, the expression of the CaM–PDE genes, including their alternative splice variants in a number of tissues, have been mapped (9,18, 19,22; Yan, Zhao, and Beavo, *unpublished observations*). In general, different tissues express different CaM–PDE isozymes. For example, while the expression of PDE1B seems to be absent from bovine heart (18), the same tissue has been shown to express a substantial level of PDE1A (9), and human and mouse heart have been shown to express substantial levels of PDE1C (23; Yan, Zhao, and Beavo, unpublished observations). Kidney papilla contains relatively high levels of PDE1A and low levels of PDE1B transcripts, but PDE1A seems to be the only CaM–PDE gene expressed in kidney medulla (9,18; Yan, Zhao, and Beavo, *unpublished observation*).

For all tissues thus far examined, the expression of different CaM–PDEs has been demonstrated to be cell-type-specific in cases in which different CaM–PDE mRNAs colocalize in the same tissue. Different alternatively spliced variants of PDE1C have also been shown to exhibit tissue-specific expression patterns (18; Zhao, Yan, and Beavo, *unpublished observation*). Our discussion here will focus mainly on the expression of PDE1 in the central nervous system (CNS) and of PDE1C in olfactory sensory neurons.

Expression of Calmodulin–Phosphodiesterases in the Central Nervous System

Mammalian brain has long been used as a source for biochemical purification of various PDE enzymes, especially the CaM–PDEs (20,24–26). Presumably, the mammalian CNS uses many types of PDE enzymes to suit the needs of extremely complex neurologic processes. With molecular techniques such as *in situ* hybridization and RNase protection assay, we have recently mapped the expression profiles of all the CaM-PDE genes in the mouse CNS (22; Yan, Zhao, and Beavo, *unpublished observation*). Figure 3 shows a representation of CaM–PDE hybridization signals in mouse brain sections. As shown in the figure, the different CaM–PDE mRNAs exhibit complex and somewhat overlapping distribution patterns. The PDE1B mRNA has a wide distribution with strong hybridization signals in the caudate–putamen, nucleus accumbens, olfactory tubercle, Purkinje cells of cerebellum, and dentate gyrus of the hippocampus. The PDE1A mRNA, on the other hand, has a more restricted pattern of distribution, with the highest level in the cerebral cortex and in the pyramidal cells of the hippocampal CA1 to CA4 regions. The expression of PDE1C is essentially limited in the granule and Purkinje cells of cerebellum, the caudate putamen, and the olfactory bulb. All CaM–PDE mRNAs are localized in neuronal cell bodies. It is not always clear why certain regions of the CNS express one particular CaM–PDE over the others. Part of the answer may be that the kinetic properties

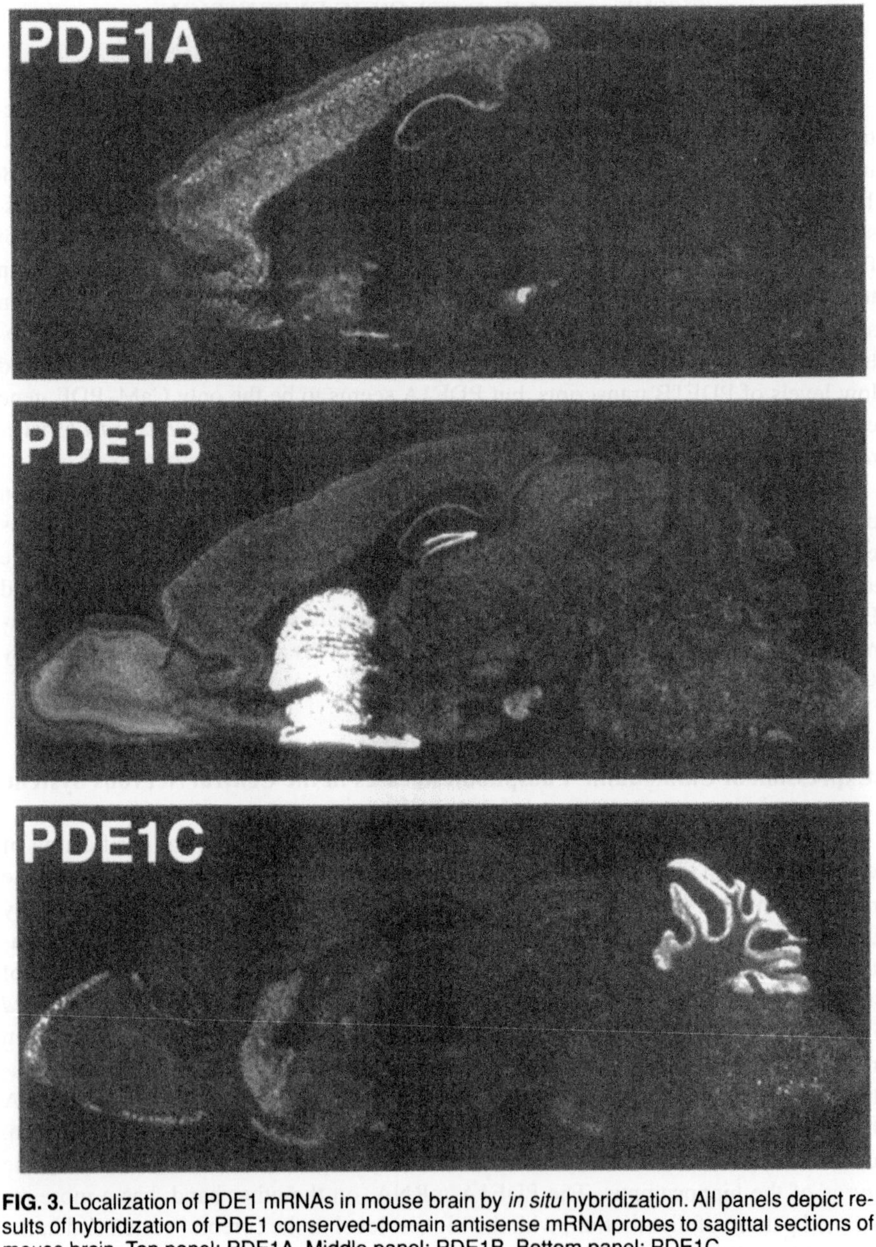

FIG. 3. Localization of PDE1 mRNAs in mouse brain by *in situ* hybridization. All panels depict results of hybridization of PDE1 conserved-domain antisense mRNA probes to sagittal sections of mouse brain. Top panel: PDE1A. Middle panel: PDE1B. Bottom panel: PDE1C.

of one CaM–PDE are better suited than those of the others to the functional needs of the region, with evolution having selected the expression of specific CaM–PDEs to coordinate with the signal-transduction pathways that regulate their activities.

The distinct distribution profiles of CaM–PDEs in the CNS suggest that they have important but different physiologic roles in the regional regulation of cyclic nucleotides. For example, the very high level of PDE1B mRNA in the striatum mirrors that of dopamine receptors, especially D1 and D2 receptors, which are prominently expressed in the caudate–putamen, nucleus accumbens, and olfactory tubercle (27, 28). The D1 receptors are known to activate adenylyl cyclase and the D2 receptors to inhibit adenylyl cyclase. In addition, binding of dopamine to D1-like receptors in the striatum has been shown to increase the level of inositol trisphosphate (IP_3) and consequently that of Ca^{2+} (29). Thus, the PDE1B enzymes presumably modulate dopamine functions by attenuating the magnitude and duration of dopamine-induced, cAMP-mediated signals in cells containing D1-like receptors. Similarly, an IP3-dependent increase in Ca^{2+} concentrations should allow PDE1B to work synergistically with dopamine to lower the cAMP level in neuronal cells containing D2 receptors. It will be very interesting to see with immunocytochemical approaches whether the PDE1B proteins colocalize with dopamine receptors in the same subcellular regions of neuronal cell bodies.

Since all three CaM–PDEs can hydrolyze cGMP very efficiently, any neurotransmitter that increases the Ca^{2+} concentration in neurons containing a high level of CaM–PDEs would, through activation of these enzymes, be expected to attenuate the magnitude and duration of the cGMP signal. A major modulator of cGMP in CNS is nitric oxide (NO). NO has been recognized as an intercellular mediator, and possibly a "retrograde messenger," in the CNS (30,31). NO is synthesized by NO synthases, most of which are Ca^{2+}/CaM dependent. The NO can then freely diffuse to adjacent neighboring cells in which it activates soluble guanylyl cyclase (30). It seems likely that one of the important functions of CaM–PDEs in the CNS is to counteract NO-activated guanylyl cyclase by regulating the cGMP level in a Ca^{2+}/CaM-dependent manner. It has recently been shown that in isolated brain synaptosome preparations, CaM–PDE activity strongly modulates cGMP responses elicited by NO (32). Another interesting observation is that Ca^{2+}/CaM-dependent NO synthetase is highly expressed in granule cells of the cerebellum, whereas the cGMP concentration within these cells, in constrast to other, adjacent neuronal cells, appears to be maintained at a very low level (33,34). Since PDE1C is the major PDE1 enzyme highly expressed in cerebellar granule cells, it may very well be the crucial enzyme needed to control the cGMP level in these cells.

A High Level of Phosphodiesterase-1C Expression in Olfactory Sensory Neurons

Recently, a novel form of CaM–PDE with very high affinity toward both cAMP and cGMP has been identified in olfactory cilia. This PDE is most likely encoded by the PDE1C gene (19,35). Cyclic nucleotides, especially cAMP, play an important role in olfactory signal transduction. They mediate the activation of an odorant-induced depolarization, most likely by directly gating a cation channel in olfactory cilia, which in turn induces a Ca^{2+} sensitive Cl^- current (36–38). Recent experimental observations suggest that odorants initially increase the intracellular Ca^{2+} concentrations in olfactory sensory neurons through an influx of Ca^{2+} and possibly also through increased elevated IP_3 formation. The increased Ca^{2+} then activates Type III adenylyl cyclase, eliciting a rapid (within ~100 ms) increase in cAMP concentration (39,40). However, the increase in cAMP is only transient, and the cAMP level rapidly (within 150 to 200 ms) returns to near basal levels. Attenuation of the cAMP signal is achieved through reduced cAMP synthesis as well as increased degradation of cAMP. The shutoff of cAMP synthesis can be achieved by the phosphorylation of odorant receptors by protein kinase A (PKA) (41,42) or β-adrenergic receptor kinase (43,44). Furthermore, it has been shown that the odorant-induced Ca^{2+} influx into olfactory sensory neurons can be decreased by Ca^{2+}/CaM binding to the cAMP-activated cation channel (45), which would in turn attenuate the activation of Type III adenylyl cyclase. Recent evidence suggests that the rapid decline of odorant-induced cAMP signal must also involve the activation of a CaM–PDE, presumably PDE1C2, in olfactory cilia (39,40). Such a regulation of odorant-induced cAMP signal may be important to animals in rapidly distinguishing the amounts of an odorant. Studies have shown that the cyclic nucleotide-gated cation channels in olfactory receptor neurons have a highly active cation conductance (36). It is estimated that the odorant stimulation mediated by increases in intracellular levels of cAMP and cGMP would require their basal levels to be submicromolar (10^{-7} or less), and would consequently require the presence of a high-affinity PDE for both cyclic nucleotides (36). We have recently demonstrated, by *in situ* hybridization and RNase protection assay, that PDE1C2, rather than PDE1A, PDE1B, or any other PDE1C splice variants, is the predominant CaM–PDE expressed in olfactory sensory neurons (19; Yan, Zhao, and Beavo, *unpublished observation*; see Fig. 4). The kinetic constants and the Ca^{2+}/CaM-stimulatable character of PDE1C2 also fit very well with the apparent physiologic requirements of olfactory signal transduction. Figure 5 shows a diagram summarizing the model of olfactory signal-transduction events proposed above.

Although we have extensively discussed high levels of CaM–PDE gene expression in the CNS, this is by no means intended to discount the probable functional significance of more moderate expression of CaM–PDEs in other parts of the brain. Other types of PDEs may also exert their functions in the same tissues or cells in which CaM–PDEs are expressed.

REGULATION OF CALMODULIN–PHOSPHODIESTERASE ACTIVITIES BY CA^{2+}/CALMODULIN AND PHOSPHORYLATION

Studies of the structure and function of PDE1A have identified two CaM-binding domains in PDE1A1 and PDE1A2 (10). A putative CaM-binding domain is located near the N-terminus of PDE1A, while a second functional CaM-binding domain lies

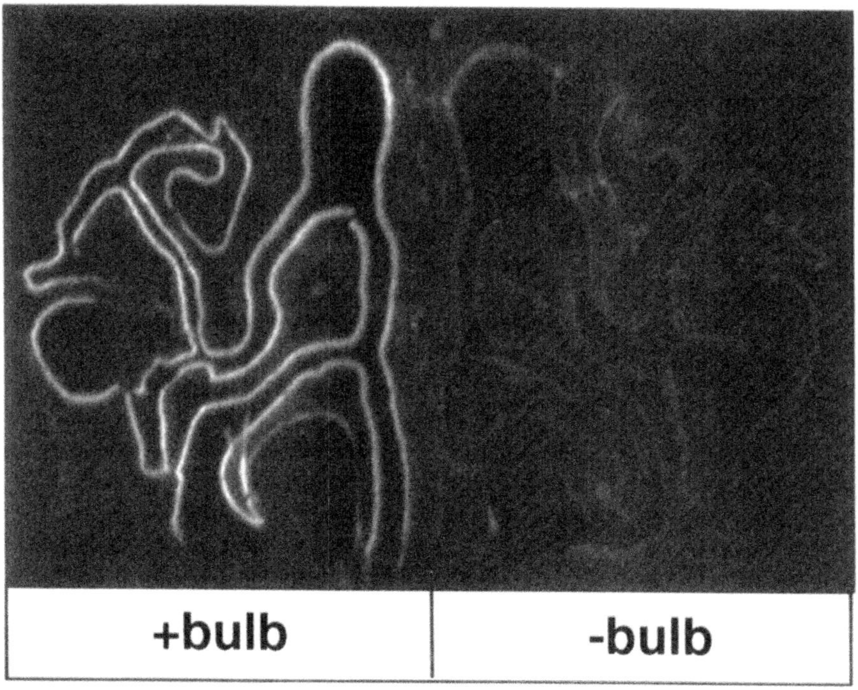

FIG. 4. Demonstration of PDE1C mRNA expression in olfactory sensory neurons. The illustration shows the effect of unilateral bulbectomy, which is known to cause the selective degeneration of olfactory sensory neurons, on the expression of PDE1C. The illustration is a dark-field autoradiographic digital image of a coronal section of a rat nasal cavity, showing *in situ* hybridization with an ^{35}S-labeled antisense riboprobe derived from the rat PDE1C2 cDNA clone. The disappearance of hybridization signals on the side affected by the unilateral bulbectomy indicates that the expression of PDE1C is within the neuronal cells. The side unaffected by the bulbectomy shows very intense hybridization signals as judged by the exposure time (only 2 days) under an NTB2 emulsion. (From ref. 19, with permission.)

approximatedly 100 residues carboxy-terminal (C-terminal) to the first domain. Although the second CaM-binding domain is highly conserved among all known CaM–PDEs, the CaM-binding domain in the N-terminal region of the PDE1A isozymes is only partly conserved. It is likely that differences in the N-terminal regions among all of the CaM-PDEs determine the differences in these enzymes' binding affinities to Ca^{2+}/CaM complexes. It is already known that the PDE1A1 enzyme has a 10-fold greater affinity than PDE1A2 for Ca^{2+}/CaM complex, and the only structural difference between the two isozymes is in their N-terminal CaM-binding domain (10). We have recently also shown that PDE1C2 displays a 5-fold greater affinity for Ca^{2+}/CaM complexes than do all the other PDE1C isozymes (Yan, Zhao, and Beavo, *unpublished observations*). An inhibitory domain located between the two CaM-binding domains, which is highly conserved and seems to be essential to maintaining the CaM–PDEs in a less active state, has also been recently identified (10; Fig. 2). The current model proposes that binding of Ca^{2+}/CaM to the CaM–PDEs reverses the effect of the inhibitory domain, thereby activating the enzymes (10).

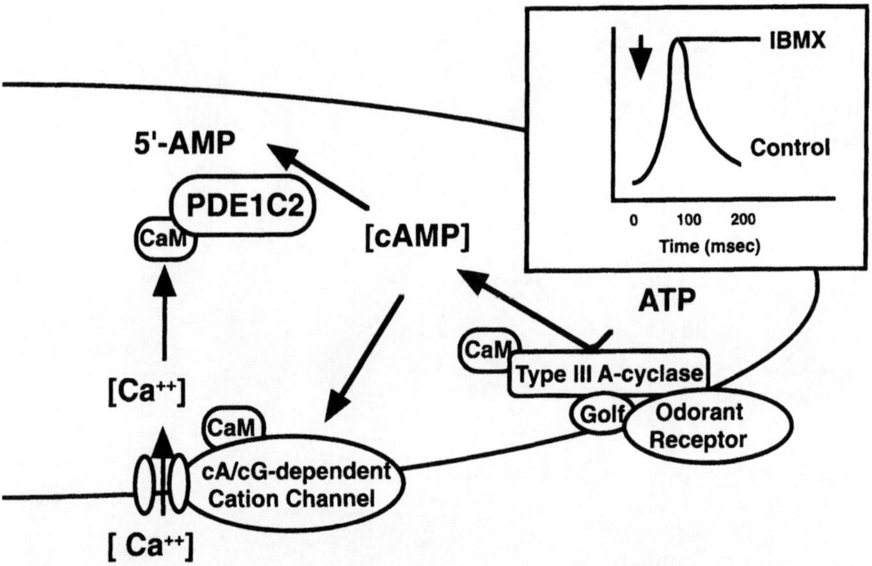

FIG. 5. Role for PDE1C in olfactory signal transduction. Model for possible role of PDE1C in rapid attenuation of a cAMP signal in response to odorant stimulation. Inset: theoretical time course for cAMP levels, which is similar to actual data obtained by Boekhoff and Breer (39,42). A-cyclase, adenylyl cyclase. The arrow inside the inset indicates the time when odorant is added. Golf, a Gs highly enriched with olfactory tissue; IBMX, 3-isobutyl-1-methylxanthine.

Studies first performed by Sharma, Wang, and their colleagues indicate that the PDE1A isozymes are phosphorylated by cAMP-dependent protein kinase (PKA), and that the PDE1B isozyme is phosphorylated by CaM-kinase II (CaMKII) (24, 46–48). Even though the phosphorylation of the two CaM–PDEs is achieved through different kinases, the effect is to render both enzymes less responsive to Ca^{2+} and therefore less active. Both enzymes can be dephosphorylated by CaM-dependent phosphatase (47,48). Although both Ser^{120} and Ser^{138} of PDE1A are shown to be phosphorylated by PKA *in vitro*, phosphorylation of Ser^{120} is responsible for most of the change in this enzyme's CaM-binding affinity (49).

The negative effect of phosphorylation on the activities of CaM–PDE may have important physiologic implications. One can easily envision that when a cell receives an outside signal (e.g., a neurotransmitter or odorant stimulation), the phosphorylation of CaM–PDEs at the initial entry of Ca^{2+} would allow a brief period in which the cells could build up the level of their cyclic nucleotides to meet their physiologic needs, such as regulating neurotransmitter release, gating ion channels, or activating transcription factors. The continued increase of Ca^{2+} would activate CaM-dependent phosphatases, which in turn would dephosphorylate CaM–PDEs. The activated CaM–PDEs would then bring down the level of cyclic nucleotides, thus reseting the entire system. Interestingly, high levels of CaM–PDEs have been localized in postsynaptic densities islolated from caudate–putamen neurons (50,51), and CaMKII has been shown to be involved in synaptic plasticity (52). Considering the high level of PDE1B expression in caudate–putamen neurons, it would be interesting to test whether phosphorylation of PDE1B by CaMKII is an important component in mediating synaptic plasticity.

CONCLUSION

Recent molecular cloning and characterization studies have revealed that the gene structure for and expression patterns of Ca2+/CaM–PDE are very complex. Probably the most important physiologic function of CaM–PDEs is to mediate the intracellular cross-talk between Ca^{2+} and cyclic nucleotides. The fluctuation in intracellular Ca^{2+} concentration dictates the activity changes of many CaM-dependent enzymes including CaM–PDEs, and consequently determines changes in the intracellular cyclic nucleotide level. Recent studies of the structure and function of CaM–PDEs have provided insight into how the activities of these isozymes are regulated. Future studies should help us to understand which physiologic processes are modulated by CaM–PDEs, and what other signal-transduction components interact with CaM–PDEs.

ACKNOWLEDGMENTS

A. Z. Z. is a recipient of a Dick and Julia McAbee Diabetes Postdoctoral Research Fellowship. C. Y. is a recipient of a fellowship for Advanced Predoctoral Training in Pharmacology/Toxicology from the Pharmaceutical Research and Manufacture of America Foundation. The work described in this chapter was also supported by an American Heart Association Washington Affiliate Grant-in-Aid to W. K. S., and by National Institutes of Health Grants DK21723 and HL44948 to J. A. B.

REFERENCES

1. Nagamine Y, Reich E. Gene expression and cAMP. *Proc Natl Acad Sci U S A* 1985;82: 4606–10.
2. Edelman AM, Raese JD, Lazar MA, Barchas JD. Tyrosine hydroxylase: studies on the phosphorylation of a purified preparation of the brain enzyme by the cyclic AMP-dependent protein kinase. *J. Pharm Exp Ther* 1981 ;216: 647–53.
3. Costa MRC, Casnellie JE, Catterall WA. Selective phosphorylation of the a subunit of the sodium channel by c-AMP-dependent protein kinase. *J. Biol Chem.* 1982; 257: 7918–21.
4. Nestler EJ, Greengard P. Protein phosphorylation in the brain. *Nature* 1983; 305: 583–8.
5. Kandel ER, Schwartz JH. Molecular biology of learning: modulation of transmitter release. *Science* 1982; 218: 433–43.
6. Schoffelmeer AN, Wardeh G, Mulder AH. Cyclic AMP facilitates the electrically evoked release of radiolabelled noradrenaline, dopamine and 5-hydroxytryptamine from rat brain slices. *Naunyn-Schmiedebergs Arch Pharmacol* 1985; 330: 74–6.
7. Bentley JK, Beavo JA. Regulation and function of cyclic nucleotides. *Curr Opin Cell Biol* 1992; 4: 233–40.

8. Beavo JA. Cyclic nucleotide phosphodiesterases: functional implications of multiple isoforms. *Physiol Rev* 1995;75:725–8.
9. Sonnenburg WK, Seger D, Beavo JA. Molecular cloning of a cDNA encoding the "61-kDa" calmodulin-stimulated cyclic nucleotide phosphodiesterase. Tissue-specific expression of structurally related isoforms. *J Biol Chem* 1993; 268: 645–52.
10. Sonnenburg WK, Seger D, Kwak KS, Huang J, Charbonneau H, Beavo JA. Identification of inhibitory and calmodulin-binding domains of the "59-kDa" and "61-kDa" calmodulin stimulated phosphodiesterases. *J Biol Chem* 1995;270:30989–31000.
11. Rossi P, Giorgi M, Geremia R, Kincaid RL. Testis-specific calmodulin-dependent phosphodiesterase. A distinct high affinity cAMP isoenzyme immunologically related to brain calmodulin-dependent cGMP phosphodiesterase. *J Biol Chem* 1988; 263: 15521–7.
12. Sharma RK, Wang JH. Purification and characterization of bovine lung calmodulin-dependent cyclic nucleotide phosphodiesterase. An enzyme containing calmodulin as a subunit. *J Biol Chem* 1986; 261: 14160–6.
13. Ho HC, Wirch E, Stevens FC, Wang JH. Purification of a Ca^{2+}-activatable cyclic nucleotide phosphodiesterase from bovine heart by specific interaction with its Ca^{2+}-dependent modulator protein. *J Biol Chem* 1977; 252: 43–50.
14. Hansen RS, Beavo JA. Purification of two calcium/calmodulin dependent forms of cyclic nucleotide phosphodiesterase by using conformation-specific monoclonal antibody chromatography. *Proc Natl Acad Sci USA* 1982; 79: 2788–92.
15. Sharma RK, Adachi AM, Adachi K, Wang JH. Demonstration of bovine brain calmodulin-dependent cyclic nucleotide phosphodiesterase isozymes by monoclonal antibodies. *J Biol Chem* 1984; 259: 9248–54.
16. Charbonneau H, Kumar S, Novack JP, et al. Evidence for domain organization within the 61-kDa calmodulin-dependent cyclic nucleotide phosphodiesterase from bovine brain. *Biochemistry* 1991; 30: 79931–940.
17. Novack JP, Charbonneau H, Bentley JK, Walsh KA, Beavo JA. Sequence comparison of the 63-, 61-, and 59-kDa calmodulin-dependent cyclic nucleotide phosphodiesterases. *Biochemistry* 1991; 30: 7940–7.
18. Bentley JK, Kadlecek A, Sherbert CH, et al. Molecular cloning of cDNA encoding a "63"-kDa calmodulin-stimulated phosphodiesterase from bovine brain. *J Biol Chem* 1992; 267:18676–82.
19. Yan C, Zhao AZ, Bentley JK, Loughney K, Ferguson K, Beavo JA. Molecular cloning and characterization of a calmodulin-dependent phosphodiesterase enriched in olfactory sensory neurons. *Proc Natl Acad Sci USA* 1995; 92: 9677–81.
20. Wang JH, Sharma RK, Mooibroek MJ. Calmodulin-stimulated cyclic nucleotide phosphodiesterases. In: Beavo J, Houslay MD, eds. *Cyclic nucleotide phosphodiesterases: structure, regulation and drug action*. Chichester, UK: John Wiley & Sons, 1990; 19–60.
21. Charbonneau, H. Structure-function relationships among cyclic-nucleotide phosphodiesterases. In: Beavo J, Houslay MD, eds. *Cyclic nucleotide phosphodiesterases: structure, function, regulation and drug action*. Chichester, UK: John Wiley & Sons, 1990; 267–98.
22. Yan C, Bentley JK, Sonnenburg WK, Beavo JA. Differential expression of the 61 kDa and 63 kDa calmodulin-dependent phosphodiesterases in the mouse brain. *J Neurosci* 1994; 14: 973–84.
23. Loughney K, Martins T, Harris EAS, et al. Isolation and characterization of cDNAs corresponding to two human calcium, calmodulin stimulated 3',5' cyclic nucleotide phosphodiesterase genes. *J Biol Chem* 1995.(In press).
24. Sharma RK, Wang JH. Differential regulation of bovine brain calmodulin-dependent cyclic nucleotide phosphodiesterase isoenzymes by cyclic AMP-dependent protein kinase and calmodulin-dependent phosphatase. *Proc Natl Acad Sci USA* 1985; 82: 2603–7.
25. Sharma RK, Wang TH, Wirch EH, Wang JH. Purification and properties of bovine brain calmodulin-dependent cyclic nucleotide phosphodiesterase. *J Biol Chem* 1980; 255: 5916–23.
26. Sharma RK, Taylor WA, Wang JH. Use of calmodulin affinity chromatography for purification of specific calmodulin-dependent enzymes. *Methods Enzymol* 1983; 102: 210–9.
27. Weiner DM, Levey AI, Sunahara RK, et al. D1 and D2 dopamine receptor mRNA in rat brain. *Proc Natl Acad Sci USA* 1991; 88: 1859–63.
28. Surmeier DJ, Eberwine J, Wilson CJ, Cao Y, Stephani A, Kitai ST. Dopamine receptor subtypes colocalize in rat striatonigral neurons. *Proc Natl Acad Sci USA* 1992; 89: 10178–82.
29. Mahan LC, Burch RM, Monsma FJJ, Sibley DR. Expression of striatal D1 dopamine receptors cou-

pled to inositol phosphate production and Ca2+ mobilization in Xenopus oocytes. *Proc Natl Acad Sci USA* 1990; 87: 2196–200.
30. Snyder SH. Nitric oxide and neurons. *Curr Opin Neurobiol* 1992; 2: 323–7.
31. Snyder SH, Bredt DS. Nitric oxide as a neuronal messenger. *Trends Pharmacol Sci* 1991; 12: 125–8.
32. Mayer B, Klatt P, Bohme E, Schmidt K. Regulation of neuronal nitric oxide and cyclic GMP formation by Ca^{2+}. *J Neurochem* 1992; 59: 2024–9.
33. Garthwaite G, Garthwaite J. Cyclic GMP and cell death in rat cerebellar slices. *Neuroscience* 1988; 26: 321–6.
34. Bredt DS, Hwang PM, Snyder SH. Localization of nitric oxide synthase indicating a neural role for nitric oxide. *Nature* 1990; 347: 768–70.
35. Borisy FF, Ronnett GV, Cunningham AM, Juilfs D, Beavo J, Snyder SH. Calcium/calmodulin-activated phosphodiesterase expressed in olfactory receptor neurons. *J Neurosci* 1992; 12: 915–23.
36. Nakamura T, Gold GH. A cyclic nucleotide-gated conductance in olfactory receptor cilia. *Nature* 1987; 325: 442–4.
37. Frings S, Lindemann B. Current recording from sensory cilia of olfactory receptor cells in situ. I. The neuronal response to cyclic nucleotides. *J Gen Physiol* 1991; 97: 1–16.
38. Lowe G, Nakamura T, Gold GH. Adenylate cyclase mediates olfactory transduction for a wide variety of odorants. *Proc Natl Acad Sci USA* 1989;86:5641–5.
39. Breer H, Boekhoff I, Tareilus E. Rapid kinetics of second messenger formation in olfactory transduction. *Nature* 1990; 345: 65–8.
40. Breer H. Molecular reaction cascades in olfactory signal transduction. *J Steroid Biochem Mol Biol* 1991; 39: 621–5.
41. Boekhoff I, Schleicher S, Strotmann J, Breer H. Odor-induced phosphorylation of olfactory cilia proteins. *Proc Natl Acad Sci USA* 1992; 89: 11983–7.
42. Boekhoff I, Breer H. Termination of second messenger signaling in olfaction *Proc Natl Acad Sci USA* 1992; 89: 471–4.
43. Dawson TM, Arriza JL, Jaworsky DE, et al. Beta-adrenergic receptor kinase-2 and beta-arrestin-2 as mediators of odorant-induced desensitization. *Science* 1993; 259: 825–9.
44. Schleicher S, Boekhoff I, Arriza J, Lefkowitz RJ, Breer H. A beta-adrenergic receptor kinase-like enzyme is involved in olfactory signal termination. *Proc Natl Acad Sci USA* 1993; 90: 1420–4.
45. Chen T-Y, Yau K-W. Direct modulation by Ca^{2+}-calmodulin of cyclic nucleotide-activated channel of rat olfactory receptor neurons. *Nature* 1994; 368: 545–8.
46. Sharma RK, Wang JH. Calmodulin and Ca2+-dependent phosphorylation and dephosphorylation of the 63-kDa subunit-containing bovine brain calmodulin-stimulated cyclic nucleotide phosphodiesterase isozyme. *J Biol Chem* 1986; 261: 1322–8.
47. Hashimoto Y, Sharma RK, Soderling TR. Regulation of Ca^{2+}/calmodulin-dependent cyclic nucleotide phosphodiesterase by the autophosphorylated form of Ca^{2+}/calmodulin-dependent protein kinase II. *J Biol Chem* 1989; 264: 10884–7.
48. Sharma RK. Phosphorylation and characterization of bovine heart calmodulin-dependent phosphodiesterase. *Biochemistry* 1991;30:5963–8.
49. Florio VA, Sonnenburg WK, Johnson R, et al. Phosphorylation of the 61-kDa calmodulin-stimulated cyclic nucleotide phosphodiesterase at serine 120 reduces its affinity for calmodulin. *Biochemistry* 1994; 33: 8948–54.
50. Ariano MA, Adinolfi AM. Subcellular localization of cyclic nucleotide phosphodiesterase in the caudate nucleus. *Exp Neurology* 1977; 55: 84–94.
51. Ariano MA, Appleman MM. Biochemical characterization of postsynaptically localized cyclic nucleotide phosphodiesterase. *Brain Res* 1979; 177: 301–9.
52. Hanson PI, Schulman H. Neuronal Ca^{2+}/calmodulin-dependent protein kinases. *Annu Rev Biochem* 1992; 61: 559–601.

Signal Transduction in Health and Disease, Advances in Second Messenger and Phosphoprotein Research, Vol. 31, edited by J. Corbin and S. Francis.
Lippincott–Raven Publishers, Philadelphia © 1997.

20

Specificity and Complexity of Receptor–G-Protein Interaction

Thomas Gudermann, Frank Kalkbrenner, Edgar Dippel, Karl-Ludwig Laugwitz, and Günter Schultz

Institute of Pharmacology, Free University of Berlin, D-14195 Berlin, Germany

INTRODUCTION

Specialized cells in a living organism communicate with each other through extracellular molecules such as hormones, neurotransmitters, and growth factors. Most of these molecules do not enter the cell, but bind to receptors at the cell surface, thereby initiating signaling cascades that finally reach the nucleus. A given cell may respond to one stimulus in a precisely defined manner, but may react toward others less specifically. Conversely, a certain signaling molecule may give rise to a focused response in one cell while eliciting a plethora of effects in a different cell. Although considerable progress has been made in recent years toward defining elements involved in transmembrane signaling processes, the issue of selectivity *versus* diversity in a cell's response to extracellular stimuli remains poorly understood.

The majority of intercellular signaling molecules interact with receptors that represent one element of a three-protein transmembrane signaling system whose single components interact sequentially and reversibly. Agonist binding to a specific receptor results in activation of heterotrimeric guanine-nucleotide-binding proteins (G proteins), which act as transducers and signal amplifiers. G proteins modulate the activity of one or more effector systems (1–5). G-protein-coupled receptors form a large and functionally diverse superfamily (6–8), and have been identified in organisms evolutionarily as distant from one another as yeasts and humans. So far, more than 300 G-protein-coupled receptors have been cloned; the total number of such proteins is assumed to exceed 1,000.

G proteins are heterotrimers composed of α, β, and γ subunits, and are classified according to their α subunit. To date, 23 distinct α subunits (including splice variants), encoded by 17 different genes, as well as five β (9) and 10 γ subunits (10), are known. β and γ subunits are tightly associated and are regarded as a single functional unit. Both guanosine triphosphate (GTP)-bound α subunits and βγ dimers are signaling molecules in their own right, and modulate the activity of specific effector systems such as enzymes, ion channels, and transporters (11).

SPECIFICITY OF RECEPTOR–G-PROTEIN INTERACTION

Considering that hundreds of G-protein-coupled receptors transduce signals through a fairly limited repertoire of G proteins, the question of specificity governing the coupling of receptors to G proteins needs to be addressed. What are the molecular determinants that funnel highly selective signals through the G-protein "bottleneck"? One mechanism for imparting selectivity on signaling processes is the compartmentalization of signal-transduction components in distinct subcellular domains. There is functional and structural evidence that certain subsets of receptors, G proteins, and effectors reside within different cellular compartments, which may or may not have access to each other (12). In the rod photoreceptor cell, for example, all components needed for transduction of the light signal are contained in the outer segment, and an asymmetric distribution of some receptors and G proteins has been described in neuronal growth cones (13), rat Sertoli cells (14), and Madin–Darby canine kidney cells (15,16). Furthermore, association of G proteins with elements of the cytoskeleton (17), and their sequestration into specialized cell-membrane structures (18) involved in transcytosis (caveolae), have been described.

A given cell is endowed with a certain complement of receptors, effectors, and G-protein α, β, and γ subunits. Expression of different G-protein subunits may result in a vast number of distinct αβγ heterotrimers if all subunits can associate randomly. There is evidence that the α subunit is not the sole determinant of selectivity in receptor–G-protein interactions. The formation of interactive complexes between transducin βγ and rhodopsin has been reported (19), and both the isoprenoid moiety and the carboxy-terminal (C-terminal) primary amino-acid sequence of γ subunits have been highlighted as specific determinants of receptor–G-protein interaction (20,21). Differences between distinct βγ complexes were reported with regard to β-adrenergic-receptor kinase-mediated receptor phosphorylation (22). These results were interpreted in terms of β subunits determining coupling to receptors and γ subunits specifying effector interaction. In addition, Neubig and colleagues (23) showed that a peptide derived from the C-terminal portion of the third cytoplasmic loop of the α_2-adrenergic receptor binds specifically to both the N-terminal region in the G-protein α subunit and to a site in the β subunit.

In recent years, antisense technologies have attracted much attention, since they allow selective suppression of distinct components of a given signal-transduction cascade, and subsequent study of the functional corollaries of these components. In studies of the hormonal regulation of voltage-gated Ca^{2+} channels, antisense oligonucleotides designed to anneal to specific messenger ribonucleic acid (mRNA) sequences for G-protein α, β, and γ subunits were microinjected into nuclei of pituitary GH_3 cells (24). These investigations provided compelling evidence that selectivity of receptor–G-protein interaction is encoded by specific G-protein αβγ heterotrimers. Inhibition of Ca^{2+} channels by activated somatostatin receptors is mediated by the $\alpha_{o2}\beta_1\gamma_3$ heterotrimer, whereas M_4 muscarinic receptor-induced inhibition is mediated by the $\alpha_{o1}\beta_3\gamma_4$ heterotrimer (25–27) (Fig. 1). This extraordinarily high degree of coupling selectivity of a receptor to heterotrimeric G proteins is retained in other cell systems as well (Fig. 1). Ca^{2+}-channel inhibition by muscarinic agonists in pheochromacytoma (PC-12) cells and by somatostatin in rat insulinoma (RINm5F) cells is mediated by the same G-protein heterotrimers as in GH_3 cells (Fig. 1). By analyzing the G-protein coupling pattern of the galanin receptor, we observed that Ca^{2+}-channel inhibition mediated by activation of this receptor occurs via two different G-protein heterotrimers. Only one α subunit, namely α_{o1}, but two β subunits, β_2 and β_3, and two different γ subunits, γ_2 and γ_4, participate in the inhibition of voltage-gated Ca^{2+} channels in GH_3 and RINm5F cells (28) (Fig. 1). The cellular $\beta\gamma$ expression pattern, in conjunction with the extent of inhibition observed, led us to the assumption that the galanin receptor in GH_3 and RINm5F cells preferentially couples to $\alpha_{o1}\beta_2\gamma_2$ and less efficiently to the G-protein heterotrimer $\alpha_{o1}\beta_3\gamma_4$ (28). Using our antisense approach, we were able to show that inhibition of voltage-gated Ca^{2+} channels by the activated α_2-adrenergic receptor endogenously expressed in PC-12 cells is mediated by a specific heterotrimer that is also utilized by the activated M_4 muscarinic receptor in PC-12 and GH_3 cells (Fig. 1).

Because the lack of effect of a certain oligonucleotide on a hormonal effect can be due either to the lack of involvement of that protein in the measured process or to an insufficient reduction of the protein level which would have functional conse-

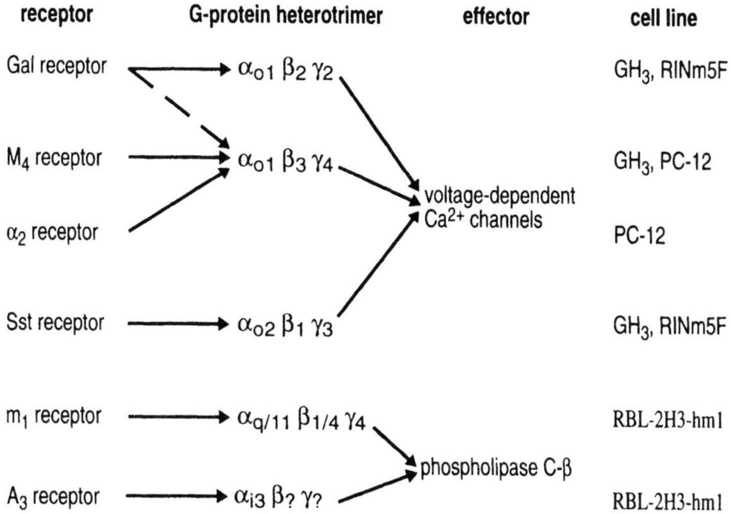

FIG. 1. Summary of functionally active G-protein heterotrimers detected by intranuclear injection of antisense oligonucleotides directed against G-protein subunits. Gal receptor, galanin receptor; M_4 receptor, M_4 muscarinic receptor; α_2 receptor, α_2 adrenoreceptor; Sst receptor, somatostatin receptor; m1 receptor, m1 muscarinic receptor; A_3 receptor, A_3 adenosine receptor.

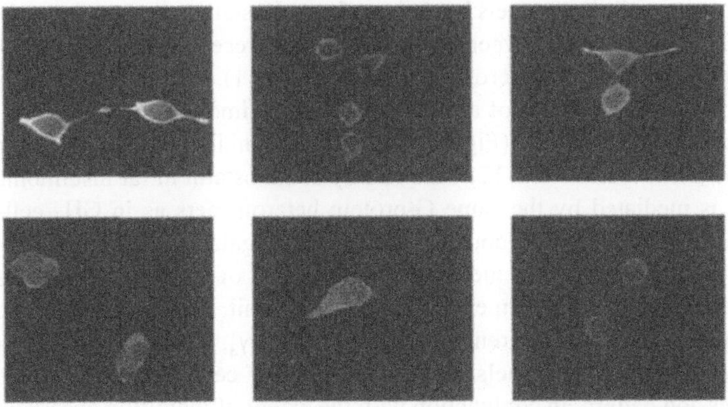

FIG. 2. Inhibition of $G\alpha_o$-protein expression in RINm5F cells injected with $G\alpha_{o1}$ or $G\alpha_{o2}$ antisense oligonucleotides. Upper panel: cells not injected with antisense oligonucleotides or injected with α_{o1} or α_{o2} antisense oligonucleotides. Forty-eight hours after injection, cells were stained with $G\alpha_{o1}$ antiserum (AS 248, 1:40). Lower panel: RINm5F cells either not injected or injected with α_{o1} or α_{o2} oligonucleotides. Forty-eight hours after injection, cells were stained with $G\alpha_{o2}$ antiserum (AS 201, 1:40). The first antibodies were detected with fluorescent isothiocyanate (FITC)-marked antirabbit IgG (1:2,000).

quences, we studied the effect of injection of α_{o1} and α_{o2} antisense oligonucleotides on $G\alpha_{o1}$ and $G\alpha_{o2}$ protein expression with immunofluorescence analysis (Fig. 2). Injection of particular antisense oligonucleotides led to a marked suppression of immunofluorescence signals when cells were stained with specific antisera at 48 hr after the injection (Fig. 2). In order to control for specificity of the phenomena observed, the effects of sequential superfusion with galanin and carbachol (which act via M_4 muscarinic receptors) were studied in GH_3 cells that had been injected with antisense oligonucleotides (Fig. 3). Suppression of individual hormonal effects by β or γ antisense oligonucleotide injection in cells stimulated with either galanin or carbachol showed that this kind of experimental approach can be used to selectively affect the signal-transduction pathway induced by one hormone without disturbing related effects elicited by a second hormone (Fig. 3). Nonspecific effects were seen only when antisense oligonucleotides were injected at concentrations of 50 μM or higher.

The antisense approach has recently been extended to study of the regulation of another effector system, that involving phospholipase C (PLC) β isoforms, by monitoring the release of Ca^{2+} from intracellular stores with a single-cell fluorimetric imaging system. In rat basophilic leukemia cells stably expressing the human m1 muscarinic receptor (RBL-hm1), carbachol was shown to stimulate PLC via $\alpha_{q/11}\beta_{1/4}\gamma_4$ (29) (Fig. 1). It will be of particular interest to clarify whether G_q and G_{11} use common or distinct βγ dimers for coupling the RBL-hml receptor to PLC β.

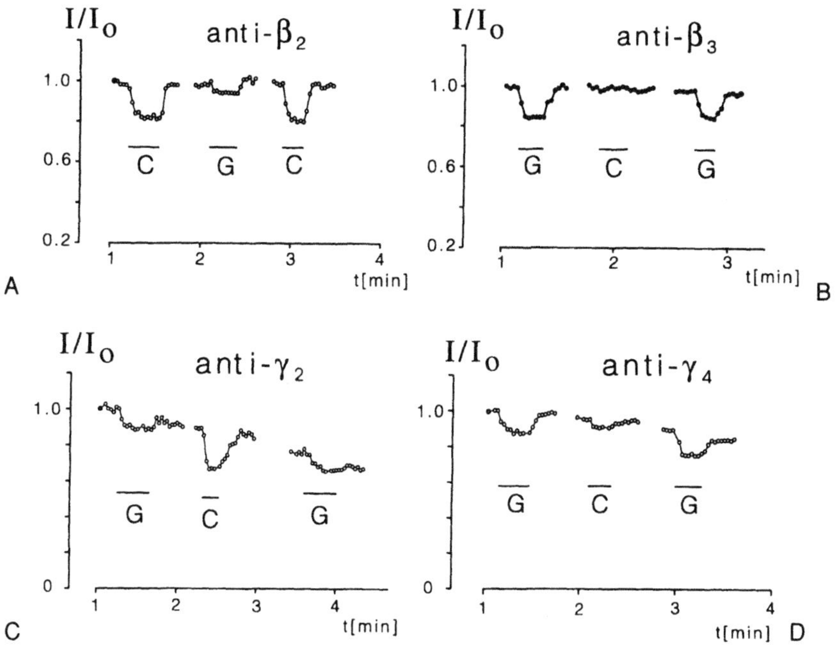

FIG. 3. Carbachol- and galanin-induced inhibition of $B\alpha^{2+}$ current through voltage-gated Ca^{2+} channels (I_{Ba}) in GH_3 cells injected with oligonucleotides annealing to the mRNAs of G-protein β and γ subunits. Shown are time courses of I_{Ba} normalized to control current (I_o) evoked by repeated depolarization in representative cells injected with anti-β_2 (A), anti-β_3 (B), anti-γ_2 (C), and anti-γ_4 oligonucleotides (D). Each cell was superfused with galanin (500 nM) (G) and carbachol (10 μM) (C) at 44 hr after injection of antisense oligonucleotides. *Bars* denote the time during which galanin (G) or carbachol (C) was present. For composition of heterotrimers coupling the M_4 muscarinic and galanin receptors to inhibition of voltage-dependent Ca^{2+} channels in GH_3 cells, see Fig. 1.

RBL cells endogenously express A_3 adenosine receptors, which stimulate PLC β isoforms via pertussis-toxin-sensitive G proteins. With the antisense approach described above, the A_3 receptor was shown to specifically couple to G_{i3} in order to stimulate PLC (Dippel E, Kalkbrenner F, Wittig B, Schultz G; in preparation). It remains to be determined whether G-protein heterotrimer selectivity is due to interactions of specific heterotrimers with each receptor or to the preference of a receptor for a particular G-protein subunit (e.g., Gα), which in turn prefers distinct partners in forming a heterotrimer.

DIVERSITY OF RECEPTOR–G-PROTEIN COUPLING

In order to channel incoming signals into different pathways, cells use divergent coupling patterns at the receptor/G protein and G protein/effector interfaces. Activation of one G protein may give rise to a bifurcating signal, since βγ dimers can propagate part of the information, in addition to the signaling capabilities, of the GTP-bound α subunit (5). Coupling of a variety of receptors primarily expressed in neuronal and neuroendocrine cells to various G proteins belonging to a single family (i.e., $G_{i/o}$) has been demonstrated. Activation of this type of receptor has been shown to result in inhibition of adenylyl cyclase (via $G\alpha_i$ and/or βγ), activation of receptor-operated K^+ currents (via βγ derived from $G_{i/o}$), and inhibition of voltage-gated Ca^{2+} channels (via $G\alpha_o$). Several cloned $G_{i/o}$-coupled receptors (e.g., M_2 and M_4 muscarinic, α_2-adrenergic, 5-hydroxytryptamine$_{1A}$ [5-HT$_{1A}$] serotonin, D_2 dopamine, $SSTR_1$ somatostatin receptors, etc.) have been described that mediate inhibition of adenylyl cyclase and also stimulate PLC β isoforms (summarized in [30]). In this type of dual coupling, adenylyl cyclase inhibition occurs at low agonist concentrations and is independent of receptor density. In contrast, PLC activation requires considerably higher agonist concentrations and is directly correlated to receptor abundance. Activated α subunits are thought to mediate inhibition of adenylyl cyclase, whereas PLC β isoforms (β2 and β3; see [31]) are assumed to be activated by βγ subunits released from activated G_i.

One of the earliest observations of divergent coupling at the receptor–G-protein level originated from studies on the action of vasopressin in liver, angiotensin II in liver and adrenal glomerulosa, and thyrotropin-releasing hormone (TRH) in GH_4C_1 cells (summarized in [1]). In all cases, pertussis-toxin-insensitive activation of PLC and inhibition of adenylyl cyclase by a pertussis-toxin-sensitive mechanism were observed. The TRH receptor has been shown to interact with G_q and G_{11} to activate PLC (32), and regulation of voltage-gated Ca^{2+} channels by TRH is partly mediated by G_i (33). Thus, the hormone receptors indicated above appear to activate G proteins of two distinct families, G_q and G_i.

Another aspect of dual coupling is realized by primarily G_s-coupled receptors, which have also been found to stimulate PLC. Among these are three glycoprotein hormone receptors, namely the receptors for TSH (34), luteinizing hormone (LH) (35,36), and follicle-stimulating hormone (FSH) (37), as well as receptors for parathyroid hormone (PTH)/PTH-related peptide (PTHrP) (38) and calcitonin (39). In analogy to the situation with G_i-coupled receptors, PLC activation by G_s-coupled receptors requires high receptor density and high agonist concentrations, as opposed to stimulation of adenylyl cyclase by the same agonist. Studies of the LH receptor in bovine corpora lutea and in L cells permanently expressing the recombinant receptor revealed that G_s and G_i proteins are activated by the agonist-bound receptor (40) (Fig. 4). In contrast, overexpression of PTHrP and the calcitonin receptors together with G-protein α subunits in COS-7 cells provided strong evidence that dual signaling of these receptors is due to activation of distinct G proteins belonging to the G_s and G_q families (41) (Fig. 4).

Photolabeling of G proteins in platelet membranes in response to thromboxane A_2 showed increased incorporation of [α-^{32}P]GTP azidoanilide into G proteins of the G_q and G_{12} families, whereas in the same membranes, the activated thrombin receptor couples to G proteins of three different families: G_q, G_i, and G_{12} (42) (Fig. 5). To date, the champion among promiscuous receptors is the human thyrotropin receptor.

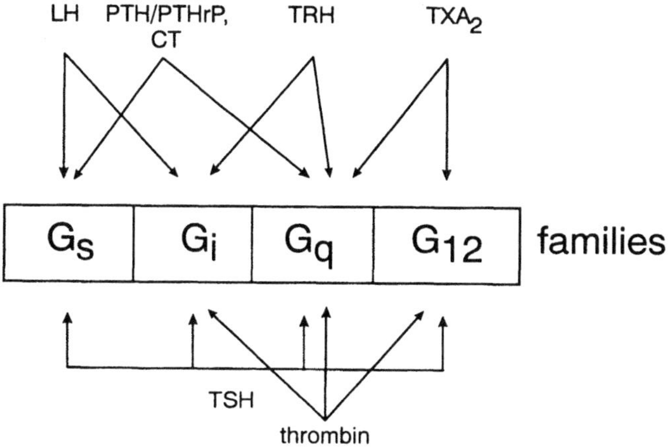

FIG. 4. Examples of dual or multiple coupling of receptors to different G-protein families. CT, calcitonin; LH, luteinizing hormone; PTH/PTHrP, parathyroid hormone/parathyroid hormone-related peptide; TRH, thyrotropin-releasing hormone; TXA_2, thromboxane A_2.

When human thyroid membranes are stimulated with bovine TSH, the thyrotropin receptor is able to couple to at least 10 different G proteins belonging to all four G-protein families (i.e., G_s, G_i, G_q, G_{12}) (43) (Fig. 4). Such a high level of promiscuity in receptor–G-protein interaction is unprecedented, and is indicative of complex multifunctional signaling by the TSH receptor. However, it remains to be seen whether this coupling pattern can also be observed in isolated thyrocytes or in transfected cells expressing the recombinant TSH receptor. Our knowledge about physiologic roles of the pertussis-toxin-insensitive G proteins of the G_{12} family is still vague, and it may be argued that the lack of known receptors exclusively coupling to $G_{12/13}$ constitutes one major reason for this situation.

CONCLUSIONS

Experimental results obtained over the past few years have shown that the conceptualization of signal-transduction pathways in a linear fashion (i.e., one receptor coupling to one G protein that in turn activates one effector) is by no means sufficient to provide adequate explanations for the diversity of effects mediated by receptor–G-protein interactions. In such a scenario, specificity can be conferred only by highly restricted interactions at each step of the pathway. G-protein-mediated signal transduction should preferably be looked upon as a highly organized, complex signaling mechanism that allows some cross-talk due to diverging and converging transduction steps at the ligand–receptor, receptor–G-protein, and G-protein–effector interfaces. In most instances, one heptahelical receptor interacts with several distinct G proteins and elicits multiple intracellular signals. The interaction of any sin-

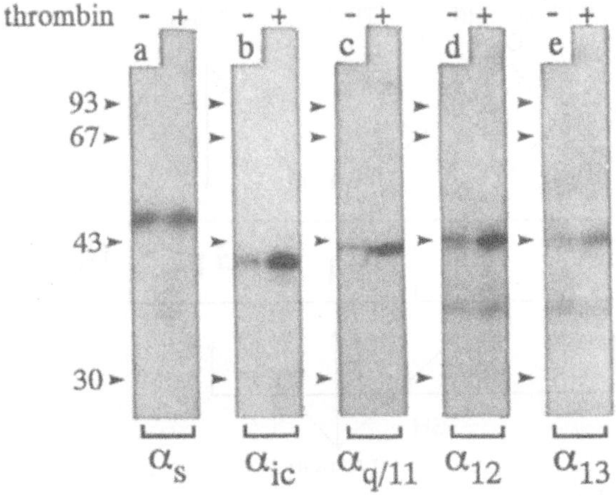

FIG. 5. G-protein α subunits in human platelet membranes photolabeled in the presence of thrombin. Human platelet membranes (200 µg of protein per tube) were photolabeled with [α-^{32}P]GTP azidoanilide in the absence (-) or presence (+) of thrombin (1 unit/ml). Solubilized membranes were incubated with the following antisera: a, AS 348 (α_s); b, AS 266 (α_i common); c, AS 370 ($\alpha_{q/11}$); d, AS 233 (α_{12}); and e, AS 343 (α_{13}). Precipitated proteins were subjected to sodium dodecyl sulfate–polyacrylamide gel electrophoresis (SDS–PAGE), and radioactively labeled bands were visualized by autoradiography. Molecular masses (kDa) of standard proteins are indicated on the left. (Modified from ref. 42, with permission.)

gle G protein with a given receptor in a certain cell, however, appears to be governed by a high degree of selectivity imparted by specific G-protein heterotrimers.

ACKNOWLEDGMENTS

We would like to thank Drs. M. I. Simon and S. Offermanns for making their results available to us before publication. Our own research reported herein was supported by the Deutsche Forschungsgemeinschaft, Fonds der Chemischen Industrie, Boehringer Ingelheim Fond, and Ernst Schering Research Foundation.

REFERENCES

1. Birnbaumer L, Abramowitz J, Brown AM. Receptor-effector coupling by G proteins. *Biochim Biophys Acta* 1990; 1031:163–224.
2. Simon MI, Strathman MP, Gautam N. Diversity of G proteins in signal transduction. *Science* 1991; 252:802–8.
3. Hepler JR, Gilman AG. G proteins. *Trends Biochem Sci* 1992; 17:383–7.
4. Spiegel AM, Shenker A, Weinstein LS. Receptor-effector coupling by G proteins: implications for normal and abnormal signal transduction. *Endocr Rev* 1992; 13:536–65.
5. Neer EJ. Heterotrimeric G proteins: organizers of transmembrane signals. *Cell* 1995; 80:249–57.
6. Dohlman HG, Thorner J, Caron MG, Lefkowitz RJ. Model systems for the study of seven-transmembrane-segment receptors. *Annu Rev Biochem* 1991; 60:653–88.

7. Strader CD, Fong TM, Tota MR, Underwood D. Structure and function of G protein-coupled receptors. *Annu Rev Biochem* 1994;63:101–32.
8. Gudermann T, Nürnberg B, Schultz G. Receptors and G proteins as primary components of transmembrane signal transduction. Part 1. G-protein-coupled receptors: structure and function. *J Mol Med* 1995; 73:51–63.
9. Nürnberg B, Gudermann T, Schultz G. Receptors and G proteins as primary components of transmembrane signal transduction. Part 2. G proteins: structure and function. *J Mol Med* 1995; 73:123–32.
10. Ray K, Kunsch C, Bonner LM, Robishaw JD. Isolation of cDNA clones encoding eight different human G protein γ subunits, including three novel forms designated the γ_4, γ_{10}, and γ_{11} subunits. *J Biol Chem* 1995; 270:21765–71.
11. Clapham DE, Neer EJ. New roles for G protein $\beta\gamma$ dimers in transmembrane signaling. *Nature* 1993; 365:403–6
12. Neubig R. Membrane organization in G-protein mechanisms. *FASEB J* 1994; 8:939–46.
13. Strittmatter SM, Valenzuela D, Kennedy TE, Neer EJ, Fishman MC. G_o is a major growth cone protein subject to regulation by GAP-43. *Nature* 1990: 344:833–41.
14. Dym M, Lamsam-Casalotti S, Jia MC, Kleinman HK, Papadopoulos V. Basement membrane increases G-protein levels and follicle-stimulating hormone responsiveness of Sertoli cell adenylyl cyclase activity. *Endocrinology* 1991; 128:1167–76.
15. Keefer JR, Limbird LE. The α_{2A}-adrenergic receptor is targeted directly to the basolateral membrane domain of Madin–Darby canine kidney cells independent of coupling to pertussis toxin-sensitive GTP-binding proteins. *J Biol Chem* 1993; 268:11340–7.
16. Keefer JR, Kennedy ME, Limbird LE. Unique structural features important for stabilization versus polarization of the α_{2A}-adrenergic receptor on the basolateral membrane of Madin-Darby canine kidney cells. *J Biol Chem* 1994; 269:16425–33.
17. Popova JS, Johnson GL, Rasenick MM. Chimeric $G\alpha_s/G\alpha_{i2}$ proteins define domains on $G\alpha_s$ that interact with tubulin for β-adrenergic activation of adenylyl cyclase. *J Biol Chem* 1994; 269:21748–54.
18. Sargiacomo M, Sudol M, Tang ZL, Lisanti MP. Signal transducing molecules and glycosyl-phosphatidylinositol-linked proteins from a caveolin-rich insoluble complex in MDCK cells. *J Cell Biol* 1993; 122:789–807.
19. Phillips WJ, Cerione RA. Rhodopsin/transducin interactions. I. Characterization of the binding of the transducin-β-complex to rhodopsin using fluorescence spectroscopy. *J Biol Chem* 1992; 267: 17032–9.
20. Kisselev O, Gautam N. Specific interactions with rhodopsin is dependent on the γ subunit type in a G protein. *J Biol Chem* 1993;268:24519–22.
21. Kisselev O, Ermolaeva MV, Gautam N. A farnesylated domain in the G protein γ subunit is a specific determinant of receptor coupling. *J Biol Chem* 1994;21399–402.
22. Müller S, Hekman M, Lohse MJ. Specific enhancement of β-adrenergic receptor kinase activity by defined G-protein β and γ subunits. *Proc Natl Acad Sci USA* 1993; 90:10339–43.
23. Taylor JM, Jacob-Mosier GG, Lawton RG, Remmers AE, Neubig RR. Binding of an α_2-adrenergic receptor third intracellular loop peptide to $G\beta$ and the amino terminus of $G\alpha$. *J Biol Chem* 1994; 269: 27618–24.
24. Kleuss C, Schultz G, Wittig B. Microinjection of antisense oligonucleotides to assess G-protein function. *Methods Enzymol* 1994; 237:345–55.
25. Kleuss C, Hescheler J, Ewel C, Rosenthal W, Schultz G, Wittig B. Assignment of G-protein subtypes to specific receptors inducing inhibition of calcium currents. *Nature* 1991; 353:43–8.
26. Kleuss C, Scherübl H, Hescheler J, Schultz G, Wittig B. Different β-subunits determine G-protein interaction with transmembrane receptors. *Nature* 1992; 358:424–6.
27. Kleuss C, Scherübl H, Hescheler J, Schultz G, Wittig B. Selectivity in signal transduction determined by γ-subunits of heterotrimeric G proteins. *Science* 1993; 259:832–4.
28. Kalkbrenner F, Degtiar VE, Schenker M, et al. Subunit composition of Go proteins functionally coupling galanin receptors to voltage-gated calcium channels. *EMBO J.* 1995;14:4728–37.
29. Dippel E, Kalkbrenner F, Wittig B, Schultz G. A heterotrimeric G-protein complex couples the muscarinic m1 receptor to phospholipase C-β. *Proc Natl Acad Sci USA* 1996;43:1391–6.
30. Gudermann T, Kalbrenner F, Schultz G. Diversity and selectivity of receptor-G protein interaction. *Annu Rev Pharmacol Toxicol* 1996; 36:429–59.
31. Exton JH. Phosphoinositide phospholipases and G proteins in hormone action. *Annu Rev Physiol* 1994; 56:349–69.

32. Aragay AM, Katz A, Simon MI. The $G\alpha_q$ and $G\alpha_{11}$ proteins couple the thyrotropin-releasing hormone receptor to phospholipase C in GH_3 rat pituitary cells. *J Biol Chem* 1992; 267:24983–8.
33. Gollasch M, Kleuss C, Hescheler J, Wittig B, Schultz G. G_{i2} and protein kinase C are required for thyrotropin-releasing hormone-induced stimulation of voltage-dependent Ca^{2+} channels in rat pituitary GH_3 cells. *Proc Natl Acad Sci USA* 1993; 90:6265–9.
34. Van Sande J, Raspé J, Perret J, et al. Thyrotropin activates both the cAMP and the PIP_2 cascade in CHO cells expressing the human cDNA of the TSH receptor. *Mol Cell Endocrinol* 1990; 74:R1–R6.
35. Gudermann T, Birnbaumer M, Birnbaumer L. Evidence for dual coupling of the murine luteinizing hormone receptor to adenylyl cyclase and phosphoinositide breakdown and Ca^{2+} mobilization. *J Biol Chem* 1992; 267:4479–88.
36. Gudermann T, Nichols C, Levy FO, Birnbaumer M, Birnbaumer L. Ca^{2+} mobilization by the LH receptor expressed in *Xenopus* oocytes independent of 3′, 5′-cyclic adenosine monophosphate formation: evidence for parallel activation of two signaling pathways. *Mol Endocrinol* 1992; 6:272–8.
37. Quintana J, Hipkin RW, Sanchez-Yagüe J, Ascoli M. Follitropin (FSH) and a phorbol ester stimulate the phosphorylation of the FSH receptor in intact cells. *J Biol Chem* 1994; 269:8772–9.
38. Abou-Samra AB, Jüppner H, Force T, et al. Expression cloning of a common receptor for parathyroid hormone and parathyroid hormone-related peptide from rat osteoblast-like cells: a single receptor stimulates intracellular accumulation of both cAMP and inositol trisphosphates and increases intracellular free calcium. *Proc Natl Acad Sci USA* 1992; 89:2732–6.
39. Chabre O, Conklin BR, Lin H, et al. A recombinant calcitonin receptor independently stimulates 3′, 5′-cyclic adenosine monophosphate and Ca^{2+}/inositol phosphate signaling pathways. *Mol Endocrinol* 1992;6:551–6.
40. Herrlich A, Kühn B, Grosse R, Schmid A, Schultz G, Gudermann T. Involvement of G_s and G_i proteins in dual coupling of the luteinizing hormone receptor to adenylyl cyclase and phospholipase C. *J Biol Chem* 1996;271:16764–72.
41. Offermanns S, Iida-Klein A, Segre GV, Simon MI. $G\alpha_q$ family members couple PTH/PTHrP and calcitonin receptors to phospholipase C in COS-7 cells. (Submitted).
42. Offermanns S, Laugwitz K-L, Spicher K, Schultz G. G proteins of the G_{12} family are activated via thromboxane A_2 and thrombin receptors in human platelets. *Proc Natl Acad Sci USA* 1994; 91: 504–8.
43. Laugwitz K-L, Allgeier A, Offermanns S, et al. The human thyrotropin receptor: a heptahelical receptor capable of stimulating members of all four G-protein families. *Proc Natl Acad Sci USA*. (In press).

Signal Transduction in Health and Disease, Advances in Second Messenger and Phosphoprotein Research, Vol. 31, edited by J. Corbin and S. Francis.
Lippincott–Raven Publishers, Philadelphia © 1997.

21

G-Protein-Coupled Receptors and Their Regulation

Activation of the MAP Kinase Signaling Pathway by G-Protein-Coupled Receptors

Louis M. Luttrell, *Tim van Biesen, †Brian E. Hawes, Walter J. Koch, ‡Kathleen M. Krueger, §Kazushige Touhara, and Robert J. Lefkowitz

*Department of Medicine, Duke University Medical Center, Durham, North Carolina 27710; *Department of Neurological and Urological Diseases Research, Abbott Laboratories, Abbott Park, Illinois 60064-3500; †CNS/Cardiovascular Department, Schering Plough Research Institute, Kenilworth, New Jersey 07033; ‡Bioscience and Biotechnology Group, Chemical Science and Technology Division, Los Alamos National Laboratory, Los Alamos, New Mexico 87545; and §Department of Neurobiochemistry, The University of Tokyo, Tokyo 113 Japan*

The processes of cell growth, division, and differentiation are regulated by a complex set of signals derived from several sources including the cell substratum, the cell's immediate neighbors, and distant cells, from which information is conveyed in the form of growth factors and hormones. Many of these externally derived signals are conveyed by cell-surface receptors that specifically bind extracellular ligands and transmit information internally either through a series of direct protein–protein interactions, often involving protein phosphorylation, or through the enzymatic generation of small second-messenger molecules. These signals ultimately influence intermediary metabolism, regulate gene transcription, and stimulate the deoxyribonucleic acid (DNA) replication and cytoskeletal changes necessary for mitogenesis. The Ras–mitogen-activated protein (MAP) kinase signaling pathway (1–2) represents one major point of convergence at which an array of complex and sometimes conflicting signals regulating growth and differentiation are integrated. The Ras–MAP kinase pathway has been elucidated largely through the study of transforming viral oncogenes and their cellular homologues, and the characterization of related pathways that regulate fruit fly, nematode, and yeast development.

The best understood pathway of Ras–MAP kinase activation is that mediated by the tyrosine kinase growth-factor receptors, including the receptors for epidermal growth factor (EGF), fibroblast growth factor, (FGF), platelet-derived growth factor (PDGF), insulin, and insulin-like growth factor 1 (IGF-1). In the case of the receptor for EGF, the binding of ligand leads to receptor dimerization and phosphorylation of tyrosine residues in the intracellular portion of the receptor. Binding of cytoplasmic src-homology 2 (SH2)-domain-containing (3,4) adapter proteins, including shc and Grb2 (5,6), which recognize specific phosphotyrosine residues on the receptor, results in recruitment of the Ras-guanine-nucleotide exchange factor Sos1 to the plasma membrane, where it catalyzes Ras guanosine triphosphate (GTP)-for-guanosine diphosphate (GDP) exchange, inducing Ras activation. Activation of Ras results in activation of the serine/threonine kinase Raf (MAP kinase kinase kinase) via direct Ras–Raf interaction. Activated Raf phosphorylates and activates the multifunctional threonine/tyrosine MEK kinases (MAP kinase/extracellular signal-regulated kinase or MAP kinase kinase) (7–9), which phosphorylate the MAP kinases (*erk* gene products) (10). The MAP kinases, and one of their substrates, the cell-cycle-regulated ribosomal S6 protein kinase RSK (11), are both cytoplasmic and nuclear in localization, and may directly convey signals that regulate gene transcription by phosphorylating nuclear transcription factors.

Although the signal-transduction pathways remain incompletely described, recent data indicate that many G-protein-coupled receptors also have the ability to activate

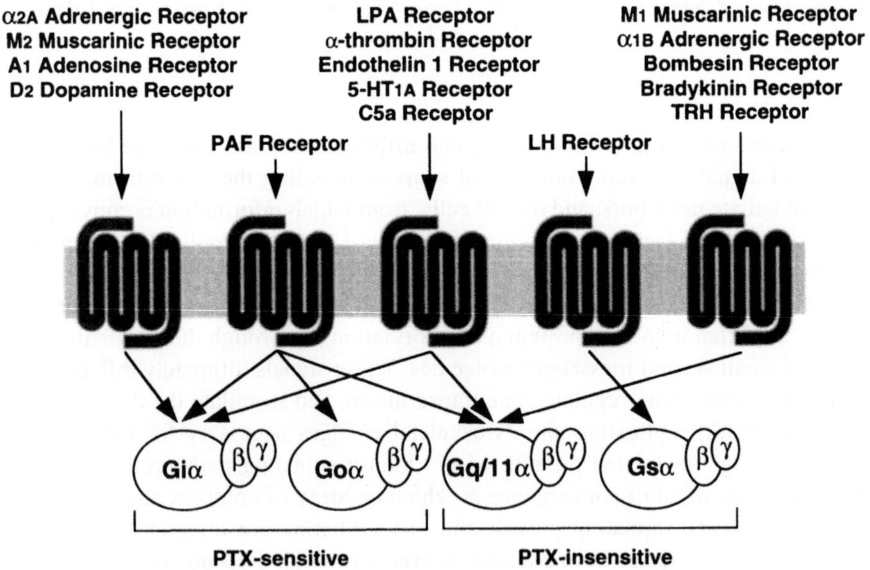

FIG. 1. G-protein-coupled receptors that activate the MAP kinase pathway interact with several different subsets of heterotrimeric G proteins. In appropriate model systems, receptors that activate effectors primarily via G_i family members, both G_i and $G_{q/11}$ family members, primarily $G_{q/11}$ family members, and Gs have been reported to stimulate MAP kinase activity (see text for references). Such heterogeneity suggests that multiple pathways may exist for G-protein-coupled, receptor-mediated MAP kinase activation.

the Ras–MAP kinase pathway. Figure 1 represents a partial list of G-protein-coupled receptors that have been reported to activate MAP kinases. Among them are receptors for several substances present either in the general circulation or released as neurotransmitters, such as catecholamines (12–15), acetylcholine (13,15–18), dopamine (16), thyrotropin-releasing hormone (TRH) (19), and luteinizing hormone (LH) (16); substances produced locally by vascular endothelium and activated platelets, such as adenosine (16), angiotensin II (20), bombesin (16), endothelin I (21), lysophosphatidic acid (LPA) (13, 22–25), complement C5a (26), and α-thrombin (27–30); or exogenously derived substances, such as formyl-methionyl peptides (31,32). Thus, a diverse array of signaling molecules acting via G-protein-coupled receptors may, like receptor tyrosine kinases, play significant roles in endocrine or paracrine regulation of cell proliferation in both physiologic and pathophysiologic states. Proliferative responses to local tissue injury, which contribute to microvascular and macrovascular complications of diabetes mellitus and atherosclerosis, may to a significant degree be regulated by G-protein-coupled receptors.

EVIDENCE FOR DISTINCT PATHWAYS OF G-PROTEIN-COUPLED RECEPTOR-MEDIATED MAP KINASE ACTIVATION

Among the list of G-protein-coupled receptors that activate the MAP kinase pathway are receptors that interact with distinct subsets of heterotrimeric G proteins, including the pertussis toxin-insensitive Gq/11 and pertussis toxin-sensitive Gi and Go families. G_s-coupled receptors have also been reported to activate MAP kinase (16,33,34), although activation of adenyl cyclase opposes MAP kinase activation in many systems (35,36). This diversity suggests that, in appropriate cell types, heterogeneous upstream signals generated by G-protein-coupled receptors must ultimately converge upon the MAP kinase pathway.

Activation of protein kinase C (PKC) isoforms, such as following exposure to phorbol esters, is known to result in MAP kinase activation, possibly via direct PKC α-mediated phosphorylation and activation of the Raf1 kinase (37). Stimulation of phospholipase Cβ (PLCβ) isoforms via G-protein-coupled receptors, resulting in diacylglycerol generation and inositol phosphate-dependent Ca^{2+} flux, might be expected to have similar effects. Indeed, G-protein-coupled receptors that couple primarily to the pertussis toxin-insensitive $G_{q/11}$ family of G proteins, such as the α_{1B} adrenergic receptor (α_{1B}-AR) and M_1 muscarinic acetylcholine receptor (M_1-AChR), appear to employ this pathway (15). As shown in Fig. 2A, pertussis toxin-insensitive activation of MAP kinase by transiently expressed M_1-AChR is attenuated in PKC-depleted cells, as well as in cells coexpressing a dominant negative mutant Raf protein (ΔN Raf), but not a dominant negative Ras protein (N17 Ras) (38).

Several lines of evidence, however, indicate that G-protein-coupled receptors also stimulate the MAP kinase pathway via a distinct, Ras-dependent mechansim. LPA, the simplest naturally occurring phospholipid, potently stimulates mitogenesis in quiescent fibroblasts (22,39). LPA activates its own G-protein-coupled receptor(s)

(40), producing several biologic effects, including the stimulation of PLC and PLD, inhibition of adenyl cyclase, and stimulation of DNA synthesis (22). Its effects on phospholipid metabolism are pertussis toxin-insensitive, and neither necessary nor sufficient for the mitogenic response. The inhibition of adenyl cyclase is pertussis toxin-sensitive and also apparently unrelated to the mitogenic response. Rather, LPA-induced mitogenesis in Rat1 and CCL39 fibroblasts depends upon the rapid, pertussis toxin-sensitive activation of MAP kinase via a pathway involving Ras and Raf activation (23, 24), which is sensitive to inhibitors of tyrosine protein kinases such as genistein and herbimycin A. The platelet-derived protease α-thrombin, a potent mitogen for various types of fibroblasts (41), activates a unique G-protein-coupled receptor (42), which is capable of coupling to Gq and Gi when the receptor is proteolytically cleaved to reveal an internal "tethered" ligand. Like LPA, α-thrombin induces rapid MAP kinase activation via a pertussis toxin-sensitive, G_i-mediated mechanism (23,27–30). The activation of MAP kinase, but not phosphatidylinositol hydrolysis or Ca^{2+} mobilization, is dependent upon the activation of Ras and is blocked by genistein.

Stable expression of the Gi-coupled α_{2A}-AR in CCL39 cells confers clonidine-stimulated, pertussis-toxin-sensitive stimulation of DNA synthesis (43), which is associated with Ras activation and is independent of phospholipase activation or inhibition of adenyl cyclase (12). Stable expression of the Gi-coupled M_2-AChR in Rat 1a fibroblasts confers carbachol-stimulated, pertussis toxin-sensitive activation of Ras, Raf, MEK, and MAP kinase (18). As shown in Fig. 2A, activation of MAP kinase via the α_{2A}-AR exhibits a pattern of sensitivity to inhibitors that is distinct from that of the Gq-coupled M_1-AChR. The pertussis toxin-sensitive α_{2A}-AR signal is insensitive to PKC depletion, but is markedly attenuated by the dominant negatives N17 Ras and ΔN Raf (15). Interestingly, the α_{2A}-AR signal is inhibited by genistein at concentrations lower than those required to inhibit pertussis toxin-insensitive, Ras-dependent MAP kinase activation via the EGF receptor tyrosine kinase. C5a receptors expressed in HEK-293 cells mediate pertussis toxin-sensitive, Ras/Raf-dependent activation of MAP kinase. Coexpression of G16α subunits, which confers upon the C5a receptor the ability to stimulate pertussis toxin-insensitive PLCβ activation, results in augmented MAP kinase activation that is reversed by PKC nhibitors (26). Thus, at least two distinct mechanisms of MAP kinase activation by G-protein-coupled receptors can be discerned; one that is PKC-dependent and mediated by receptors coupling to pertussis toxin-insensitive G proteins, and another that is Ras-dependent, sensitive to inhibitors of tyrosine protein kinases, and mediated by receptors coupling to pertussis toxin-sensitive G proteins. As shown schematically in Fig. 2B, the two pathways apparently converge upstream of MAP kinase, at the level of Raf kinase.

FIG. 2. Distinct patterns of sensitivity to inhibitors characterize Gq-coupled and Gi-coupled receptor-mediated MAP kinase activation. **A:** COS-7 or CHO cells were transiently transfected with hemagluttinin-tagged p44 MAP kinase ($p44^{HAmapk}$), and MAP kinase activation was measured following stimulation of coexpressed M_1-AChR or α_{2A}-AR, coexpression of Gβ1γ2 subunits, or stimulation of endogenous EGF receptors. The effects of PKC depletion, pertussis-

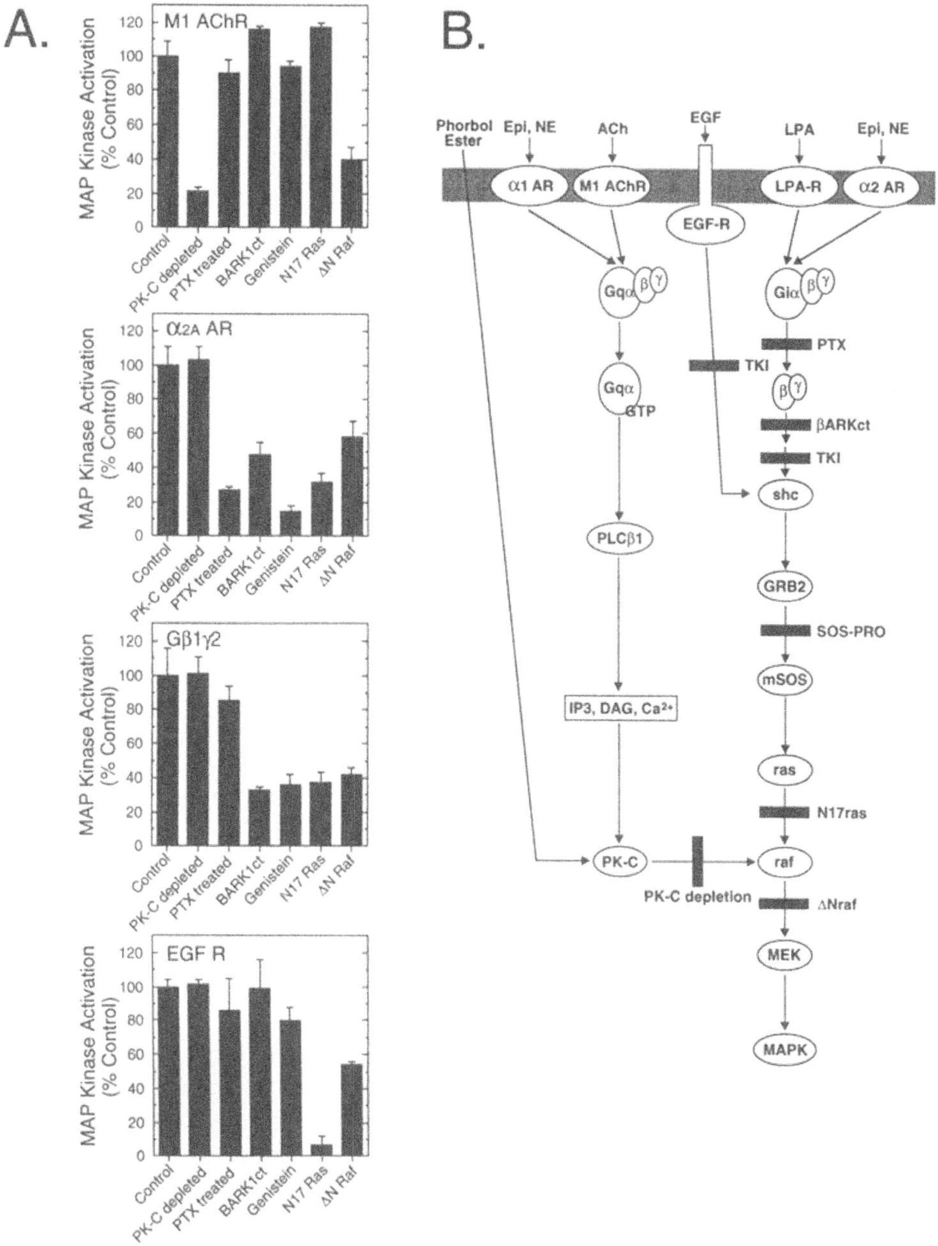

toxin treatment (PTX), βARK1ct-peptide expression (βARKct), exposure to the tyrosine protein kinase inhibitors genistein or herbimycin A (TKI), and expression of dominant negative mutants of Ras (N17Ras) and Raf (ΔNraf), were determined. Data shown represent the percent of control MAP kinase activation detected in the presence of each of the inhibitors. MAP kinase activity was measured by phosphorylation of myelin basic protein (MBP) following immunoprecipitation of $p44^{HAmapk}$ from basal and stimulated cells. **B:** Model of putative PKC-dependent and Ras-dependent MAP kinase activation pathways mediated by receptor tyrosine kinases, Gq-coupled, and Gi-coupled receptors, based upon the observed sensitivity of each pathway to inhibitors.

Agonist activation of G-protein-coupled receptors results in the dissociation of heterotrimeric G proteins into Gα-GTP and Gβγ heterodimers. Appreciation of the role of Gβγ subunits in regulating a variety of effectors of G-protein-coupled receptors has recently grown (44). To determine whether the effectors required for MAP kinase activation are regulated by Gα or Gβγ subunits, activated mutant Gα subunits or Gβγ heterodimers have been employed. In Rat 1a fibroblasts, stable expression of activated $G\alpha_{i2}$ mutants causes neoplastic transformation and constitutive activation of the MAP kinase pathway (45). However, in COS-7 cells, expression of activated $G\alpha_q$ and $G\alpha_s$, but not $G\alpha_i$, is sufficient to activate MAP kinase (16). Expression of Gβγ subunits in COS-7 cells leads directly to MAP kinase activation (13,15,16,46), suggesting that in this system, MAP kinase activation by Gi-coupled receptors depends upon the release of free Gβγ subunits. As shown in Fig. 2A, Gβγ-subunit-mediated MAP kinase activation, like that mediated by α_{2A}-AR, is sensitive to genistein, N17 Ras, and ΔN Raf.

Cellular expression of a specific Gβγ-subunit-binding peptide derived from the carboxyl-terminus of the β-adrenergic receptor kinase 1 (βARK1ct) (47,48) specifically antagonizes Gβγ subunit-mediated PLC activation (49) and conditional stimulation of Type II adenyl cyclase (49,50). As shown in Fig. 2A, expression of the βARK1ct peptide inhibits α_{2A}-AR- and direct Gβγ-subunit-mediated MAP kinase activation, with no effect on either M_1-AChR- or EGF-receptor-mediated activation. Similar results have been obtained by coexpression of transducin Gα subunits (16). Thus, expression of Gβγ-subunit scavengers inhibits Ras-dependent MAP kinase activation via pertussis-toxin-sensitive G proteins, but not PKC dependent MAP kinase activation via pertussis-toxin-insensitive G proteins.

THE ROLE OF TYROSINE PROTEIN PHOSPHORYLATION IN G-PROTEIN-COUPLED, RECEPTOR-MEDIATED MAP KINASE ACTIVATION

Activation of Ras via receptor tyrosine kinases requires receptor-catalyzed tyrosine protein phosphorylation, which forms the basis for SH2 domain-directed recruitment of the Grb2/Sos1 complex to the plasma membrane. Growing evidence suggests that Ras-dependent MAP kinase activation via G-protein-coupled receptors also requires the tyrosine phosphorylation of "docking" proteins, which serve as platforms for assembly of a Ras activation complex. Both PKC-dependent and PKC-independent mechanisms for G-protein-coupled, receptor-mediated tyrosine phosphorylation apparently exist. Stimulation of endogenous angiotensin II, bombesin, LPA, and endothelin receptors includes tyrosine protein phosphorylation via a PKC-dependent pathway (25,51–53), although the LPA receptor does not require PKC activity for MAP kinase activation (25). Tyrosine phosphorylation of the shc adapter protein by two predominantly Gq-coupled receptors, those for thyrotropin-releasing hormone (TRH) (19) and endothelin-1 (21), has been described. In both cases, shc phosphorylation is associated with Grb2 binding. In the case of the TRH receptor,

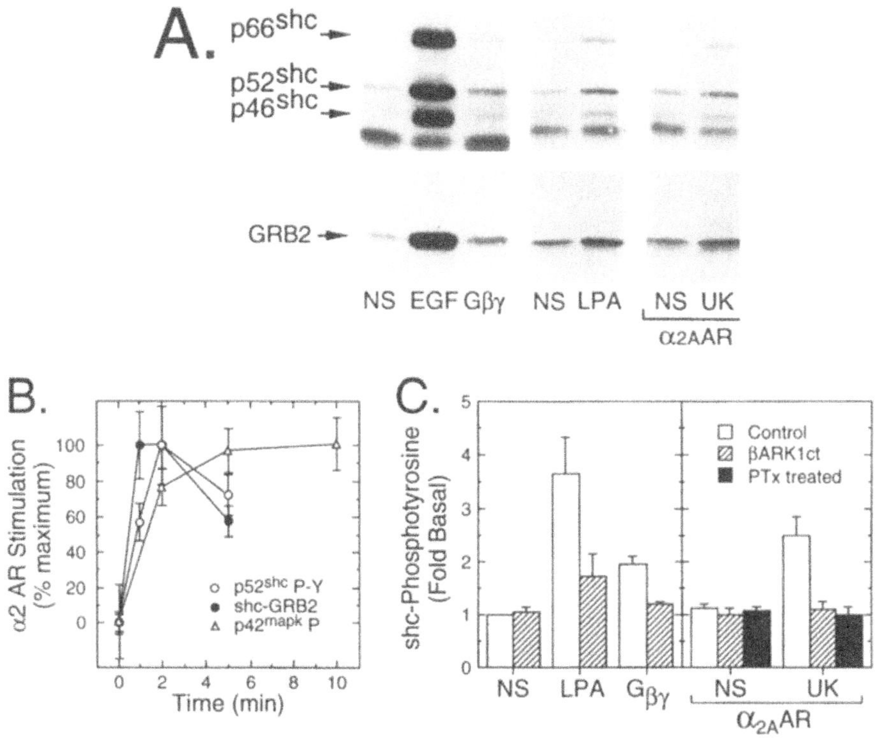

FIG. 3. Stimulation of shc tyrosine phosphorylation and shc-Grb2 complex formation follows activation of receptor tyrosine kinases and Gi-coupled receptors or overexpression of Gβ1γ2 subunits in COS-7 cells. **A:** Shc phosphotyrosine content was determined by antiphosphotyrosine immunoblotting of shc immunoprecipitates from COS-7 cells following stimulation of endogenous EGF and LPA receptors or transiently overexpressed α_{2A}-AR, or following overexpression of Gβ1γ2 subunits (*upper panel*). Shc-Grb2 complex formation was determined by anti-Grb2 immunoblotting of shc immunoprecipitates (*lower panel*). **B:** Time courses of shc tyrosine phosphorylation and shc-Grb2 complex formation were determined as described following stimulation of transiently overexpressed α_{2A}-AR in COS-7 cells. Time course of endogenous p42 MAP kinase phosphorylation was determined by electrophoretic-mobility-shift measurement. **C:** Effects of coexpression of the Gβγ-sequestrant βARKct peptide and pertussis-toxin treatment on Gi-coupled receptor and Gβ1γ2-subunit-stimulated shc tyrosine phosphorylation were determined as described. (From ref. 54, with permission.)

the tyrosine phosphorylation is unrelated to PKC activation (19). As shown in Fig. 3, stimulation of endogenous LPA receptors and transiently expressed α_{2A}-AR, or expression of Gβ1γ2 subunits, also induces PKC-independent shc tyrosine phosphorylation and shc/Grb2 complex formation in COS-7 cells (54). The receptor-mediated responses are rapid, reaching a peak within 1 to 2 min after stimulation, and are inhibited both by treatment with pertussis toxin and coexpression of the Gβγ subunit sequestrant βARKct peptide, indicating that in this system the phosphorylations are mediated by the Gβγ subunits of pertussis toxin-sensitive G proteins.

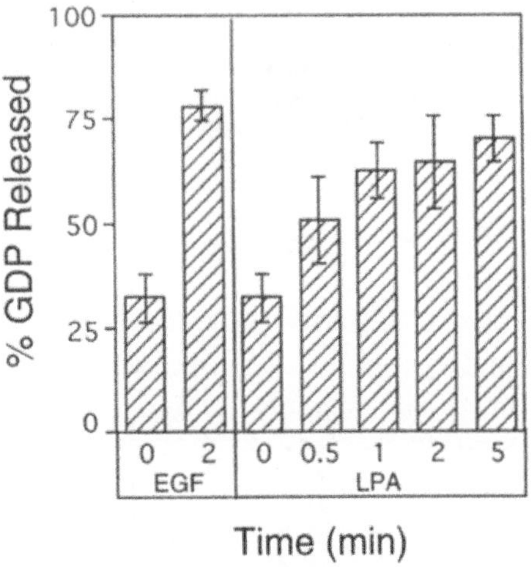

FIG. 4. Association of guanine-nucleotide-exchange factor activity with shc follows stimulation of endogenous EGF and LPA receptors. Shc immunoprecipitates from COS-7 cells overexpressing mSos1 were assayed for the presence of Ras-guanine-nucleotide-exchange factor activity using [^{3}H]GDP loaded K-Ras as substrate. Data are presented as percent of labelled GDP released from the substrate pool. (From ref. 54, with permission.)

Rapid association of Ras-guanine-nucleotide-exchange factor activity with shc can be directly demonstrated in shc immunoprecipitates from LPA-stimulated cells, as shown in Fig. 4. The inhibition of Gi-coupled-receptor- and Gβγ-subunit-mediated MAP kinase activation by coexpression of dominant negative mutants of Sos1 (Δm-Sos1 and SOS-PRO) (54,55) demonstrates that Grb2-mediated recruitment of Sos1 is a required step in the signaling pathway. Expression of enzymatically inactive

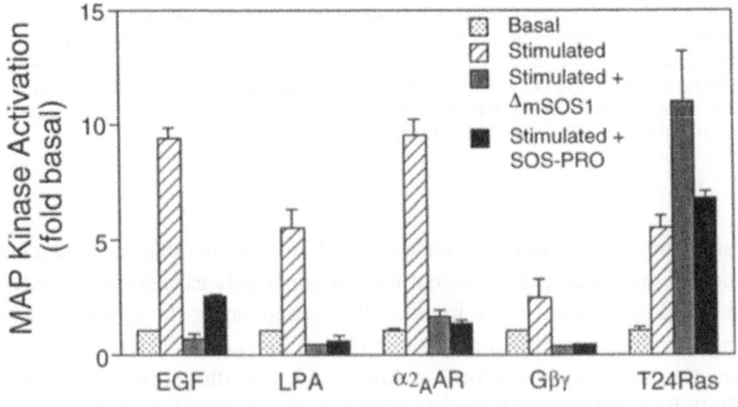

FIG. 5: Expression of the dominant negative ΔmSos1 and Sos-Pro mutants of mSos1 inhibits receptor-mediated MAP kinase activation. COS-7 cells were transiently transfected with p44^{HAmapk} plus the indicated mSos1 construct. MAP kinase activation was determined following stimulation through either endogenous EGF and LPA receptors or coexpressed α_{2A}-AR, Gβ1γ2 subunits, or the constitutively activated mutant Ras, T24Ras. (From ref. 54, with permission.)

Sos1 mutants results in the formation of functionally inactive Grb2/Sos mutant complexes, which disrupt the recruitment of wild-type Sos1 into shc/Grb2/Sos1 complexes (54). As shown in Fig. 5, expression of ΔmSos1 or SOS-PRO inhibits endogenous EGF receptor- and LPA receptor-mediated MAP kinase activation and that mediated by coexpressed α_{2A}-AR or Gβ1γ2 subunits, with no effect on activation via constitutively activated mutant Ras (T24 Ras). Thus, Gβγ subunit-mediated tyrosine phosphorylation of the shc adaptor protein, and Grb2-mediated recruitment of Ras-guanine-nucleotide-exchange factor activity, are apparently required for Ras-dependent MAP kinase activation via Gi-coupled receptors.

Although it is clear that some G-protein-coupled receptors can stimulate tyrosine protein phosphorylation and initiate SH2/3 domain-directed assembly of a Ras-activation complex in a manner similar to that effected by receptor tyrosine kinases, the identity of the responsible kinase(s) and their mechanism of activation remains unclear. The antigen receptors on T and B cells, as well as the receptors for growth hormone, erythropoietin, and several cytokines, do not possess intrinsic tyrosine kinase activity, but stimulate tyrosine phosphorylation through association with src-family tyrosine kinases such as Lck, Lyn, and Fyn (56). Similar recruitment of nonreceptor tyrosine kinases to perform functions analogous to those performed by the intrinsic tyrosine kinase domain of the receptor tyrosine kinases might also be mediated by G-protein-coupled receptors. Pertussis toxin-sensitive activation of Src, Fyn, and Yes via the α-thrombin receptor, α_{2A}-AR, and M_2-AChR, as well as pertussis toxin-insensitive activation via the M_1AChR, have been reported (57). Recently, activation of the N-formyl peptide (fMLP) chemoattractant receptor of phagocytic leukocytes has been shown to stimulate the formation of a protein signaling complex containing Lyn kinase, shc, Grb2, and phosphatidylinositol-3-kinase (PI 3-kinase) (58). In human neutrophils, the fMLP receptor mediates pertussis toxin-sensitive, PKC independent activation of Ras and Raf (32).

Other known nonreceptor tyrosine kinases are also candidates for mediating tyrosine protein phosphorylation via G-protein-coupled receptors. The recently cloned Fak family tyrosine protein kinase, PYK2, has been shown to cause shc phosphorylation and Grb2/Sos1 recruitment in neuronal PC12 cells (59). PYK2 is activated by stimuli that increase intracellular Ca^{2+} or activate PKC, including activation of the Gq-coupled bradykinin receptor. Although PYK2 is apparently limited in tissue distribution, PYK2-related kinases may provide a mechanism for linking G-protein-coupled receptor-mediated PLC and PKC activation to the Ras–MAP kinase pathway. The pleckstrin-homology (PH)-domain-containing family of tyrosine protein kinases, which includes Btk, Tsk, TecA, and BMX, represents another candidate. Although these kinases are also restricted in tissue distribution, *in vitro* activation of two of them, Btk and Tsk, by Gβγ subunits has recently been reported (60). Interestingly, the B-cell-specific tyrosine protein kinase, Btk, has been reported to interact with the SH3 domains of the src family kinases Fyn, Lyn, and Hck (61). This interaction with src-family kinases may contribute to Btk activation by stimulating Btk autophosphorylation (62).

Other proteins are likely to contribute to the assembly of a mitogenic signaling complex in response to G-protein-coupled receptor activation. Tyrosine phosphorylation of the ubiquitous phosphotyrosine phosphatase PTP1D is increased in CCL39 cells following α-thrombin stimulation (63,64). Expression of catalytically inactive PTP1D inhibited α-thrombin-induced gene transcription in these cells. Tyrosine-phosphorylated PTP1D can act as a "docking protein" for Grb2, and might therefore function in a scaffolding role analogous to that of shc. Further, the phosphatase might directly activate src family kinases by catalyzing dephosphorylation of their carboxyl-terminal regulatory tyrosine.

PI-3-kinases also play an important, although poorly understood, role in intracellular signaling. Wortmannin, a potent inhibitor of PI-3-kinases, blocks fMLP-stimulated, pertussis toxin-sensitive phosphatidylinositol-3,4,5-trisphosphate accumulation in neutrophils (65). The drug also blocks direct Gβγ-subunit-stimulated shc phosphorylation in COS cells (66), suggesting that PI-3-kinases play an important upstream role in MAP kinase activation. Gβγ subunit-responsive PI-3-kinases (67, 68), such as the recently cloned p110γ PI-3-kinase (69), may be among the initial effector molecules involved in Gβγ subunit-mediated MAP kinase activation. In addition, heterodimeric complexes of PI-3-kinase p110 catalytic and p85 adaptor subunits associate either with tyrosine-phosphorylated shc (70) or with Grb2 (71), allowing for the potential recruitment of PI-3-kinase activity to shc complexes as a consequence of Gβγ subunit-mediated shc phosphorylation.

THE ROLE OF PLECKSTRIN HOMOLOGY DOMAINS IN G-PROTEIN-COUPLED RECEPTOR-MEDIATED MAP KINASE ACTIVATION

Pleckstrin-homology (PH) domains, so named because of their original description as internal repeats in pleckstrin, the major PKC substrate in platelets, are approximately 110-amino-acid regions of protein-sequence homology that are found in a variety of proteins involved in signal transduction and growth control, among them several that participate in the function of Ras, including Ras-GRF, Sos1, and Ras-GAP; the growth-factor-binding protein Grb7; the insulin-receptor substrate IRS-1; serine/threonine kinases including Rac-α, Rac-β, and the β-adrenergic receptor kinases βARK1 and βARK2; tyrosine kinases including Btk, Tec, and TSK; and PLC species including PLCγ and PLCδ (72). The three-dimensional structures of the pleckstrin, spectrin, and dynamin PH domains are essentially superimposable, indicating that despite low primary-sequence homology, PH domains represent modular protein domains of defined tertiary structure. This structure, comprising seven antiparallel β-sheets forming a β-barrel and a carboxyl-terminal amphiphilic α-helix, bears considerable similarity to proteins that bind small hydrophobic molecules, such as FK506 binding protein and retinol binding protein (RBP).

The function of PH domains is unknown. The PH domains of several proteins bind preferentially to lipid vesicles containing phosphatidylinositol-4-phosphate, phosphatidylinositol-4,5-bisphosphate or phosphatidylinositol-1,4,5-trisphosphate (73), suggesting that they represent binding sites for inositol phospholipids (72). In addition, some PH-domain-containing fusion proteins apparently undergo specific protein–protein interactions. GST-fusion proteins containing the PH domains of nine proteins, βARK1, Ras-GRF, Ras-GAP, PLCγ, IRS-1, Racβ, Atk, spectrin, and OSBP reversibly bind Gβγ subunits *in vitro* with varying affinity (74), as do the PH domains of Btk (75) and Dbl, but not Sos1 (76). Deletion mapping of the Gβγ subunit-binding regions of βARK1, Btk, and Ras-GRF (74,75) indicates that the sequences required for Gβγ subunit-binding span only the carboxyl-terminal α-helix of the PH domain and sequences just distal to it. Detailed mutational analysis of the βARK1 PH domain reveals that the Gβγ-subunit- and phosphatidylinositol-4,5-bisphosphate-binding properties reside in adjacent, but distinct, parts of the molecule (77).

A physiologic role for PH domains in mitogenic signaling is suggested by reports that deletion of the PH domain from either IRS-1 (78) or Sos1 (79) impairs, but does not abolish, their function. The PH domain in the C-terminus of βARK1 and βARK2 corresponds approximately to that region of the protein responsible for agonist-induced, Gβγ subunit-dependent translocation of the kinase from the cytosol to the plasma membrane, where it initiates the process of homologous desensitization (48,49). Optimal βARK1 translocation *in vitro* requires both Gβγ subunits and inositol phospholipids (80), indicating that this region of the molecule coordinates both protein–protein and protein–phospholipid interactions. The similarities between the βARK1 and Btk PH domains suggest that the PH-domain-containing tyrosine kinases might be similarly regulated. Indeed, cellular expression of PH-domain-containing peptides derived from βARK1, IRS-1, Ras-GAP, Ras-GRF (81), and Btk (75) antagonizes Gi-coupled receptor- and Gβγ-mediated Ras-GTP exchange and MAP kinase activation, suggesting that PH-domain-containing peptides derived from several proteins can behave as Gβγ subunit-binding domains when expressed in intact cells.

The finding that disruption of endogenous shc/Grb2/Sos1 blocks Gβγ subunit-mediated MAP kinase activation (54) indicates that Gβγ subunits cannot mediate MAP kinase activation via a direct interaction with the Sos1 PH domain, underscoring the point that Gβγ subunit-binding is not a universal property of PH-domain-containing proteins. In the case of βARK1 and probably other proteins as well, juxtaposition of partially overlapping lipid- and protein-recognition domains serves to recruit the protein and orient it at the proper place on the membrane. Although little is known about the lipid-binding specificity of PH domains (72), recent reports that pleckstrin phosphorylation in human platelets is inhibited by the PI-3-kinase inhibitor wortmannin (82) and stimulated by phosphatidylinositol-3,4,5-trisphosphate (83) suggest that one role of PI-3-kinase may be to modulate the function of PH domain-containing proteins.

SUMMARY

G-protein-coupled receptors that mediate cellular responses to a variety of humoral, endothelial-, or platelet-derived substances are able to stimulate MAP kinase activity. In transfected model systems, G-protein-coupled receptors that couple to pertussis toxin-insensitive G proteins of the Gq/11 family mediate this activation predominantly via a PKC-dependent mechanism. In contrast, activation of MAP kinase by receptors that couple to pertussis toxin-sensitive Gi proteins is PKC-independent and requires downstream activation of the low-molecular-weight G protein, Ras. This pathway can be inhibited by coexpression of peptides that sequester Gβγ subunits, and is mimicked by overexpression of Gβγ subunits. This Ras-dependent MAP kinase activation requires tyrosine phosphorylation of "docking proteins," including the shc adapter protein, and depends upon recruitment of Grb2/Sos1 complexes to the plasma membrane, thus resembling the pathway of MAP kinase activation employed by the receptor tyrosine kinases. Other molecules, including PI-3-kinases and phosphotyrosine phosphatases, probably also contribute to Gβγ-subunit-mediated assembly of a mitogenic signaling complex. Identification of the G-protein-coupled, receptor-regulated tyrosine kinase(s), and the means by which the mitogenic signaling complex is assembled at the plasma membrane, remain subjects of further study.

ACKNOWLEDGMENTS

This work was supported in part by National Institutes of Health Grant HL16037 to R.J.L. L.M.L. is the recipient of a National Institutes of Health Clinical Investigator Development Award. T.vB. is supported by a postdoctoral fellowship from the Alberta Heritage Foundation for Medical Research.

REFERENCES

1. Egan S, Weinberg RA. The pathway to signal achievement. *Nature* 1993; 365: 781–3.
2. Medema RH, Bos JL. The role of p21^{ras} in receptor tyrosine kinase signaling. *Crit Rev Oncogen* 1993; 4: 615–61.
3. Pawson T, Gish GD. SH2 and SH3 domains: from structure to function. *Cell* 1992; 71: 359–62.
4. Kuriyan J, Cowburn D. Structures of SH2 and SH3 domains. *Curr Opin Struct Biol* 1993; 3: 828–37.
5. Lowenstein EJ, Daly RJ, Batzer AG, et.al. The SH2 and SH3 domain-containing protein Grb2 links receptor tyrosine kinases to Ras signaling. *Cell* 1992; 70: 431–42.
6. Rozakis-Adcock M, Fernley R, Wade J, Pawson T, Bowtell D. The SH2 and SH3 domains of mammalian Grb2 couple the EGF receptor to the Ras activator mSos1. *Nature* 1993;363:83–5.
7. Kyriakis JM, App H, Zhang X, Bannerjee P, Brautigan DL, Rapp UR, Avruch J. Raf-1 activates MAP kinase kinase. *Nature* 1992; 358: 417–21.
8. Haung W, Alessandrini A, Crews CM, Erikson RL. Raf-1 forms a stable complex with Mek1 and activates Mek1 by serine phosphorylation. *Proc Natl Acad Sci USA* 1993; 90: 10947–51.
9. Moodie SA, Willumsen BM, Weber MJ, Wolfman A. Complexes of Ras-GTP with Raf1 and mitogen-activated protein kinase kinase. *Science* 1993; 260: 1658–61.
10. Ahn NG, Seger R, Krebs EG. The mitogen-activated protein kinase activator. *Curr Opin Cell Biol* 1992; 4: 992–9.
11. Blenis J. Signal transduction via the MAP kinases: proceed at your own RSK. *Proc Natl Acad Sci USA* 1993; 90: 5889–92.
12. Alblas J, van Corven EJ, Hordijk PL, Milligan G, Moolenaar WH. Gi-mediated activation of the p21^{ras}-mitogen-activated protein kinase pathway by α_2-adrenergic receptors expressed in fibroblasts. *J Biol Chem* 1993; 268: 22235–8.

13. Koch WJ, Hawes BE, Allen LF, Lefkowitz RJ. Direct evidence that Gi-coupled receptor stimulation of mitogen-activated protein kinase is mediated by Gβγ activation of p21^{ras}. *Proc Natl Acad Sci USA* 1994; 91: 12706–10.
14. Flordellis CS, Berguerand M, Gouache P, et.al. α_2 Adrenergic receptor subtypes expressed in Chinese hamster ovary cells activate differentially mitogen-activated protein kinase by a p21^{ras} independent pathway. *J Biol Chem* 1995; 270: 3491–4.
15. Hawes BE, van Biesen T, Koch WJ, Luttrell LM, Lefkowitz RJ. Distinct pathways of Gi- and Gq-mediated mitogen-activated protein kinase activation. *J Biol Chem* 1995; 270: 17148–53.
16. Faure M, Voyno-Yasenetskaya TA, Bourne HR. cAMP and βγ subunits of heterotrimeric G proteins stimulate the mitogen-activated protein kinase pathway in COS-7 cells. *J Biol Chem* 1994; 269: 7851–4.
17. Crespo P, Xu N, Daniotti JL, Troppmair J, Rapp UR, Gutkind JS. Signaling through transforming G protein-coupled receptors in NIH 3T3 cells involves c-Raf activation. Evidence for a protein kinase C-independent pathway. *J Biol Chem* 1994; 269: 21103–9.
18. Winitz S, Russell M, Quian N-X, Gardner A, Dwyer L, Johnson GL. Involvement of Ras and Raf in the Gi-coupled acetylcholine muscarinic M2 receptor activation of mitogen-activated protein (MAP) kinase kinase and MAP kinase. *J Biol Chem* 1993; 268: 19196–9.
19. Ohmichi M, Sawada T, Kanda Y, et al. Thyrotropin-releasing hormone stimulates MAP kinase activity in GH3 cells by divergent pathways. *J Biol Chem* 1994; 269: 3783–8.
20. Linseman DA, Benjamin CW, Jones DA. Convergence of angiotensin II and platelet-derived growth factor receptor signaling cascades in vascular smooth muscle cells. *J Biol Chem* 1995; 270: 12563–8.
21. Cazaubon SM, Ramos-Morales F, Fischer S, Schweighoffer F, Strosberg AD, Couraud P-O. Endothelin induces tyrosine phosphorylation and GRB2 association of Shc in astyrocytes. *J Biol Chem* 1994; 269: 24805–9.
22. Moolenaar WH. Lysophosphatidic acid, a multifunctional phospholipid messenger. *J Biol Chem* 1995; 270: 12949–52.
23. van Corven EJ, Hordijk PL, Medema RH, Bos JL, Moolenaar WH. Pertussis toxin-sensitive activation of p21^{ras} by G protein-coupled receptor agonists in fibroblasts. *Proc Natl Acad Sci USA* 1993; 90: 1257–61.
24. Howe LR, Marshall CJ. Lysophosphatidic acid stimulates mitogen-activated protein kinase activation via a G-protein-coupled pathway requiring p21^{ras} and p74^{raf-1}. *J Biol Chem* 1993; 268: 20717–20.
25. Hordijk PL, Verlaan I, van Corven EJ, Moolenaar WH. Protein tyrosine phosphorylation induced by lysophosphatidic acid in Rat-1 fibroblasts. *J Biol Chem* 1994; 269: 645–51.
26. Buhl AM, Osawa S, Johnson GL. Mitogen-activated protein kinase activation requires two signal inputs from the human anaphylatoxin C5a receptor. *J Biol Chem* 1995; 270: 19828–32.
27. Vouret-Craviari V, van Obberghen-Schilling E, Scimeca JC, van Obberghen E, Pouyssegur J. Differential activation of p44mapk (ERK 1) by α-thrombin and thrombin-receptor peptide agonist. *Biochem J* 1993; 289: 209–14.
28. Gardner AM, Vaillancourt RR, Johnson GL. Activation of mitogen-activated protein kinase/extracellular signal-regulated kinase kinase by G protein and tyrosine kinase oncoproteins. *J Biol Chem* 1993; 268: 17896–901.
29. LaMorte VJ, Kennedy ED, Collins LR, et al. A requirement for Ras protein function in thrombin-stimulated mitogenesis in astrocytoma cells. *J Biol Chem* 1993; 268: 19411–5.
30. Winitz S, Gupta SK, Qian N-X, Heasley LE, Nemenoff RA, Johnson GL. Expression of a mutant Gi2 α subunit inhibits ATP and thrombin stimulation of cytoplasmic phospholipase A_2-mediated arachidonic acid release independent of Ca^{2+} and mitogen-activated protein kinase regulation. *J Biol Chem* 1994; 269: 1889–95.
31. Grinstein S, Furuya W. Chemoattractant-induced tyrosine phosphorylation and activation of microtubule-associated protein kinase in human neutrophils. *J Biol Chem* 1992; 267: 18122–5.
32. Worthen GS, Avdi N, Buhl AM, Suzuki N, Johnson GL. FMLP activates Ras and Raf in human neutrophils. Potential role in activation of MAP kinase. *J Clin Invest* 1994; 94: 815–23.
33. Faure M, Bourne HR. Differential effects of cAMP on the MAP kinase cascade: evidence for a cAMP-insensitive step that can bypass Raf-1. *Mol Biol Cell* 1995; 6: 1025–35.
34. Erhardt P, Troppmair J, Rapp UR, Cooper GM. Differential regulation of Raf-1 and B-Raf and Ras-dependent activation of mitogen-activated protein kinase by cyclic AMP in PC-12 cells. *Mol Cell Biol* 1995; 15: 5524–30.
35. Cook SJ, McCormick F. Inhibition by cAMP of Ras-dependent activation of Ras. *Science* 1993; 262: 1069–72.

36. Hordijk PL, Verlaan I, Jalink K, van Corven EJ, Moolenaar WH. cAMP abrogates the p21^{ras}-mitogen-activated kinase pathway in fibroblasts. *J Biol Chem* 1993; 269: 3534–8.
37. Kolch W, Heidecker G, Kochs G, et al. Protein kinase C activates Raf-1 by direct phosphorylation. *Nature* 1993; 364: 249–52.
38. Feig LA, Cooper GM. Inhibition of NIH3T3 cell proliferation by a mutant Ras protein with preferential affinity for GDP. *Mol Cell Biol* 1988; 8: 3235–43.
39. van Corven EJ, Groenink A, Jalink K, Eicholtz T, Moolenaar WH. Lysophosphatidate-induced cell proliferation: identification and dissection of signaling pathways mediated by G proteins. *Cell* 1989; 59: 45–54.
40. van der Bend RL, Brunner J, Jalink K, van Corven EJ, Moolenaar WH, van Blitterswijk WJ. Identification of a putative membrane receptor for the bioactive phospholipid, lysophosphatidic acid. *EMBO J.* 1992; 11: 2495–501.
41. Chen LB, Buchannan JM. Mitogenic activity of blood components. I. Thrombin and prothrombin. *Proc Natl Acad Sci USA* 1975; 72: 131–5.
42. Vu TK, Hung D, Wheaton VI, Coughlin SR. Molecular cloning of a functional thrombin receptor reveals a novel proteolytic mechanism of receptor activation. *Cell* 1991; 64: 1057–62.
43. Seuwen K, Magnaldo I, Kobilka BK, et al. α2-adrenergic agonists stimulate DNA synthesis in Chinese hamster lung fibroblasts transfected with a human α2-adrenergic receptor gene. *Cell Regul* 1990; 1: 445–51.
44. Clapham DE, Neer EJ. New roles for G protein βγ-dimers in transmembrane signalling. *Nature* 1993; 365: 159–61.
45. Gupta SK, Gallego C, Johnson GL, Heasley LE. MAP kinase is constitutively activated in Gip2 and src transformed Rat 1a fibroblasts. *J Biol Chem* 1992; 267: 7987–90.
46. Ito A, Satoh T, Kaziro Y, Itoh H. G protein βγ subunit activates Ras, Raf, and MAP kinase in HEK 293 cells. *FEBS Lett* 1995; 368: 183–87.
47. Pitcher JA, Inglese JI, Higgins JB, et al. Role of βγ subunits of G proteins in targeting the β-adrenergic receptor kinase to membrane-bound receptors. *Science* 1992; 257: 1264–7.
48. Koch WJ, Inglese JI, Stone WC, Lefkowitz RJ. The binding site for the βγ subunits of heterotrimeric G proteins of the β-adrenergic receptor kinase. *J Biol Chem* 1993; 268: 8256–60.
49. Koch WJ, Hawes BE, Inglese JI, Luttrell LM, Lefkowitz RJ. Cellular expression of the carboxyl terminus of a G protein-coupled receptor kinase attenuates Gβγ-mediated signaling. *J Biol Chem* 1994; 269:6193–7.
50. Inglese JI, Luttrell LM, Iniguez-Lluhi J, Touhara K, Lefkowitz RJ. A functionally active subdomain of the β-adrenergic receptor kinase: an inhibitor of Gβγ-mediated stimulation of type II adenylyl cyclase. *Proc Natl Acad Sci USA* 1994; 91: 3637–41.
51. Simonson MS, Herman WH. Protein kinase C and protein tyrosine kinase activity contribute to mitogenic signaling by endothelin I. Cross-talk between G protein-coupled receptors and pp60^{c-src}. *J Biol Chem* 1993; 268: 9347–57.
52. Page L, Decker SJ, Saltiel AR. Bombesin and epidermal growth factor stimulate the mitogen activated protein kinase pathway through different pathways in Swiss 3T3 cells. *Biochem J* 1993; 289: 283–7.
53. Molloy CJ, Taylor DS, Weber H. Angiotensin II stimulation of rapid protein tyrosine phosphorylation and protein kinase activation in rat aortic smooth muscle cells. *J Biol Chem* 1993; 268: 7338–45.
54. van Biesen T, Hawes BE, Luttrell DK, et al. Receptor-tyrosine-kinase- and Gβγ-mediated MAP kinase activation by a common signaling pathway. *Nature* 1995; 376: 781–4.
55. Sakuae M, Bowtell D, Kasuga M. A dominant negative mutant of mSOS1 inhibits insulin-induced Ras activation and reveals Ras-dependent and -independent insulin signaling pathways. *Mol Cell Biol* 1995; 15: 379–88.
56. Satoh T, Nakafuku M, Kaziro Y. Function of Ras as a molecular switch in signal transduction. *J Biol Chem* 1992; 267: 24149–52.
57. Chen Y-H, Pouyssegur J, Courtneidge SA, Van Obberghen-Schilling E. Activation of Src family kinase activity by the G-protein-coupled thrombin receptor in growth-responsive fibroblasts. *J Biol Chem* 1994; 269: 27372–7.
58. Ptasznik A, Traynor-Kaplan A, Bokoch GM. G-protein-coupled chemoattractant receptors regulate Lyn tyrosine kinase-Shc adapter protein signaling complexes. *J Biol Chem* 1995; 270: 19969–73.
59. Lev S, Moreno H, Martinez R, et al. Protein tyrosine kinase PYK2 involved in Ca^{2+}-induced regulation of ion channel and MAP kinase functions. *Nature* 1995; 376: 737–45.

60. Langhans-Rajasekaran SA, Wan Y, Huang X-Y. Activation of Tsk and Btk tyrosine kinases by G protein βγ subunits. *Proc Natl Acad Sci USA* 1995; 92: 8601–5.
61. Cheng G, Ye Z-S, Baltimore D. Binding of Bruton's tyrosine kinase to Fyn, Lyn, or Hck through a Src homology 3 domain-mediated interaction. *Proc Natl Acad Sci USA* 1994; 91: 8152–5.
62. Mahajan S, Fargnoli J, Burkhardt AL, Kut SA, Saouaf SJ, Bolen JB. Src family protein tyrosine kinases induce autoactivation of Bruton's tyrosine kinase. *Mol Cell Biol* 1995; 15: 5304–11.
63. Rivard N, McKenzie FR, Brondello J-M, Pouyssegur J. The phosphotyrosine phosphatase PTP1D but not PTP1C is an essential mediator of fibroblast proliferation induced by tyrosine kinase and G-protein-coupled receptors. *J Biol Chem* 1995; 270: 11017–24.
64. Li RY, Gaits F, Ragab A, Ragab-Thomas JMF, Chap H. Tyrosine phosphorylation of an SH2-containing protein tyrosine phosphatase is coupled to platelet thrombin receptor via a pertussis toxin-sensitive heterotrimeric G-protein. *EMBO J* 1995; 14: 2519–26.
65. Ui M, OkadaT, Hazeki K, Hazeki O. Wortmannin as a unique probe for an intracellular signalling protein, phosphoinositide 3-kinase. *TIBS* 1995; 20: 303–7.
66. Touhara K, Hawes BE, van Biesen T, Lefkowitz RJ. G protein βγ subunits stimulate phosphorylation of Shc adapter protein. *Proc Natl Acad Sci USA* 1995; 92: 9284–7.
67. Stephens L, Smrcka A, Cooke FT, Jackson TR, Sternweis PC, Hawkins PT. A novel phosphoinositide-3-kinase activity in myeloid-derived cells is activated by G protein βγ subunits. *Cell* 1994; 77: 83–93.
68. Thomason PA, James SR, Casey PJ, Downes CP. A G-protein βγ-subunit-responsive phosphoinositide 3-kinase activity in human platelet cytosol. *J Biol Chem* 1994; 269: 16525–8.
69. Stoyanov B, Volinia S, Hanck T, et al. Cloning and characterization of a G protein-activated human phosphoinositide-3 kinase. *Science* 1995; 269: 690–3.
70. Harrison-Findik D, Susa M, Varticovski L. Association of phosphatidylinositol 3-kinase with SHC in chronic myelogenous leukemia cells. *Oncogene* 1995; 10: 1385–91.
71. Wang J, Auger KR, Jarvis L, Shi Y, Roberts TM. Direct association of Grb2 with the p85 subunit of phosphatidylinositol 3-kinase. *J Biol Chem* 1995; 270: 12774–80.
72. Ferguson KF, Lemmon MA, Sigler PB, Schlessinger J. Scratching the surface with the PH domain. *Nature Struct Biol* 1995; 2: 715–8.
73. Harlan JE, Hajduk PJ, Yoon HS, Fesik SW. Pleckstrin homology domains bind to phosphatidylinositol-4,5-bisphosphate. *Nature* 1994; 371: 168–70.
74. Touhara K, Inglese JI, Pitcher JA, Shaw G, Lefkowitz RJ. Binding of G protein βγ-subunits to pleckstrin homology domains. *J Biol Chem* 1994; 269: 10217–20.
75. Tsukada S, Simon MI, Witte ON, Katz A. Binding of βγ subunits of heterotrimeric G proteins to the pleckstrin homology domain of Bruton tyrosine kinase. *Proc Natl Acad Sci USA* 1994; 91: 11256–60.
76. Mahadevan D, Thanki N, Singh J, et al. Structural studies on the PH domains of Dbl, Sos1, IRS-1, and βARK1 and their differential binding to Gβγ subunits. *Biochemistry* 1995; 34: 9111–7.
77. Touhara K, Koch WJ, Hawes BE, Lefkowitz RJ. Mutational analysis of the pleckstrin homology domain of the β-adrenergic receptor kinase. *J Biol Chem* 1995; 270: 17000–5.
78. Myers MG Jr, Grammer TC, Brooks J, et al. The pleckstrin homology domain in insulin receptor substrate-1 sensitizes insulin signaling. *J Biol Chem* 1995; 270: 11715–8.
79. McCollam L, Bonfini L, Karlovich CA, et al. Functional roles for the pleckstrin and Dbl homology regions in the Ras exchange factor son-of sevenless. *J Biol Chem* 1995; 270: 15954–7.
80. Pitcher JA, Touhara K, Payne ES, Lefkowitz RJ. Pleckstrin homology domain-mediated membrane association and activation of the β-adrenergic receptor kinase requires coordinate interaction with Gβγ subunits and lipid. *J Biol Chem* 1995; 270: 11707–10.
81. Luttrell LM, Hawes BE, Touhara K, van Biesen T, Koch WJ, Lefkowitz RJ. Effect of cellular expression of pleckstrin homology domains on Gi-coupled receptor signaling. *J Biol Chem* 1995; 270: 12984–9.
82. Yatomi Y, Hazeki O, Kume S, Ui M. Suppression by wortmannin of platelet responses to stimuli due to inhibition of pleckstrin phosphorylation. *Biochem J* 1992; 285: 745–51.
83. Zhang J, Falck JR, Reddy KK, Abrams CS, Zhao W, Rittenhouse SE. Phosphatidylinositol-(3,4,5)-trisphosphate stimulates phosphorylation of pleckstrin in human platelets. *J Biol Chem*, 1995; 270: 22804–10.

Signal Transduction in Health and Disease, Advances in Second Messenger and Phosphoprotein Research, Vol. 31, edited by J. Corbin and S. Francis.
Lippincott–Raven Publishers, Philadelphia © 1997.

22

Rab3A–Rabphilin-3A System in Neurotransmitter Release

Takuya Sasaki, *Hiromichi Shirataki, †Hiroyuki Nakanishi, and Yoshimi Takai

*Department of Molecular Biology and Biochemistry, Osaka University Medical School, Suita 565 Osaka Japan; *Department of Cell Physiology, National Institute for Physiological Sciences, Okazaki 444, Japan; and †Takai Biotimer Project, ERATO, Kobe 651-22, Japan*

INTRODUCTION

There is a small G-protein superfamily occurring in species from yeast to mammals and consisting of more than 50 members (Fig. 1). The small G proteins indicated in boxes in the figure were discovered in our laboratory. According to their structures, these small G proteins fall into at least four groups, Ras, Rho, Rab, and other groups. The Ras subfamily regulates gene expression through an MAP kinase cascade and possibly through other pathways. The Rho group mainly regulates the cytoskeleton through reorganization of actin filaments, as well as regulating gene expression through another MAP kinase cascade involving a Jun N-terminal kinase (JNK) kinase. The Rab group regulates intracellular vesicle transport, including exocytosis, endocytosis, and transcytosis. The Arf family and Sar1, belonging to other groups, also regulate intracellular vesicle transport. The Ran family regulates protein transport into the cell nucleus. Thus, it is well established that small G proteins have various important cell functions. In this chapter we focus on Rab3A, which was found in our laboratory, and describe its function and mode of action.

PURIFICATION, MOLECULAR CLONING, AND CHARACTERIZATION OF Rab3A

When we attempted to purify Ras from a bovine brain-membrane fraction, we separated at least 15 G proteins, which had molecular weight values similar to those of Ras (~20,000 *Mr*) in several column chromatographic analyses (1–5). We named

	Ras	Rho	Rab			Others
Mammal	Ha-Ras	RhoA	Rab1A	Rab9	Rab23	Arf1
	Ki-Ras	RhoB	Rab1B	Rab10	Rab24	Arf2
	N-Ras	RhoC	Rab2	Rab11	Rab25	Arf3
	R-Ras	RhoG	Rab3A	Rab12	Rab26	Arf4
	Ral	Rac1	Rab3B	Rab13	Ram	Arf5
	Rap1A	Rac2	Rab3C	Rab14		Arf6
	Rap1B	Cdc42	Rab3D	Rab15		Sar1a
	Rap2	Tc10	Rab4	Rab16		Sar1b
	Tc21		Rab5A	Rab17		Ran
			Rab5B	Rab18		Rad
			Rab5C	Rab19		
			Rab6	Rab20		
			Rab7	Rab21		
			Rab8	Rab22		
Yeast	Ras1	Rho1	Ypt1	Ypt6		Arf1
	Ras2	Rho2	Ypt2	Ypt7		Arf2
	Rsr1	Rho3	Ypt3	Ypt8		Arf3
		Rho4	Ypt4	Sec4		Sar1
		Cdc42	Ypt5			

FIG. 1. Small-G-protein superfamily.

these G proteins small-*M*r G proteins (we later renamed them small G proteins) and purified to homogeneity one of them with an *M*r of about 24,000, which was estimated by sodium dodecylsulfate–polyacrylamide gel electrophoresis (SDS–PAGE) (1). We first named this protein Smg 25 (*S*mall *M*r *G* protein with an Mr of *25* Kda). The purified sample of Smg 25 showed both guanosine diphosphate (GDP) and guanosine triphosphate (GTP)-binding and guanosine triphosphatase (GTPase) activities, as described for the G proteins that had been reported at that time, including heterotrimeric G proteins (G_s, G_i, G_o, and G_t) and Ras. In collaboration with Teranishi's group (Mitsubishi Kasei Research Institute, Yokohama, Japan), we determined the partial amino-acid sequence of Smg 25, and derived a complementary deoxyribonucleic acid (cDNA) from a bovine-brain cDNA library on the basis of this information (6). We isolated three highly homologous cDNAs and determined their primary structures. We named them Smg 25A, -25B, and -25C (6). They encoded proteins with calculated molecular weight values of 24,954 *M*r, 24,766 *M*r, and 25,975 *M*r, and with amino-acid contents of 220, 219, and 227, respectively. The Smg 25A cDNA encoded Smg 25 purified from bovine brain. At about the same time, Tavitian's group (7) isolated new small-G proteins from a rat-brain cDNA library containing several oligonucleotide mixtures corresponding to the conservative region of Ras. One of these proteins, named Rab3, was identical with our Smg 25A (8). The nomenclatures for the proteins were unified, and Smg 25A, -25B, and -25C were renamed Rab3A, -3B, and -3C, respectively. Later, Lodish et al. (9) cloned a Rab3-like molecule from

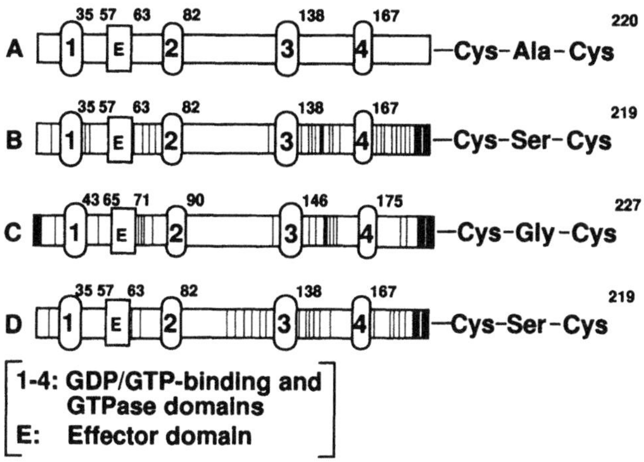

FIG. 2. The structure of Rab3 family members.

the differentiated 3T3-l cell-specific subtractive library and named it Rab3D. As shown in Fig. 2, all of these four Rab3s have consensus amino-acid sequences for GDP- and GTP-binding and GTPase activities, as described for other G proteins. Moreover, they have a unique carboxy-terminal (C-terminal) structure: Cys-X-Cys (where X is any amino acid). By 1989, Ras had been shown to undergo posttranslational modifications with lipids (10). Glomset and Gelb's group at the University of Washington in Seattle, in collaboration with our group (11), identified the posttranslationally modified structure of Rab3A; the two cysteine residues were geranylgeranylated, and the C-terminal cysteine residue was further carboxymethylated.

TISSUE AND INTRACEREBRAL DISTRIBUTION OF Rab3A

Northern and Western blot analyses revealed that Rab3A was detected only in cells with regulated secretion, including neuronal cells, exocrine cells, and endocrine cells, and not in cells with constitutive secretion, including hepatocytes and lymphocytes (12,13). In brain, Rab3A was concentrated mainly in the synaptic vesicle fraction in the presynaptic nerve terminal (14,15). These results, together with the fact that yeast Sec4 and Ypt1 are involved in vesicle transport (16,17), suggested that Rab3A is involved in Ca^{2+}-dependent exocytosis.

ISOLATION, MOLECULAR CLONING, AND CHARACTERIZATION OF A Rab REGULATOR, Rab GDI

During the course of studying Rab3A, a model for the mode of action of the Rab-family members in vesicle transport was proposed by Bourne (18). This model was

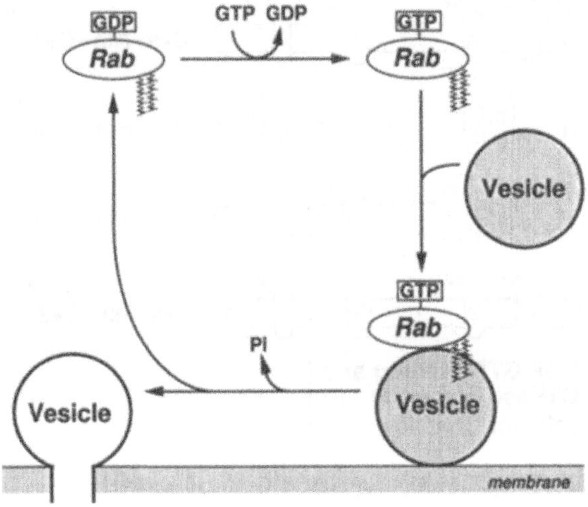

FIG. 3. A model for the mode of action of the Rab family members in vesicle transport.

based on the mode of action of elongation factor Tu in protein synthesis. As schematically shown in Fig. 3, the GDP-bound form of Rab is located in the cytosol. When it is converted to the GTP-bound form, the GTP-bound form binds to a vesicle, which is then transported to an acceptor membrane. According to this model, GTP hydrolysis occurs before the fusion of the vesicle with the membrane. After the hydrolysis, the GDP-bound form of Rab is translocated to the cytosol. In this model, two types of cycling of Rab between the GDP-bound and GTP-bound forms and the soluble and membrane fractions are essential for the action of Rab. However, these cycling mechanisms had not been demonstrated.

Before this model was proposed, we had begun to isolate a regulatory protein of Rab3A that stimulates conversion of the GDP-bound form to the GTP-bound form. It was named GDP/GTP exchange protein (GEP), because at that time the Cdc25 gene had been suggested genetically to encode a GEP for Ras in the yeast *Saccharomyces cerevisiae* (19), but no GEPs had been biochemically identified. We succeeded in isolating a GEP for Rab3A (20). This GEP did not stimulate the GDP/GTP exchange reaction, but rather inhibited it. Because the rate-limiting step of this reaction is the dissociation of GDP, we originally named this protein Smg 25A GDI (*G*DP *d*issociation *i*nhibitor), and it was later renamed Rab GDI because the protein turned out to be active not only on Rab3A, but also on all the Rab family members studied (21–23).

The purified sample of Rab GDI showed *M*r values of about 54,000, 82,000, and 65,000 as estimated by SDS–PAGE, gel filtration, and sucrose density-gradient ultracentrifugation, respectively (20). All of these results suggested that Rab GDI is a monomeric protein. Rab GDI has been recovered in the cytosol fraction of all tissues thus far examined. In collaboration with Teranishi's group (24), we then isolated a cDNA of Rab GDI from a bovine-brain cDNA library and determined its primary structure. The protein showed a calculated *Mr* of 50,565, and consisted of 447 amino acids. Northern and Western blot analyses indicated that it was widely distributed in all the tissues examined (24,25). However, there were two transcripts, of 2.3 kb and 3.1 kb, in brain but not in other tissues. It was later clarified that Rab GDI constitutes a family consisting of at least three members, Rab GDIα, -β, and -γ, as described below. The originally isolated member, corresponding to the larger, 3.1-kb transcript, was named Rab GDIα, and it was later clarified that Rab GDIα is specifically expressed in neuronal cells. During the course of these studies, a novel type of regulator for Ras, which stimulated its GTPase activity, and was named Ras GTPase-activating protein (GAP), was first identified in *Xenopus* oocytes by McCormick's group (26). The cDNA for Ras GAP was cloned from a bovine-brain cDNA library by Gibb's group (27).

UNIQUE BIOCHEMICAL PROPERTIES OF Rab GDI AND ITS FUNCTION

The purified sample of Rab GDI showed unique biochemical properties, as shown in Fig. 4. It inhibited the dissociation of GDP only from the lipid-modified form of Rab3A, and was inactive on the lipid-unmodified form (28). Similarly, Rab GDI in-

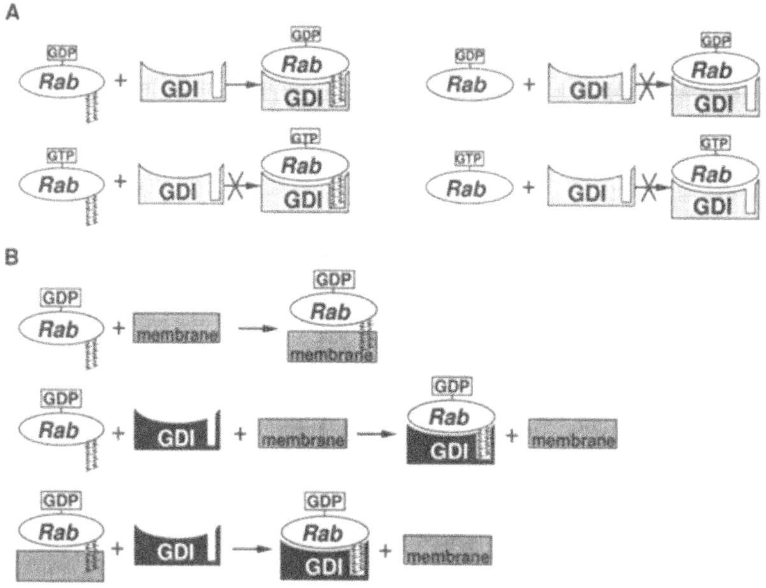

FIG. 4. Properties of Rab GDI.

hibited the binding of GTP to the lipid-modified, GDP-bound form of Rab3A, but not that to the lipid-unmodified, GDP-bound form. Moreover, Rab GDI interacted with the lipid-modified, GDP-bound form of Rab3A to form a stable ternary complex of GDP–Rab3A–Rab GDI. Rab GDI did not interact with the lipid-unmodified, GDP-bound form, the lipid-modified, GTP-bound form, or the lipid-unmodified, GTP-bound form. The lipid-modified, GDP-bound and GTP-bound forms of Rab3A were associated with all types of membranes, such as erythrocyte ghosts, synaptic vesicles, and synaptic plasma membranes, as far as tested in a cell-free system, presumably through the interactions between the geranylgeranyl moieties of Rab3A and membrane phospholipid (29). The lipid-unmodified form was not associated with these membranes. We assume that this binding of Rab3A is a nonspecific artifact, as described below; the binding of the GDP-bound form was inhibited by the addition of Rab GDI, with formation of a ternary complex of GDP–Rab3A–Rab GDI (Fig. 4B), whereas the binding of the GTP-bound form was not inhibited. Moreover, Rab GDI induced the dissociation of the lipid-modified, GDP-bound form of Rab3A, which was prebound to the membranes, by forming the stable ternary complex. Rab GDI did not show such an action for the lipid-modified, GTP-bound form of Rab3A. These biochemical properties indicated that Rab GDI has at least two activities, one of which is to regulate the GDP/GTP exchange reaction of Rab3A, and the other of which is to regulate the reversible binding of Rab3A to membranes. On the other hand, Zerial and colleagues at the European Molecular Biology Laboratory in Heidelberg, Germany, in collaboration with our group (22), found that Rab GDI was active not only on Rab3A but also on all the Rab family members thus far examined. Moreover, Novick's group at Yale University, in collaboration with our group (21,23), found that Rab GDI was cross-reactive with the yeast Rab family members including Sec4 and Ypt1. It was for these reasons that we changed the original name of Smg 25A GDI to Rab GDI. On the basis of these properties, we proposed the mode of action of Rab GDI in the Rab-family-member regulated vesicle transport as shown in Fig. 5.

The GDP-bound form of Rab is complexed with Rab GDI and stays in the cytosol. After it is released from Rab GDI and is converted to the GTP-bound form, or after it is converted to the GTP-bound form and is released from Rab GDI, the GTP-bound form is associated with the vesicle, which is consequently transported to the acceptor membrane. Before fusion of the vesicle with the membrane, the GTP-bound form is converted to the GDP-bound form. After the GDP-bound form is produced from the membrane, it is complexed with Rab GDI and is translocated from the membrane to the cytosol. Wollheim's group at the University of Geneva, Switzerland, in collaboration with our group (30), showed that Rab3A was complexed with Rab GDI in the cytosol of resting, insulin-secreting cells. Afterward, Novick's group (31) found a yeast counterpart of Rab GDI, named GDI1, and clarified genetically that yeast Rab GDI indeed regulates the Sec4 function in intact cells. Moreover, Balch's group (32), using a cell-free system for vesicle transport between endoplasmic reticulum and Golgi complexes, showed that Rab GDI indeed functions in this system. Our model is currently accepted in general.

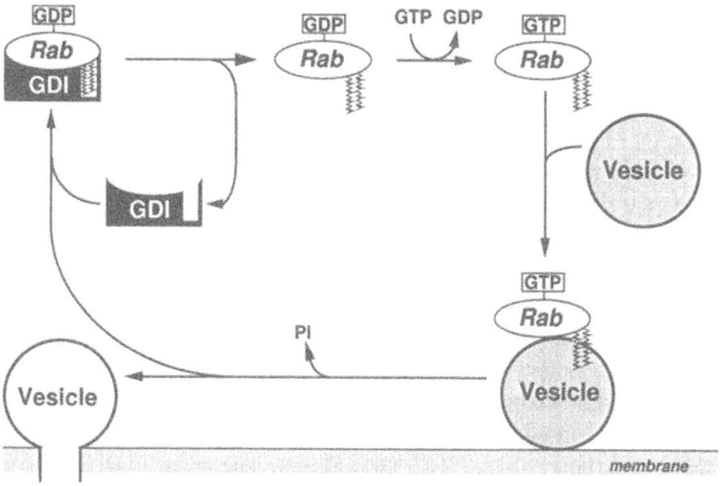

FIG. 5. Our model for the mode of action of Rab GDI in Rab-family-member-regulated vesicle transport.

IDENTIFICATION OF RAB GDI ISOFORMS

As mentioned earlier, Rab GDI transcripts of two different sizes were expressed in brain but not in other tissues (24). This result suggested that there are isoforms of Rab GDI. In accord with this assumption, at least three isoforms of Rab GDI have been thus far isolated. The originally identified isoform, named Rab GDIα, was relatively specifically expressed in brain, whereas the new isoform, named Rab GDIβ, isolated by Sano's group at Kobe University, in collaboration with our group (33), was ubiquitously expressed in all of the tissues examined. The third isoform, named Rab GDIγ, was isolated by Czech's group (34), but its tissue distribution was not precisely examined. Although the tissue distributions of Rab GDIα and -β were different, their biochemical properties, including their substrate specificity and requirement for the lipid modification of their substrates, were similar (33,35). It was also shown that the proportion of Rab complexed with either Rab GDIα or -β depends on the relative abundance of each isoform in the cytosol (36). However, it is unknown why there are many isoforms for Rab GDI.

PURIFICATION AND CHARACTERIZATION OF RAB3A GEP

The conversion from the GDP-bound to the GTP-bound form of Rab3A is regulated by GEP. Thus far, two GEPs for Rab3A have been reported: one is a GEP named MSS4, isolated by De Camelli's group (37) as a mammalian counterpart of yeast DSS4, a GEP for yeast Sec4 and Ypt1 (38); the other is a GEP named Rab3A GRF, which was partially purified by Macara's group (39) from the cytosol fraction of rat brain. MSS4 was a protein with a calculated *M*r of 13,928 and 123 amino acids. De Camelli's group (37) showed that recombinant MSS4 was active not only on Rab3A but also on other Rab family members, including at least Rab1, -8, -10, Sec4, and Ypt1. We found that recombinant MSS4 was equally active on both the lipid-modified and -unmodified forms of Rab3A, and that the action of MSS4 was inhibited by Rab GDI when the lipid-modified form of Rab3A was used as a substrate (40). Rab3A GRF was partially purified from the cytosol fraction of rat brain, and its properties have not been intensively studied (39).

We attempted to isolate a GEP specific for Rab3A, and recently succeeded in purifying it to a high degree from a rat-brain soluble synaptic fraction by column chromatography (41). The protein appeared as two peaks on hydroxyapatite chromatography, and we named them Rab3A GEPI and -II. Rab3A GEPI and -II showed similar physical and kinetic properties. Their *M*r values, as estimated by SDS-PAGE and gel filtration were about 200,000 and 270,000, respectively, suggesting that they are monomeric proteins. Rab3A GEPI and -II were specifically active on Rab3A, -3C, and -3D and inactive on Rab3B and the other Rab members, including at least Rab2, -5, and -11. They were active only on the lipid-modified form and inactive on the lipid-unmodified form of Rab3A. Their action was inhibited by Rab GDI. It remains to be clarified whether our Rab3A GEPs are identical with Macara's Rab3A GRF. We are currently cloning the cDNA of Rab3A GEPII.

PROPOSED PRESENCE OF RAB GDI DISPLACEMENT FACTOR

The GDP-bound form of Rab3A complexed with Rab GDI was not converted to the GTP-bound form by the action of any of the Rab3A GEPs described above, suggesting that another factor is necessary for this conversion. Pfeffer's group (42) found that the GDP-bound form of Rab9 complexed with Rab GDI was converted to the GTP-bound form when the complex was incubated with the endosome membrane, whereas the GDP-bound form of Rab9 free of Rab GDI was not converted to the GTP-bound form under similar conditions. On the basis of this observation, Pfeffer's group proposed that a factor is necessary for the dissociation of the GDP-bound form of Rab9 from Rab GDI, and they named this factor GDI displacement factor (GDF) (42). Zerial's group (43) obtained a similar result with Rab5. Although GDF has not been identified, a similar factor may be present for Rab3A.

ISOLATION, MOLECULAR CLONING, AND CHARACTERIZATION OF RABPHILIN-3A

Once the GTP-bound form of Rab3A is produced, it interacts specifically with synaptic vesicles. It is assumed that this specific localization of the GTP-bound form of Rab3A is due to the presence of a specific protein on synaptic vesicles that interacts with the GTP-bound form of Rab3A. On this assumption, we attempted to isolate this downstream target protein of Rab3A. Using a cross-linking technique, we identified one protein in the rat-brain crude membrane fraction that was specifically cross-linked by the GTP-bound form of radioiodinated Rab3A (44). This cross-linking was competitively inhibited by the GTP-bound form of nonradioiodinated Rab3A but not by the GDP-bound form. By use of this assay method, we purified this protein, determined its partial amino-acid sequence, cloned its cDNA from a rat-brain cDNA library, and determined its primary structure (44,45). We named this protein rabphilin-3A. The purified sample of rabphilin-3A was a protein with *M*r values of about 86,000 and 100,000 as estimated by SDS–PAGE and sucrose density-gradient ultracentrifugation, respectively, suggesting that the protein is a monomeric protein. The calculated *M*r of rabphilin-3A was 77,976, and it consisted of 704 amino acids. Recombinant rabphilin-3A indeed interacted preferentially with the GTP-bound form of Rab3A and poorly with the GDP-bound form. It also interacted with the GTP-bound form of Rab3B, -3C, and -3D, but did not interact with other Rab members, including Rab2, -5, or -11 (*unpublished data*). Recombinant rabphilin-3A showed weak Rab3A GAP activity and strong Rab3A GAP-inhibiting activity (46). The physiologic significance of these two activities is not known, but the Rab3A GAP-inhibiting activity of rabphilin-3A may be involved in keeping Rab3A in the GTP-bound form until Rab3A accomplishes its function.

Structural analysis indicated that rabphilin-3A has no transmembrane segment but has two C2-like domains at its C-terminal region (45). The C2 domain was originally found in protein kinase C (PKC), which is activated by Ca^{2+} and phospholipid, particularly phosphatidylserine (47). Rabphilin-3A interacted with Ca^{2+} and phospholipid, particularly phosphatidylserine, at the C2-like domains, whereas it interacted with Rab3A at its amino-terminal (N-terminal) region (48). Thus, rabphilin-3A has at least two functionally different domains, the N-terminal Rab3A-binding domain and the C-terminal Ca^{2+}- and phospholipid-binding domain (48). Before the discovery of rabphilin-3A, synaptotagmin had been shown to have two C2-like domains and to interact with Ca^{2+} and phospholipid, particularly phosphatidylserine (49). This protein was later shown to comprise a family consisting of at least eight members (50). The originally isolated member, synaptotagmin I, was shown to be specifically localized on synaptic vesicles through its single transmembrane segment, and has been suggested to serve as a Ca^{2+} sensor for neurotransmitter release (51,52). Rabphilin-3A is phosphorylated by protein kinase A (PKA) and calmodulin-dependent protein kinase II (CaMK II) (53,54). The CaMK II-catalyzed phosphorylation sites are Ser^{34}, Thr^{205}, Thr^{209}, and Thr^{537}. The function of the phosphorylation of rabphilin-3A remains to be established.

TISSUE AND INTRACEREBRAL DISTRIBUTION OF RABPHILIN-3A

Northern and Western blot analyses indicated that rabphilin-3A is expressed only in brain and not in other tissues, including exocrine cells or endocrine cells, in which Rab3A is expressed (45,55). In brain, rabphilin-3A was concentrated largely in the synaptic-vesicle fraction in the presynaptic nerve terminal (15). Rabphilin-3A interacted with synaptic vesicles in a dose-dependent and saturable manner, and this interaction was abolished by prior digestion of the vesicles with trypsin, suggesting that rabphilin-3A interacts with the vesicles through its anchoring protein (56). This result is consistent with the structural absence of a transmembrane segment in rabphilin-3A. The interaction of full-length rabphilin-3A with synaptic vesicles was competitively inhibited by the N-terminal fragment (57), suggesting that rabphilin-3A interacts with the vesicle protein at least through its N-terminal region. Therefore, it is likely that the specific localization of the GTP-bound form of Rab3A at synaptic vesicles is determined by the specific localization of rabphilin-3A, which is in turn determined by the specific localization of its anchoring protein. We are currently attempting to isolate this anchoring protein.

INVOLVEMENT OF RABPHILIN-3A IN CA^{2+}-DEPENDENT NEUROTRANSMITTER RELEASE

Because rabphilin-3A interacts with the GTP-bound form of Rab3A, is specifically located on synaptic vesicles, and interacts with Ca^{2+}, it is likely that rabphilin-3A as well as synaptotagmin plays roles in Ca^{2+}-dependent neurotransmitter release. To obtain evidence for this assumption, we took advantage of a human growth hormone (HGH) coexpression assay system (58,59). In this system, HGH was overexpressed in bovine adrenal chromaffin cells or PC12 cells. Exogenous HGH was stored in dense-core vesicles and released in response to an agonist in an extracellular Ca^{2+}-dependent manner. Holz's group at the University of Michigan, in collaboration with our group (58), showed that when the N-terminal fragment (amino acids 1 to 286) of chromaffin cell rabphilin-3A, which is highly homologous with brain rabphilin-3A except that six amino acids are inserted at the N-terminal region, was co-overexpressed with HGH in bovine adrenal chromaffin cells, the Ca^{2+}-dependent, high K^+-induced HGH release was markedly inhibited. In contrast, overexpression of the C-terminal fragment (amino acids 287 to 710), which contains a part of the N-terminal and the two C2-like domains, had no effect on the HGH release. We confirmed this result in PC12 cells, and moreover found that overexpression of the fragment containing only the two C2-like domains (amino acids 403 to 704) inhibited the Ca^{2+}-dependent, high K^+-induced HGH release (59). The exact reason for this difference in effect of these two fragments is not known, but it may be due to their different three-dimensional structures or the inhibitory action of this part of the N-terminal region on the two C2-like domains. Augustine's group at Duke University, again working in collaboration with our group (60), took advantage of the squid giant axon system, in which the presynaptic nerve terminal is large enough to be microinjected with protein samples. The N-terminal fragment of rabphilin-3A (amino acids 1 to 280) or the C-terminal fragment (amino acids 403 to 704) was microinjected into the nerve terminal and the latter was stimulated electrically. The postsynaptic action potential was then recorded. Microinjection of the N-terminal or C-terminal fragment inhibited Ca^{2+}-dependent neurotransmitter release. Electron microscopic analysis of the spatial distribution of synaptic vesicles in active zones showed that microinjection of the N-terminal or C-terminal fragment caused a selective relative accumulation of pre-"docked" vesicles at 50 to 100 nm away from the plasma membrane. Microinjection of peptides of the C2-like domains of synaptotagmin also inhibited Ca^{2+}-dependent neurotransmitter release, and caused a selective relative accumulation of "docked" vesicles within 50 nm from the plasma membrane (61). These ultrastructural data suggested that rabphilin-3A is involved in pre-docking or docking, whereas synaptotagmin is involved in fusion after docking. These two lines of evidence indicate that the Rab3A–rabphilin-3A system indeed functions in Ca^{2+}-dependent exocytosis in intact cells.

POSSIBLE FUNCTION OF RABPHILIN-3A AS A Ca^{2+} SENSOR

Synaptotagmin I is located on synaptic vesicles through its transmembrane segment, and could interact with syntaxin, which is located on the presynaptic plasma membrane (62). The exact mode of action of synaptotagmin in Ca^{2+}-dependent neurotransmitter release has not been established, but one proposed and fascinating function is that synaptotagmin serves as a Ca^{2+} sensor. According to this concept, the docking and fusion of the vesicles with the plasma membranes are mediated by the *N*-ethylmaleimide-sensitive fusion protein (NSF)–soluble NSF attachment protein (SNAP)–SNAP receptor (SNARE) system proposed by Rothman's group (63). Before Ca^{2+} influx, synaptotagmin interacts with syntaxin, which is one of the SNARES, and regulates this NSF–SNAP–SNARE system. When Ca^{2+} flows into the cytoplasm through Ca^{2+} channels, it binds to synaptotagmin, causing the release of this inhibitory action and operation of the NSF–SNAP–SNARE system. Recent biochemical observations have shown that the C2-like domains of synaptotagmin interact with syntaxin in a Ca^{2+}-dependent manner (50). Therefore, it is likely that synaptotagmin interacts with syntaxin after Ca^{2+} influx, and then regulates the NSF–SNAP–SNARE system. In this case, synaptotagmin serves as a positive regulator to stimulate Ca^{2+}-dependent exocytosis. The precise mechanism by which synaptotagmin serves as a Ca^{2+} sensor in neurotransmitter release is unknown.

It is also unknown how rabphilin-3A regulates synaptic-vesicle transport. There are two lines of evidence that the Rab system is an upstream regulator of the NSF–SNAP–SNARE system (64,65). Therefore, it is possible that the Rab3A–rabphilin-3A system also regulates the NSF–SNAP–SNARE system in neurotransmitter release. Moreover, because rabphilin-3A has two C2-like domains, as does synaptotagmin, rabphilin-3A may also serve as a Ca^{2+} sensor. Our working model for the mode of action of the Rab3A–rabphilin-3A system in neurotransmitter release is as follows: there is an acceptor protein directly or indirectly recognized by rabphilin-3A complexed with the GTP-bound form of Rab3A on the presynaptic plasma membrane. When rabphilin-3A is complexed with the GTP-bound form of Rab3A, it interacts with this acceptor protein and thereby causes the docking of the vesicle with the presynaptic plasma membrane. As described above, rabphilin-3A interacts with the vesicle at least through its N-terminal region, and rabphilin-3A interacts with at least one protein of the vesicle. Therefore, if the interaction of the

GTP-bound form of Rab3A with rabphilin-3A causes interaction with the acceptor protein of the presynaptic plasma membrane, there may be at least two interaction sites for rabphilin-3A. This assumption is consistent with the results obtained in the HGH overexpression experiment and the squid-giant-axon experiment described earlier, in which the N-terminal and C-terminal fragments of rabphilin-3A inhibit Ca^{2+}-dependent exocytosis. The Rab3A-dependent interaction of rabphilin-3A with the acceptor protein may be followed by the interaction of synaptotagmin with syntaxin. Binding of influxed Ca^{2+} to rabphilin-3A as well as to synaptotagmin may cause dissociation of these proteins from their respective interacting proteins, or cause tighter interactions of these proteins and thereby lead to the operation of the NSF–SNAP–SNARE system. The functional relationship between the Ca^{2+}-binding proteins and the NSF–SNAP–SNARE system is unknown, but it is possible that the Rab3A–rabphilin-3A system has two functions: transporting the vesicle to the presynaptic plasma membrane, and serving as a Ca^{2+} sensor. We are currently attempting to isolate an acceptor protein on the presynaptic plasma membrane.

PURIFICATION AND CHARACTERIZATION OF Rab3A GAP

It is unknown when the GTP-bound form of Rab3A is converted to the GDP-bound form by the action of Rab3A GAP. The conversion may occur before, during, or after fusion of the vesicle with the plasma membrane. Rab3A GAP was partially purified from the cytosol fraction of rat brain by Macara's group (66), but its properties have not been intensively studied. We have recently succeeded in highly purifying Rab3A GAP from the soluble synaptic fraction of rat brain and characterizing it (67). The purified sample of Rab3A GAP consisted of two subunits with *M*r values of 130,000 and 150,000 as estimated with SDS-PAGE. The protein showed an *M*r of about 300,000 by gel filtration and sucrose density-gradient ultracentrifugation, suggesting that Rab3A GAP is a heterodimeric protein. Rab3A GAP is specific for Rab3A, -3B, -3C, and -3D but not active on Rab2, -5, or -11. Moreover, it was active only on the lipid-modified form of Rab3A, being inactive on the lipid-unmodified form. It was active on the GTP-bound form of Rab3A that was free of rabphilin-3A, and was inactive on the GTP-bound form of Rab3A that was complexed with rabphilin-3A.

IMPORTANCE OF LIPID MODIFICATIONS OF Rab3A FOR PROTEIN–PROTEIN INTERACTIONS

As previously described, the lipid modifications of Rab3A are of crucial importance for its interaction with regulatory proteins, including Rab GDI, Rab3A GEP, and Rab3A GAP. Moreover, they are essential for the action of these interactive products. Rab3A undergoes both geranylgeranylation and carboxylmethylation (11). We found that the geranylgeranylation was essential for the action of Rab GDI, and that the carboxylmethylation was not essential (68). It is unknown which modification is essential for the actions of Rab3A GEP and Rab3A GAP. It is also unknown how the lipid modifications are important for the actions of these regulatory proteins. It is possible that the lipid moieties directly interact with the interacting proteins, or interact with the Rab3A molecule itself, causing a conformational change that enables Rab3A to interact with the interacting proteins. Three-dimensional structure analysis, through crystallization of the lipid-modified form of Rab3A and its interacting proteins, and through their cocrystallization, is important for understanding the function of the lipid moieties in the protein–protein interactions.

CONCLUSION

Evidence is accumulating that Rab3A and its interacting proteins play important roles in vesicle transport, particularly in Ca^{2+}-dependent neurotransmitter release. Our current working model for the mode of action of Rab3A and its interacting proteins in neurotransmitter release is shown schematically in Fig. 6. Of these proteins, it has been established that Rab GDI indeed functions *in vivo*, but the *in vivo* functions of other interacting proteins of Rab3A have not been established. To under-

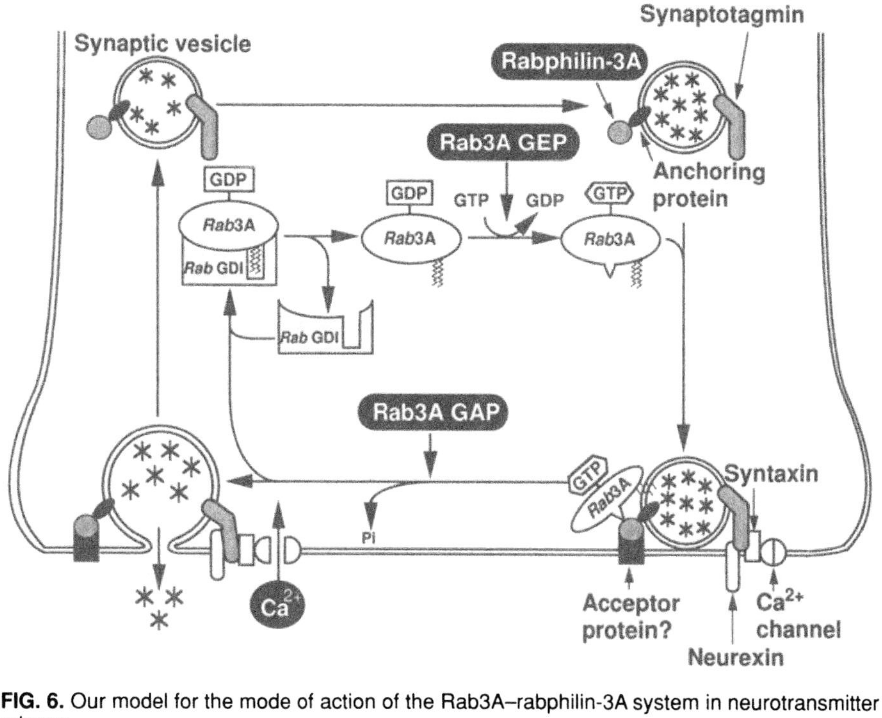

FIG. 6. Our model for the mode of action of the Rab3A–rabphilin-3A system in neurotransmitter release.

stand the mode of action of Rab3A in neurotransmitter release, it will be important to determine how the GDP-bound form is converted to the GTP-bound form and transferred to rabphilin-3A, to isolate at least two rabphilin-3A-interacting proteins, and to elucidate how Rab3A GAP is activated.

REFERENCES

1. Kikuchi A, Yamashita T, Kawata M, et al. Purification and characterization of a novel GTP-binding protein with a molecular weight of 24,000 from bovine brain membranes. *J Biol Chem* 1988;263: 2897–904.
2. Kawata M, Matsui Y, Kondo J, Hishida T, Teranishi Y, Takai Y. A novel small molecular weight GTP-binding protein with the same putative effector domain as the ras proteins in bovine brain membranes. Purification, determination of primary structure, and characterization. *J Biol Chem* 1988;263: 18965–71.
3. Yamamoto K, Kondo J, Hishida T, Teranishi Y, Takai Y. Purification and characterization of a GTP-binding protein with a molecular weight of 20,000 in bovine brain membranes. Identification as the rho gene product. *J Biol Chem* 1988;263:9926–32.
4. Yamashita T, Yamamoto K, Kikuchi A, et al. Purification and characterization of c-Ki-ras p21 from bovine brain crude membranes. *J Biol Chem* 1988;263:17181–8.
5. Hoshijima M, Kondo J, Kikuchi A, Yamamoto K, Takai Y. Purification and characterization from bovine brain membranes of a GTP-binding protein with a *Mr* of 21,000, ADP-ribosylated by an ADP-ribosyltransferase contaminated in botulinum toxin type C1—identification as the rhoA gene product. *Mol Brain Res* 1990;7:9–16.
6. Matsui Y, Kikuchi A, Kondo J, Hishida T, Teranishi Y, Takai Y. Nucleotide and deduced amino acid sequences of a GTP-binding protein family with molecular weights of 25,000 from bovine brain. *J Biol Chem* 1988;263:11071–4.
7. Touchot N, Chardin P, Tavitian A. Four additional members of the ras gene superfamily isolated by an oligonucleotide strategy: molecular cloning of YPT-related cDNAs from a rat brain library. *Proc Natl Acad Sci USA* 1987;84:8210–4.
8. Zahraoui A, Touchot N, Chardin P, Tavitian A. Complete coding sequences of the ras related rab 3 and 4 cDNAs. *Nucleic Acids Res* 1988;16:1204.
9. Baldini G, Hohl T, Lin HY, Lodish HF. Cloning of a Rab3 isotype predominantly expressed in adipocytes. *Proc Natl Acad Sci USA* 1992;89:5049–52.
10. Hancock JF, Magee AI, Childs JE, Marshall CJ. All ras proteins are polyisoprenylated but only some are palmitoylated. *Cell* 1989;57:1167–77.
11. Farnsworth CC, Kawata M, Yoshida Y, Takai Y, Gelb MH, Glomset JA. C terminus of the small GTP-binding protein smg p25A contains two geranylgeranylated cysteine residues and a methyl ester. *Proc Natl Acad Sci USA* 1991;88:6196–200.
12. Sano K, Kikuchi A, Matsui Y, Teranishi Y, Takai Y. Tissue-specific expression of a novel GTP-binding protein (smg p25A) mRNA and its increase by nerve growth factor and cyclic AMP in rat pheochromocytoma PC-12 cells. *Biochem Biophys Res Commun* 1989;158:377–85.
13. Mizoguchi A, Kim S, Ueda T, Takai Y. Tissue distribution of smg p25A, a ras p21-like GTP-binding protein, studied by use of a specific monoclonal antibody. *Biochem Biophys Res Commun* 1989;162: 1438–45.
14. Mizoguchi A, Kim S, Ueda T, Kikuchi A, Yorifuji H, Hirokawa N, Takai Y. Localization and subcellular distribution of smg p25A, a ras p21-like GTP-binding protein, in rat brain. *J Biol Chem* 1990; 265:11872–9.
15. Mizoguchi A, Yano Y, Hamaguchi H, et al. Localization of rabphilin-3A on the synaptic vesicle. *Biochem Biophys Res Commun* 1994;202:1235–43.
16. Pryer NK, Wuestehube LJ, Schekman R. Vesicle-mediated protein sorting. *Annu Rev Biochem* 1992; 61:471–516.
17. Ferro-Novick S, Novick P. The role of GTP-binding proteins in transport along the exocytic pathway. *Annu Rev Cell Biol* 1993;9:575–99.
18. Bourne HR. Do GTPases direct membrane traffic in secretion? *Cell* 1988;53:669–71.

19. Broek D, Toda T, Michaeli T, et al. The *S. cerevisiae* CDC25 gene product regulates the RAS/adenylate cyclase pathway. *Cell* 1987;48:789–99.
20. Sasaki T, Kikuchi A, Araki S, Hata Y, Isomura M, Kuroda S, Takai Y. Purification and characterization from bovine brain cytosol of a protein that inhibits the dissociation of GDP from and the subsequent binding of GTP to smg p25A, a ras p21-like GTP-binding protein. *J Biol Chem* 1990;265: 2333–7.
21. Sasaki T, Kaibuchi K, Kabcenell AK, Novick PJ, Takai Y. A mammalian inhibitory GDP/GTP exchange protein (GDP dissociation inhibitor) for smg p25A is active on the yeast SEC4 protein. *Mol Cell Biol* 1991;11:2909–12.
22. Ullrich O, Stenmark H, Alexandrov K, et al. Rab GDP dissociation inhibitor as a general regulator for the membrane association of rab proteins. *J Biol Chem* 1993;268:18143–50.
23. Garrett MD, Kabcenell AK, Zahner JE, et al. Interaction of Sec4 with GDI proteins from bovine brain, *Drosophila melanogaster* and *Saccharomyces cerevisiae*. Conservation of GDI membrane dissociation activity. *FEBS Lett* 1993;331:233–8.
24. Matsui Y, Kikuchi A, Araki S, et al. Molecular cloning and characterization of a novel type of regulatory protein (GDI) for smg p25A, a ras p21-like GTP-binding protein. *Mol Cell Biol* 1990;10: 4116–22.
25. Nonaka H, Kaibuchi K, Shimizu K, Yamamoto J, Takai Y. Tissue and subcellular distributions of an inhibitory GDP/GTP exchange protein (GDI) for smg p25A by use of its antibody. *Biochem Biophys Res Commun* 1991;174:556–63.
26. Trahey M, McCormick F. A cytoplasmic protein stimulates normal N-ras p21 GTPase, but does not affect oncogenic mutants. *Science* 1987;238:542–5.
27. Vogel US, Dixon RA, Schaber MD, et al. Cloning of bovine GAP and its interaction with oncogenic ras p21. *Nature* 1988;335:90–3.
28. Araki S, Kaibuchi K, Sasaki T, Hata Y, Takai Y. Role of the C-terminal region of smg p25A in its interaction with membranes and the GDP/GTP exchange protein. *Mol Cell Biol* 1991;11:1438–47.
29. Araki S, Kikuchi A, Hata Y, Isomura M, Takai Y. Regulation of reversible binding of smg p25A, a ras p21-like GTP-binding protein, to synaptic plasma membranes and vesicles by its specific regulatory protein, GDP dissociation inhibitor. *J Biol Chem* 1990;265:13007–15.
30. Regazzi R, Kikuchi A, Takai Y, Wollheim CB. The small GTP-binding proteins in the cytosol of insulin-secreting cells are complexed to GDP dissociation inhibitor proteins. *J Biol Chem* 1992;267: 17512–9.
31. Garrett MD, Zahner JE, Cheney CM, Novick PJ. GDI1 encodes a GDP dissociation inhibitor that plays an essential role in the yeast secretory pathway. *EMBO J* 1994;13:1718–28.
32. Peter F, Nuoffer C, Pind SN, Balch WE. Guanine nucleotide dissociation inhibitor is essential for Rab1 function in budding from the endoplasmic reticulum and transport through the Golgi stack. *J Cell Biol* 1994;1263:1393–1406.
33. Nishimura N, Nakamura H, Takai Y, Sano K. Molecular cloning and characterization of two rab GDI species from rat brain: brain-specific and ubiquitous types. *J Biol Chem* 1994;269:14191–8.
34. Shisheva A, Südhof TC, Czech MP. Cloning, characterization, and expression of a novel GDP dissociation inhibitor isoform from skeletal muscle. *Mol Cell Biol* 1994;14:3459–68.
35. Araki K, Nakanishi H, Hirano H, Kato M, Sasaki T, Takai Y. Purification and characterization of Rab GDIβ from rat brain. *Biochem Biophys Res Commun* 1995;211:296–305.
36. Yang C, Slepnev VI, Goud B. Rab proteins form *in vivo* complexes with two isoforms of the GDP-dissociation inhibitor protein (GDI). *J Biol Chem* 1994;269:31891–9.
37. Burton J, Roberts D, Montaldi M, Novick P, De Camilli P. A mammalian guanine-nucleotide-releasing protein enhances function of yeast secretory protein Sec4. *Nature* 1993;361:464–7.
38. Moya M, Roberts D, Novick P. DSS4-1 is a dominant suppressor of sec4–8 that encodes a nucleotide exchange protein that aids Sec4p function. *Nature* 1993;361:460–3.
39. Burstein ES, Macara IG. Characterization of a guanine nucleotide-releasing factor and a GTPase-activating protein that are specific for the ras-related protein p25rab3A. *Proc Natl Acad Sci USA* 1992; 89:1154–8.
40. Miyazaki A, Sasaki T, Araki K, et al. Comparison of kinetic properties between MSS4 and Rab3A GRF GDP/GTP exchange proteins. *FEBS Lett* 1994;350:333–6.
41. Wada M, Nakanishi H, Sato A, et al. Isolation and characterization of a GDP/GTP exchange protein specific for the Rab3 subfamily small G proteins. *J Biol Chem.* 1997;272;3875–8.
42. Soldati T, Shapiro AD, Svejstrup AB, Pfeffer SR. Membrane targeting of the small GTPase Rab9 is accompanied by nucleotide exchange [see comments]. *Nature* 1994;369:76–8.

43. Ullrich O, Horiuchi H, Bucci C, Zerial M. Membrane association of Rab5 mediated by GDP-dissociation inhibitor and accompanied by GDP/GTP exchange. *Nature* 1994;3681:157–60.
44. Shirataki H, Kaibuchi K, Yamaguchi T, Wada K, Horiuchi H, Takai Y. A possible target protein for smg-25A/rab3A small GTP-binding protein. *J Biol Chem* 1992;267:10946–9.
45. Shirataki H, Kaibuchi K, Sakoda T, et al. Rabphilin-3A, a putative target protein for smg p25A/rab3A p25 small GTP-binding protein related to synaptotagmin. *Mol Cell Biol* 1993;13: 2061–8.
46. Kishida S, Shirataki H, Sasaki T, Kato M, Kaibuchi K, Takai Y. Rab3A GTPase-activating protein-inhibiting activity of rabphilin-3A, a putative Rab3A target protein. *J Biol Chem* 1993;268: 22259–61.
47. Nishizuka Y. The molecular heterogeneity of protein kinase C and its implications for cellular regulation. *Nature* 1988;334:661–5.
48. Yamaguchi T, Shirataki H, Kishida S, et al. Two functionally different domains of rabphilin-3A, Rab3A p25/smg p25A-binding and phospholipid- and Ca^{2+}-binding domains. *J Biol Chem* 1993;268: 27164–70.
49. Perin MS, Fried VA, Mignery GA, Jahn R, Südhof TC. Phospholipid binding by a synaptic vesicle protein homologous to the regulatory region of protein kinase C. *Nature* 1990;345:260–3.
50. Li C, Ullrich B, Zhang JZ, Anderson RG, Brose N, Südhof TC. Ca^{2+}-dependent and -independent activities of neural and non-neural synaptotagmins. *Nature* 1995;375:594–9.
51. Brose N, Petrenko AG, Südhof TC, Jahn R. Synaptotagmin: a calcium sensor on the synaptic vesicle surface. *Science* 1992;256:1021–5.
52. Geppert M, Goda Y, Hammer RE, et al. Synaptotagmin I: a major Ca^{2+} sensor for transmitter release at a central synapse. *Cell* 1994;79:717–27.
53. Numata S, Shirataki H, Hagi S, Yamamoto T, Takai Y. Phosphorylation of rabphilin-3A, a putative target protein for Rab3A, by cyclic AMP-dependent protein kinase. *Biochem Biophys Res Commun* 1994;203:1927–34.
54. Kato M, Sasaki T, Imazumi K, et al. Phosphorylation of rabphilin-3A by calmodulin-dependent protein kinase II. *Biochem Biophys Res Commun* 1994;205:1776–84.
55. Inagaki N, Mizuta M, Seino S. Cloning of a mouse rabphilin-3A expressed in hormone-secreting cells. *J Biochem* 1994;116:239–42.
56. Shirataki H, Yamamoto T, Hagi S, et al. Rabphilin-3A is associated with synaptic vesicles through a vesicle protein in a manner independent of Rab3A. *J Biol Chem* 1994;269:32717–20.
57. Senbonmatsu T, Shirataki H, Jin-no Y, Yamamoto T, Takai Y. Interaction of rabphilin 3 with synaptic vesicles through multiple regions. *Biochem Biophys Res Commun* 1996;228:567–72.
58. Chung SH, Takai Y, Holz RW. Evidence that the Rab3a-binding protein, rabphilin3a, enhances regulated secretion. Studies in adrenal chromaffin cells. *J Biol Chem* 1995;270:16714–8.
59. Komuro R, Sasaki T, Orita S, Maeda M, Takai Y. Involvement of rabphilin-3A in Ca^{2+}-dependent exocytosis from PC12 cells. *Biochem Biophys Res Commun* 1996;219:435–40.
60. Burns M, Sasaki T, Shirataki H, Takai Y, Augustine G. submitted.
61. Bommert K, Charlton MP, DeBello WM, Chin GJ, Betz H, Augustine GJ. Inhibition of neurotransmitter release by C2-domain peptides implicates synaptotagmin in exocytosis. *Nature* 1993;363: 163–5.
62. Bennett MK, Calakos N, Scheller RH. Syntaxin: a synaptic protein implicated in docking of synaptic vesicles at presynaptic active zones. *Science* 1992;257:255–9.
63. Söllner T, Whiteheart SW, Brunner M, et al. SNAP receptors implicated in vesicle targeting and fusion. *Nature* 1993;362:318–24.
64. Lian JP, Stone S, Jiang Y, Lyons P, Ferro-Novick S. Ypt1p implicated in v-SNARE activation. *Nature* 1994;372:698–701.
65. Sogaard M, Tani K, Ye RR, et al. A rab protein is required for the assembly of SNARE complexes in the docking of transport vesicles. *Cell* 1994;78:937–48.
66. Burstein ES, Linko-Stentz K, Lu ZJ, Macara IG. Regulation of the GTPase activity of the ras-like protein p25rab3A. Evidence for a rab3A-specific GAP. *J Biol Chem* 1991;266:2689–92.
67. Fukui T, Sasaki T, Imazumi K, Matsuura Y, Nakanishi H, Takai Y. Isolation and characterization of a GTPase activating protein specific for the Rab3 subfamily of small G proteins. *J Biol Chem.* 1997; 272:4655–8.
68. Musha T, Kawata M, Takai Y. The geranylgeranyl moiety but not the methyl moiety of the smg-25A/rab3A protein is essential for the interactions with membrane and its inhibitory GDP/GTP exchange protein. *J Biol Chem* 1992;267:9821–5.

Subject Index

H

I

R

Printed and bound by CPI Group (UK) Ltd, Croydon, CR0 4YY
08/05/2025
01865014-0001